14 ⁰⁰

Advances in

BOTANICAL RESEARCH

VOLUME 1

Advances in
BOTANICAL RESEARCH

Edited by

R. D. PRESTON

The Astbury Department of Biophysics
The University, Leeds, England

VOLUME 1

1963

ACADEMIC PRESS London and New York

ACADEMIC PRESS INC. (LONDON) LTD.
BERKELEY SQUARE HOUSE
LONDON, W.1

U.S. Edition published by

ACADEMIC PRESS INC.
111 FIFTH AVENUE
NEW YORK 3, NEW YORK

Library of Congress Catalog Card Number : 62–21144

Printed in Great Britain by Adlard & Son Ltd.,
Bartholomew Press, Dorking, Surrey

CONTRIBUTORS TO VOLUME 1

G. M. ANDROES, *Physics Department, American University of Beirut, Beirut, Lebanon* (p. 327).

HANS BURSTRÖM, *Institute of Plant Physiology, Lund University, Lund, Sweden* (p. 73).

JACK DAINTY, *Biophysics Department, University of Edinburgh, Scotland* (p. 279).

G. ERDTMAN, *Palynological Laboratory, Nybodagatan, Stockholm-Solna, Sweden* (p. 149).

A. L. KURSANOV, *Institute of Plant Physiology, K. A. Timiriasev, Academy of Sciences, Moscow, U.S.S.R.* (p. 209).

C. R. METCALFE, *The Jodrell Laboratory, Royal Botanic Gardens, Kew, England* (p. 101).

ALAN WESLEY, *Department of Botany, The University, Leeds, England* (p. 1).

PREFACE

As in most other scientific disciplines, research in the various aspects of botany has grown vigorously since the war and on the way has achieved spectacular results which have received much publicity. These developments have led, and will progressively lead still further, to a corresponding growth, steadily accelerating, in the number of papers published over a wide spectrum of learned journals until it is even now virtually impossible for any worker to stay abreast of his own field and remain active in research. The classical type of review article is still a valuable help; but the reviewing journals are too few in number to give a satisfactory cover. Consisting as they normally—and properly —do, moreover, of little more than ordered sequences of abstracts of papers published over the past year or so, they can neither examine the subject concerned against the background of less recent knowledge or of matter from cognate fields nor allow authors freedom to express opinions and to speculate on the future. This undoubtedly goes far to explain the appearance over the past few years of volumes, either singly or in series, sometimes but not always confined to one field of botanical science, such as Physiology or Cytology, which deal with the retrieval of information in a different way. In these the authors are allowed freedom not only to collate facts but also to express opinions— to deal not only with the letter but also with the spirit. The present series is designed to be of this kind.

It is the intention, however, that this series should have special features at present unique to it, in our view features which will make all the articles of special merit and some of them of lasting value. The authors of this first volume have been asked to do three things. They have been asked to write about some special topic within their chosen field, of especial current interest to them and upon which they have been actively engaged; they have been asked to set their own work against the background of cognate researches in other laboratories both past and present; and they have been asked to express opinions freely and to speculate as widely as they dare upon future trends and future developments. This will remain the policy of these volumes. The articles will, moreover, range over the whole field of botanical enquiry dealing both with the more spectacular modern chemical and physical approaches and the less well publicized developments in more classical fields upon which all else depends. In this again the first volume sets the standard.

In these publicity-minded days it is by no means an easy matter for an editor to persuade scientists, already deeply immersed in paper, to write yet another article. We are all the more grateful to the authors of this volume that they accepted their tasks cheerfully and presented manuscripts punctually. They are all recognized authorities in their fields and we need say no more about them. Their articles range from the classical fields of anatomy and palaeobotany to the most modern treatments of irreversible thermodynamics and electron spin magnetic resonance. There should therefore be something here for all; though it is my sincere hope that both the clasically- and the modern-minded readers will at least dip into each other's pages so that each may appreciate the other and learn "what it is all about".

The editing of volumes such as these could not possibly be attempted without the assurance of co-operation with many whose names will not appear in these pages. To all of these I offer my sincere thanks, particularly to my colleagues for their support and encouragement and especially to my secretary, Miss Eunice D. Lister, for her untiring attention to detail and for her skill in ensuring that I have not lost a manuscript. I am especially indebted to the publishers who have throughout smoothed my path in every possible way and who have carried out their own part of the task without fuss and with quiet efficiency.

R. D. PRESTON

Leeds, 1962

CONTENTS

The Status of Some Fossil Plants

ALAN WESLEY

Growth Regulation by Metals and Chelates

HANS BURSTRÖM

Comparative Anatomy as a Modern
Botanical Discipline

With special reference to recent advances in
the systematic anatomy of monocotyledons

C. R. METCALFE

Palynology

G. ERDTMAN

Metabolism and the Transport of Organic Substances in the Phloem

A. L. KURSANOV

Water Relations of Plant Cells

JACK DAINTY

Electron Paramagnetic Resonance in Photosynthetic Studies

G. M. ANDROES

The Status of Some Fossil Plants

ALAN WESLEY

Department of Botany, The University, Leeds, England

I. INTRODUCTION

The aim of this article is to present to the general reader a survey of some of the more interesting contributions that have been made in the field of palaeobotany during the past decade or two. It will be seen that relatively few topics have been selected for treatment, but in this way it has been possible to present detailed, though not necessarily exhaustive, discussions in each case. It is felt that this may be a more useful method of approach since to have attempted a complete coverage of all the numerous and specialist minutiae of palaeobotanical research published over a short period of time would have meant devoting a disproportionate amount of space to listing discoveries and statements, with little, or no, opportunity for extended discussion of any particular example. Consequently some groups of fossil plants and certain aspects of their study receive scant, or no, attention, but this in no way signifies belittlement of the many authors whose names and valuable contributions have inevitably and regrettably to be excluded.

Though the treatment of the several topics varies somewhat according to the subject matter, an endeavour has been made to indicate current attitudes and the extent to which they may be accepted. Historical considerations cannot pass unnoticed in most cases, and reference has been made to previous knowledge where it is of relevance to new discoveries or modified interpretations. In addition, some emphasis has been placed on the difficulties which palaeobotanists continually encounter in their studies of fossil plants and which the non-specialist does not always readily appreciate.

II. Fossil Liverworts

Apart from spores, liverwort remains are poorly represented in the fossil record on account of the small size and delicate nature of their bodies. Unless fossilization has occurred under optimum conditions, preservation is usually so poor that, at best, no more than a non-committal name may be assigned to them. The chances of a fossilized thallus being the remains of a marine alga or a terrestrial hepatic are equal, especially when it is wholly sterile. In the absence of knowledge about the structure of the cells of the plant body, and rhizoids should they be present, the affinities of any thalloid fossil must always remain doubtful.

Unfortunately much of the fossil material of this type has acquired names recalling those of living genera, which has resulted in implied relationships even though such may not have been intended by the original investigators. To avoid further confusion and misconjecture, it has now become customary to follow the proposal made some years ago by Walton (1925, 1928) to name all those showing characters exclusive to the liverworts as *Hepaticites* and those which agree equally with algae as *Thallites*, unless there be some special character indicating relationship with some more narrowly limited taxon of the group (Harris, 1942a, 1961a; Lundblad, 1954). Examples of this latter type that may be noted are *Marchantites* Brongn. *emend.* Walton (Walton, 1925; Lundblad, 1955), *Metzgeriites* Steere and possibly *Jungermannites* Goepp. *emend.* Steere (Steere, 1946), and the lately described *Ricciopsis* and *Marchantiolites* (Lundblad, 1954). There is also, of course, the very fully known *Naiadita lanceolata* Buckman *emend.* Harris (Harris, 1938).

Yet the number of Pre-Tertiary forms that may be regarded as having belonged to the Hepaticae is not particularly large and amounts to probably no more than twenty, of which about fifteen are species of *Hepaticites*. There are, in addition, probably a dozen good species of

Thallites, but their affinities are unknown at present and they could equally represent algal remains.

Nothing new has been added to our knowledge of the small assemblage of Upper Carboniferous liverworts (Walton, 1925, 1928), except that Walton now finds that his original *Hepaticites willsii* should be more correctly called *Thallites* since it lacked rhizoids (1949a). These Carboniferous hepatics were thalloid plants with a habit recalling that of certain modern members of the anacrogynous Jungermanniales, though one of them, *Hepaticites kidstoni* Walton, was definitely leafy. This latter type has been compared with the acrogynous Jungermanniales by some authors, but there seems little reason for doubting Walton's opinion that it too was another anacrogynous form very much like *Treubia*.

There is a greater number of satisfactory records of liverworts from the Mesozoic but, with the exception of *Naiadita* which has been tentatively referred to the Sphaerocarpales (Harris, 1938), sterile remains alone are known. These nearly all take the form of dichotomizing thalli and are mostly referable to the anacrogynous Jungermanniales, but recent investigations by Harris (1961a) and Lundblad (1954, 1955) indicate that plants with undoubted marchantialean characters existed during the Period.

Two species of *Hepaticites* from the Jurassic rocks of Yorkshire have been assigned to the Marchantiales, though the attribution of one of them, *H. wonnacotti* (Harris, 1942a, 1961a), is rather uncertain since neither tuberculate rhizoids nor ventral scales have been found. The other species, *H. haiburnensis* (Harris, 1961a), is only known from one specimen, but it is clear that the dichotomously branched thallus possessed numerous rhizoids arising from the underside of the midrib, as well as two rows of conspicuous ventral scales (Fig. 1, B). Harris has not been able to see details of the walls of the rhizoids, or to recognize pores in the centres of the oblique polygons which he considers to represent the outlines of air chambers within the lamina. For such reasons he justifiably refrains from attempting a more precise classification of the specimen.

On the other hand, some material from the Swedish Liassic, comprising rather complete sterile thalli (*Ricciopsis florinii* and *R. scanica*), fragmentary segments with cellular structure and air-pores (*Marchantiolites porosus*), and associated *Riccia*-like spores (*Ricciisporites tuberculatus*), quite clearly represents the remains of plants that belonged to the Marchantiales (Lundblad, 1954). *Ricciopsis florinii* consisted of small, dichotomously branched thalli, which occurred "singly or in groups, of more or less circular shape, forming rosettes of crowded segments" about 2·5 cm in diameter. The segments were channelled and bore ventral rhizoids which were "tuberculate or smooth-walled, uni-cellular or multicellular, with oblique transverse walls". Ventral scales have not been observed, but the resemblance between this species and modern members of the Ricciaceae (Marchantiales) is remarkable (Fig. 1, A). The same is also true for *R. scanica* which, though no ventral scales or rhizoids have been seen, was composed of dichotomous thalli

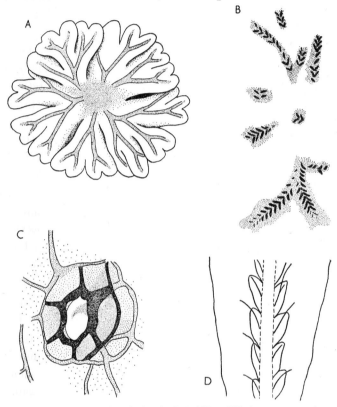

FIG. 1. Mesozoic Liverworts. A. *Ricciopsis florinii* Lundblad; reconstruction of a thallus (redrawn from Lundblad, 1954). B. *Hepaticites haiburnensis* Harris; fragments of a thallus with ventral scales (redrawn from Harris, 1961a). C. *Marchantiolites porosus* Lundblad; air-pore (redrawn from Lundblad, 1954). D. *Marchantites hallei* Lundblad; reconstruction of a segment of thallus, showing ventral scales (redrawn from Lundblad, 1955).

with rather more slender segments. Though the two forms appear to be quite distinct, Lundblad points out the possibility of their representing two different growth forms of a single species resulting from different environmental conditions or even a difference in sexuality. Amongst modern liverworts the aquatic form of *Riccia fluitans* has narrower segments than, and lacks the rhizoids and ventral scales of, the terres-trial form, and there may be considerable differences in size and shape between the male and female thalli of the dioecious *Riccia cupulifera* Duthie and Garside.

Only fragments of the thallus of *Marchantiolites* are known, and since the fossil material is of the compression type no more than the limiting layer of cells is preserved. Thus the upper and lower epidermis alone have been seen and nothing is known of the internal structure. One of these layers is pierced by a number of elevated air-pores, each of which is surrounded by a concentric ring of cells (Fig. 1, C). Ventral scales are not known, nor is it clear whether some of the rhizoids were tuberculate or not. Nevertheless the remains are undoubtedly marchantialean. The comparatively simple air-pores of the fossil do not resemble the compound barrel-shaped openings of the air chambers of the living Marchantiaceae, nor are they so simplified as in the Ricciaceae. The nearest equivalent is to be found in such living genera as *Plagiochasma*, *Grimaldia* and *Oxymitra*, but to attempt any very close comparison between *Marchantiolites* and living forms is impossible until more complete specimens become available.

Another species with air-pores is now known to have existed during the Mesozoic, but it differs considerably from the Liassic *Marchantiolites* just described. This is *Marchantites hallei* (Lundblad, 1955), based on three specimens from the Lower Cretaceous of Patagonia which were collected and figured some years previously as "*Marchantites* ? sp." by Halle (1913). The specimens represent segments of dichotomous, sterile thalli, but only one of them is actually branched. There is a median thickened zone from which arcuate lateral ribs diverge towards the margins. There are two distinct rows of ventral scales along the sides of the midrib (Fig. 1, D), and dense clusters of unicellular rhizoids attached to the proximal part of the lateral ribs. The rhizoids were apparently of two kinds, but owing to their unsatisfactory state of preservation Lundblad has not been able to make out details. The structure of the air-pores is not preserved, yet it seems that they were barrel- or cone-shaped.

The arcuate lateral ribs, the two rows of ventral scales and the large size of the air-pores are features in favour of the classification of *Marchantites hallei* within the sub-order Marchantiineae, if not in the Marchantiaceae *sens. strict.*, rather than near the ricciaceous types which have small pores and a single row of ventral scales.

While these discoveries are extremely interesting as recording the undoubted existence of the Marchantiales as far back as the older Mesozoic, they afford no data towards a better understanding of the evolution of the liverworts which must still be based purely upon comparative morphology of the living forms. They do, however, indicate that the group was already reaching a world-wide distribution towards the end of the Mesozoic Period, and possibly even earlier if the less convincing record from the Lower Jurassic of Australia (Medwell, 1954) is also taken into consideration.

III. Early Vascular Plants

Vascular plant remains have long been known in rocks of Devonian age, but whence the varied and highly organized vegetation which they represent came and what its antecedents were have been unanswered questions around which much interest, and often highly speculative discussion, has centred. In some cases, preservation of specimens is so good that it has been possible to obtain reasonably accurate reconstructions of the original plants and some indications of their relationships, but in others the specimens are so fragmentary or obscure as to offer no more evidence than that there had existed other types of plant.

The dramatic discoveries by Kidston and Lang (1917–21) of petrified remains in the Middle Old Red Sandstone of Scotland almost 50 years ago, and their recognition of the Psilophytales, still ranks as the most outstanding contribution to our knowledge of early vascular plants. The deposit at Rhynie has given a remarkable glimpse of one type of vegetation of the early Devonian period. The four higher plants have the double interest of being not only the most ancient fully-known vascular cryptogams, but also the most simply organized members of the group. These discoveries had a profound influence on morphological thought and theories of evolution of the higher plants, and the Psilophytales rapidly achieved importance as the basis of ideas which became crystallized in the Telome Theory of Zimmerman. This theory, which has received wide acceptance, derives all vascular plant groups, by a few elementary processes, from ancestors of *Rhynia*-type with simple, dichotomous, vascularized, but rootless, axes with terminal sporangia.

There is much supporting evidence for many aspects of the telomic concept, but it inevitably favours a monophyletic origin of the vascular plants. It apparently makes no provision for the unquestionable fact that the lycopsid line, represented by the Silurian *Baragwanathia longifolia* (Lang and Cookson, 1935), was already well defined and strongly established prior to the appearance of the rhyniaceous types which are conceived as ancestral to all other forms. There seems little doubt, too, that the sphenopsid and pteropsid lines were already marked out by mid-Devonian time and that the average level of differentiation of the plant association was too advanced to have originated from the contemporaneous representatives of the Psilophytales. However, as will be shown in a later paragraph, the evidence may not be quite so conflicting. Yet Leclercq (1954) inclines to the view "that the Psilophytales represent a division possibly equal in importance to that of the Lycopsides, Sphenopsides, Pteropsides, running parallel with them, instead of being their converging point", and goes on to add that "if this conception were confirmed, the Psilophytales might be considered as a resulting point instead of a starting point". The idea of polyphylesis may by no means be ruled out, and Andrews (1959, 1961; Andrews and Alt, 1956), amongst others, has recently pleaded very strongly in favour of its acceptance.

The appearance of such a relatively complex form as *Baragwanathia* during the late Silurian indicates that a fairly long period of evolution of the vascular plants had occurred prior to that moment, and that a search must be made for their real origin at least in the early part of the Period, if not before. Indeed, evidence from plant microfossils from rocks of an even earlier age suggests that vascular plants may have already been in existence in pre-Silurian time. As will be learnt from the next two sections, however, much of this evidence still awaits confirmation and is insufficiently convincing as the basis for extended evolutionary arguments.

It has sometimes been suggested that the simplicity of the Rhynie plants had resulted from reduction in response to some environmental stimulus, rather than being a manifestation of genuine primitiveness. Yet, if this be the case, what more simple form of construction for a primitive vascular plant may be expected? This writer, for one, believes that the plants are primitive both in their vegetative and sporangial construction. Anything more primitive than a simple sac-like terminal sporangium, with a wall several layers thick and no provision for dehiscence, cannot be conceived. And though they are antedated by other more complex vascular plants, is it not satisfactory to consider these plants as being the last surviving members of a family of *Psilopsida* which had existed under favourable conditions and remained unaltered since the moment of its inception until the mid-Devonian?

From this point of view, it is particularly significant that every Devonian flora contains records of genuine psilopsid remains. These are even to be found side by side with *Baragwanathia* in the Australian Silurian. This co-existence of both the psilopsid and lycopsid lines at an early period in time offers, of course, no evidence as to whether one antedated the other or whether they represent quite distinct and separate lines of evolution as polyphyletism demands. Therefore, until such evidence as may settle this point is forthcoming, it would seem that the more conventional outlook, which favours psilopsids of *Rhynia*-type as having provided the ancestral stock for all other groups of vascular plants, will remain without serious challenge.

So far reference has been made only to two long-famous early floras containing the remains of vascular plants. It should not be forgotten, however, that others exist and that much is known about the plants they contain. New material and critical investigations are continually providing abundant evidence of the great diversity, and often high complexity, of the assemblage of early plants. Some of this evidence will now be discussed, but since a comprehensive treatment would extend to many pages of text, it has been decided to highlight only relatively few examples here. This seems to be the more profitable approach since some excellent contributions have already been written by Andrews (1959, 1961), Stewart (1960) and Chaloner (1960).

A. MICROFOSSILS OF PRE-DEVONIAN AGE

Many records of plant microfossils from pre-Devonian rocks have appeared over the past ten or more years. These include spores and fragments of vascular tissue, and much importance has been attached to them by authors favouring a polyphyletic origin of the vascular plants. While the discovery of isolated tracheids would undoubtedly attest to the former presence of vascular plants, it is by no means certain that the increasingly large numbers of records of spores offer supporting evidence.

The diversity in form and ornamentation shown by these early spores indicates, of course, a varied parent flora, but these characters are no proof that the spores had been produced by vascular plants. It should be remembered that many bryophytes have highly ornamented and cutinized spore walls which in many cases may simulate those of vascular plants to an extent which makes their certain identification difficult. This has been very forcibly expressed by Knox (1939), who concluded that "Except where fossil spores are found in organic union with recognizable parent material, however, there can be no certainty as to their relationships". There seems no valid reason why this statement should not apply equally to ancient spores.

In a useful survey of many recent reports of pre-Devonian microfossils, Chaloner (1960) has clearly underlined the caution that is needed before accepting many determinations as fact. One major factor of importance derives from the risk of contamination both in the field and in the laboratory when handling such microscopic bodies. Of interest in this connection are two papers by Radforth and McGregor (1954, 1956) who, at first, recorded a variety of spores from rock samples supposedly of Devonian age which pointed to the existence of plant types much more advanced than would have been expected from the known macrofossils of that age. They later found that the samples contained a mixture of Cretaceous and Pennsylvanian (Carboniferous) elements and that no definite conclusions could be reached on such conflicting evidence.

Perhaps more disquieting is the inference to be obtained from the extensive review by Obrhel (1958) of spore reports from early Palaeozoic and Pre-Cambrian strata. It appears that a re-investigation of some supposedly Pre-Cambrian rocks in Russia, which had been reported to contain well-preserved Devonian-type spores in the lower part and Lower Carboniferous-type in the upper, revealed that spores were only actually to be found in samples containing cracks; in unfractured parts there were no spores. It is concluded, therefore, that the spores had entered the rock along cracks produced by tectonic action, and that there had been a migration of spores from younger to older rocks during the course of time. Whether such a process is of widespread occurrence is not known, but it certainly seems to be a real possibility and a very important factor that should be taken into account in the interpretation of any plant microfossil work. It could at least account for one startling report of pollen attributed to *Potamogeton* and of gymnosperm tracheids from rock of Cambrian age, as well as the controversial discoveries of spores and tracheids in rocks of a similar age by several Indian workers (Ghosh and Bose, 1947, 1952, amongst others) which have been critically reviewed by Sitholey *et al.* (1953).

B. THREE PRESUMED VASCULAR PLANTS FROM PRE-DEVONIAN ROCKS

1. Aldanophyton antiquissimum

For 20 years the flora containing *Baragwanathia longifolia* stood unchallenged as the oldest known record of authentic vascular land plants, but in 1956 Leclercq brought to the notice of Western readers the discovery by Kryshtofovich, 3 years previously in Cambrian deposits in eastern Siberia, of lycopod-like fossil remains to which he had given the name *Aldanophyton antiquissimum* (Kryshtofovich, 1953).

In a remarkably short time this evidence for the existence of vascular plants at an even earlier moment in geological time became widely adopted as fact by some writers (Leclercq, 1956; Axelrod, 1959; Darrah, 1960), and it formed an additional basis for polyphyletic arguments. Many authors have shown some caution, however, especially with regard to the supposed lycopod affinities of the fossil remains (Andrews, 1959, 1961; Andrews and Alt, 1956; Banks, 1960; Chaloner, 1960). Only recently has a critical survey of the evidence by Stewart

2§

(1959, 1960, 1961) shown that there is no proof that *Aldanophyton* was either a vascular plant or a lycopod.

The remains of *Aldanophyton* consist of some fragments of axes measuring up to 8·5 cm in length and 1·3 cm in width, which are clothed with a number of elongated, simple enations that arise from oblong thickenings and may attain a length of 9·0 mm. There is no definite arrangement of these enations or of their scars. Vascular tissue has not been described in the enations, though there is some kind of strand which extends to the base of each thickening. No cuticle has been obtained, and the reproductive structures remain unknown. In fact, extremely little is known about four very poorly preserved specimens that have served as the basis for support of arguments in favour of a polyphyletic origin of the vascular plants.

It is unfortunate that botanists should have been led into accepting what might have amounted to a spectacular discovery as a result of a not particularly accurate translation into English (Leclercq, 1956) of Kryshtofovich's original Russian text. As Stewart has clearly pointed out, there are some points of fundamental disagreement between this translation and another originating from an independent source. Thus the sentence in the translation quoted by Leclercq, which reads "In some places it is possible to trace a thin stripe, a vascular bundle right out to the base of the thickenings . . .", becomes rendered in the other version as "In places a thin rod-conducting bundle may be traced as far as the base of the enations". Further, in the latter translation, there is the very important statement that "no obvious cellular structure could be determined" in the traces of carbonaceous substance. Therefore, proof of the presence of vascular tissue, and hence that *Aldanophyton* was a vascular plant, is still wanting.

Even assuming that it should eventually be proved to have been a vascular plant, it seems that the very criterion for assigning it to the *Lycopsida* has still to be demonstrated i.e. that the enations are true microphylls with a vascular strand running their entire length. As it is, the "strands" (or whatever one may like to call them) are reported as extending only to the base of the enations. This condition existed in the genuine psilopsid *Asteroxylon*, though the strand was composed of tracheids in that plant.

Stewart has suggested that the fossil could well have been some kind of alga or non-vascular green land plant. These are possibilities that should by no means be ruled out for there are certain modern members of both the Laminariales and the Fucales and also of the Musci which exhibit a remarkably high degree of external and internal differentiation —to such an extent, indeed, that a rapid glance might suggest their belonging to higher groups in the plant kingdom.

2. Hepaticaephyton and Musciphyton

At the IXth International Botanical Congress, held in Montreal in 1959, Greguss announced the joint discovery with Kozlowski of a new group of ancient vascular plants from the Ordovician of Poland (Kozlowski and Greguss, 1959; Greguss, 1959). This group, the *Propsilophyta*, contains two genera, *Musciphyton* and *Hepaticaephyton*, and is said by the authors to have been more primitive than the *Psilopsida*.

Only small fragments, less than 1·0 cm in length were obtained, some of them branching in a monopodial fashion (*Musciphyton*), others dichotomously (*Hepaticaephyton*), but Greguss and Kozlowski claim to have seen sporangia with peristome teeth and spores. There were no leaves or stomata, but the remains are said to possess conducting elements with simple pitting in their walls. Greguss (1961) has recently published illustrations in substantiation of these reports. From the phylogenetic standpoint, it is claimed that these diminutive fossils point to a very early origin of the land plants from unicellular algae, and the existence of both monopodial and dichtomous branching is regarded as rendering unlikely the derivation of the one from the other.

If all these claims should be confirmed, Greguss and Kozlowski will have made a very important contribution to palaeobotany. The recent illustrations (Greguss, 1961: Pl. 2, 3, 5, 6, 7), however, are not at all convincing, and there must be some doubt as to whether the authors have been dealing with genuine palaeozoic fossil remains. It is truly remarkable that what must have been delicate and perishable structures should have survived in a marine limestone and still have retained their original plasticity. There is the very real possibility that the remains may actually be the roots of comparatively recent plants that have penetrated into minute cracks and crevices of the rock, and be thus contaminants of the genuine microfossil assemblage.

C. A NEW INTERPRETATION OF THE MORPHOLOGY
OF *Rhynia* AND *Horneophyton*

The one puzzling and unsolved feature about the Rhynie plant bed was the absence of any traces of the gametophytes of the vascular plants which it contained. This is truly remarkable for, though Kidston and Lang supposed them to have been delicate and perishable structures, other equally delicate material, such as rhizoids, fungal hyphae and blue-green algae, is well preserved in the chert.

A possible solution to this problem has recently been offered by Merker (1958, 1959, 1961), but though his arguments may stimulate thought, few will be convinced by his new interpretation of certain

aspects of the basic morphology of *Rhynia* and *Horneophyton*. It seems
unlikely, too, that Kidston and Lang, who were such authoritative and
thorough investigators, would have missed any evidence that may have
been available in the material at their disposal.

It will be remembered that *Rhynia* essentially consisted of a horizontal underground creeping rhizome with rhizoids and erect dichotomous aerial branches with terminal sporangia. A vascular cylinder
was continuous throughout the plant. *Horneophyton* differed only in that
there was no creeping rhizomatous portion, but instead a swollen base
with rhizoids. This swollen base had no vascular strand.

Merker suggests that the creeping underground parts of *Rhynia* were
not in actual fact the rhizomes of the sporophyte, but rather the gameto-

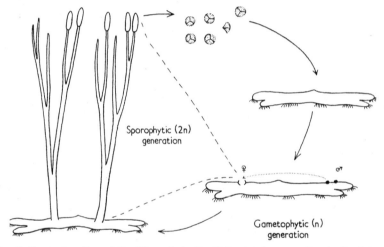

FIG. 2. Reconstruction of the life-cycle in the *Psilophytales* according to the interpretation of Merker (1961).

phyte, and that the aerial part alone represents the sporophytic generation. Thus the sporophyte is considered to have been borne on a massive
and persistent gametophyte which served as a supporting organ (Fig 2,
left-hand drawing), and from its rhizoids the water source for the erect,
sub-aerial sporophyte. He notes that Kidston and Lang commented on
the relative rarity of their "rhizomes" and suggests that had this been
their true nature, it would be expected that they would have been as
long lived as the sub-aerial axes, or even more so. From his interpretation of *Rhynia*, Merker believes that the swollen base of *Horneophyton*
becomes a plausible gametophyte.

In support of his thesis, Merker illustrates what he regards as a young
sporophyte growing out from a deep-seated archegonium in the creeping
part, as well as what is interpreted as a possible embryo. Both photo-

graphic and line illustrations of the former are given, and there is certainly a tract of somewhat different cells passing to the exterior through the outer part of the rhizome. But of the supposed embryo, only a line drawing is given which may not accurately represent the actual specimen. He was unable to discover any convincing sex organs, though there are illustrations of marginal cavities (which could represent their original location) and some peculiarities of the orientation of the cells at the periphery of the creeping parts which might indicate the proximity of such organs. There is also a line drawing of an illustration by Kidston and Lang, which purports to show a collar of gametophytic tissue surrounding the base of an aerial branch. This feature is far less clear in the original illustration than one would be led to believe from the new drawing, and there is no obvious discontinuity between the tissues of the aerial branch and the creeping part as might be expected if they were to represent parts of two different generations.

Before accepting Merker's ideas, it will be necessary for someone to make a very thorough search through all the extant material of the Rhynie plants. As it is it seems that insufficient specimens have yet been re-examined. Nevertheless this re-interpretation allows a closer comparison of the Psilophytales with the bryophytes, but it hardly adds support to the Homologous Theory of the origin of the sporophyte.

D. *Calamophyton* AND OTHER EARLY SPHENOPSIDS

Most of our essential knowledge of the Devonian genus *Calamophyton* has been derived from the original studies of Kräusel and Weyland (1926, 1929), but recently Leclercq and Andrews have demonstrated a greater complexity of organization in their *Calamophyton bicephalum* than in the type species *C. primaevum* K. and W. (Leclercq and Andrews, 1959, 1960; Andrews, 1959, 1961).

Calamophyton was probably shrubby in habit, and consisted of upright stems about 2·0 cm in diameter which branched first in a more or less digitate manner into as many as seven members. These in their turn branched monopodially, or with a mixture of monopodial and dichotomous forking. It is still not known, however, how these branch systems were borne on the parent plant.

On the basal parts there were stoutish, unforked sterile appendages, but higher on the plant these became replaced by more slender leaves which forked once or twice and tended to be inserted in whorls. There now seems to be little evidence for regarding the leaves as narrow and wedge-shaped with notched tips as was implied by the two somewhat misleading figures that have so often been reproduced in text-books (Kräusel and Weyland, 1926, Fig. 26). Some of the shoots were fertile,

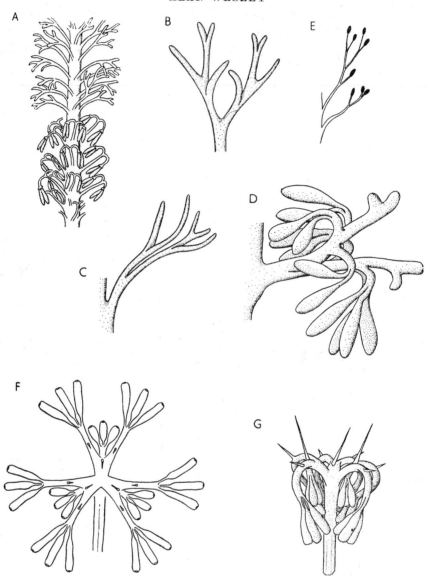

Fig. 3. Early sphenopsids. A. *Calamophyton renieri* Leclercq; part of a fertile axis (redrawn from Leclercq, 1940). B and C. Restorations of two leaves of *Calamophyton bicephalum* Leclercq and Andrews (redrawn from Leclercq and Andrews, 1960). D. *Calamophyton bicephalum*; restoration of a sporangiophore (redrawn from Leclercq and Andrews, 1960). E. *Protohyenia janovii* Ananiev; two sporangiophores showing the erect sporangia (redrawn from Ananiev, 1957). F. *Eviostachya hoegi* Stockmans; schematic representation of the apical trichotomy of a sporangiophore flattened into a horizontal plane (redrawn from Leclercq, 1957). G. *Eviostachya hoegi*; reconstruction of a sporangiophore with the sporangia in their natural position (redrawn from Leclercq, 1957).

and in place of sterile appendages there were numerous bi-forked sporangiophores with each arm ending in a recurved sporangium. Though there was a marked tendency for the sporangiophores to be aggregated together on certain branches, there were no well-organized strobili.

In the Belgian *C. renieri* (Leclercq, 1940) the leaves were more filiform and more frequently divided, and each branch of the sporangiophore usually bore two sporangia (Fig. 3, A).

Using the method of "degagement" (the careful chipping away with needles of small fragments of rock, allowing plant organs to be excavated and traced in any direction through the matrix), which Leclercq has very successfully applied in many of her researches on Devonian plants, it has been revealed that the sporangiophore of *C. bicephalum* was a three-dimensional structure and even more complex. It characteristically branched into an upper and a lower segment, both of which bore three short side stalks that terminated in a pair of recurved sporangia. There was thus a total of twelve sporangia on each sporangiophore (Fig. 3, D). The sterile appendages, fairly rigid and also three-dimensional, were slender dichotomous structures which divided two to four times (Fig. 3, B and C).

"Degagement" of some sterile leaves of a Belgian specimen of *C. primaevum* by Leclercq and Andrews revealed that these, too, forked at least twice in a three-dimensional fashion (Leclercq and Andrews, 1960, Fig. 5, Pl. III, Figs. 15 and 16; Andrews, 1961, Fig. 9–2A incorrectly called *C. bicephalum*). This led them to consider the three-dimensional branching as possibly of generic significance. It is most unfortunate that Kräusel and Weyland's type material of this species was destroyed during the 1939–45 war, since a re-investigation might have disclosed that its sterile and fertile appendages were both perhaps more complex than has generally been supposed.

Careful perusal of the literature dealing with the closely related *Hyenia* Nath. shows that the general accounts of this genus are somewhat out of date. In at least two modern texts (Foster and Gifford, 1959; Zimmerman, 1959) reproductions of the original reconstruction of *Hyenia elegans* (Kräusel and Weyland, 1926, Pl. 16) are still offered. A preferred, and more correct, reconstruction has been extant, however, in Germany for some time, appearing in its original form (Kräusel, 1950) or in a slightly modified version (Gothan and Weyland, 1954; Mägdefrau, 1942). This reconstruction, first published by Kräusel and Weyland in 1935 and based on information contained in an earlier paper of theirs (1932a), shows the plant to have possessed a stout creeping rhizome from which the aerial shoots arose.

The upright shoots were usually unbranched and are thought to have been wholly fertile or sterile (Leclercq, 1940; Høeg, 1945; Andrews,

1959). In the light of modern investigations, it is now known that another species of *Hyenia* had more complex sporangiophores than the type-species, and Leclercq and Andrews (1960) have questioned whether *Hyenia* and *Calamophyton* should be retained as separate genera. It seems that, at present, general habit is the only distinguishing character that may be used with assurance.

In a first contribution on some Belgian specimens called *Hyenia elegans*, Leclercq showed that the sporangiophores were not of such simple form as those of the type material (1940). Instead of the entire sporangiophore tip being fertile, with two or three sporangia on each of the recurved ends of the single bifurcation, the axis of the sporangiophore was prolonged into a long filiform appendage which was simple or forked several times. In some cases, there were similar structures on the convex side of the recurved sporangial stalk. As a result of investigations of another Devonian sphenopsid, Leclercq (1957) doubted whether her first report of the dichotomous construction of the sporangiophores of this material was correct. Re-examination of the specimens showed that it was not (Leclercq, 1959, 1961). It now appears that the construction of the sporangiophores was even more complicated, and the material has accordingly been re-named *Hyenia complexa*.

In *H. complexa* there were three pairs of sporangiophores inserted in a whorl, with successive whorls superimposed. Each sporangiophore divided into three lateral stalks, prolonged by spiniform appendages and bearing two anatropous sporangia. There was a total, therefore, of six sporangia on each sporangiophore.

Remains of what is considered to be an even earlier articulate plant have been described from the Lower Devonian of western Siberia (Ananiev, 1957; Andrews, 1959, 1961; Leclercq, 1959, 1961). These, to which the name *Protohyenia janovii* Ananiev has been assigned, consist of unbranched fragments of sterile and fertile shoots. The sterile shoots bore slender appendages which forked several times and probably functioned as leaves. On the fertile branchlets the appendages were of similar size and form, but bore pairs of sporangia at the tip of each segment. These sporangia, which were oval and about 3·0 mm long, were erect (Fig. 3, E), not recurved as in the later *Calamophyton*, *Hyenia* and *Eviostachya* Stockmans. In Ananiev's reconstruction, the shoots are shown attached to a creeping rhizome much as in *Hyenia*, but it is believed that this has yet to be confirmed.

Eviostachya hoegi, a reproductive structure of Upper Devonian age, was first described by Stockmans (1948), but its complete organization has only recently been worked out in great detail by Leclercq (1957). It consisted of a main axis carrying a proximal whorl of six bracts, above which were as many as twelve whorls of sporangiophores, the whole aggregated to form a strobilus. Strobili up to 5·5 cm in length and 0·8 cm in diameter are known, but they were usually smaller. There were three pairs of sporangiophores at each insertion, an individual pair being supplied with a trace from an angle of the triangular stele of the axis. Each sporangiophore was an elaborate structure consisting of a stalk which divided at its apex into three branches, each of these being further divided into three forming two lateral and one central branch. These nine ultimate branches were recurved and each was terminated by a cluster of three sporangia making a total of twenty-seven sporangia on each sporangiophore (Fig. 3, F). Due to the difference in length of the ultimate branches the sporangia were disposed in two circles (Fig. 3, G). Fairly long, undivided, spine-like emergences projected outwards from the convex sides of the branches. The plant which bore the strobili has yet to be discovered, but the stelar pattern points to affinities with the Sphenophyllales, and it is thought to have been homosporous since spores of only one size have been found in the sporangia.

These discoveries have thrown some light on two aspects of the early history of the *Sphenopsida*. In the first instance, comparison between *Eviostachya*, *Calamophyton bicephalum* and *Hyenia complexa* reveals a feature in the architecture of the sporangiophore which is common to all. This is the trichotomous mode of branching, an unsuspected character since earlier investigations on other species of *Calamophyton* and *Hyenia* had led to the assumption that the sporangiophores divided dichotomously. Since the type material of both *Calamophyton primaevum* and *Hyenia elegans* is no longer available, it will never be known whether the forkings of their sporangiophores were dichotomous, as original studies suggested, or trichotomous as in the more recently instituted species. *Calamophyton renieri* had dichotomously forked sporangiophores, so there seems to be no special reason for doubting their presence in *C. primaevum* and *H. elegans*.

That certain of the early sphenopsids possessed dichotomous sporangiophores seems fairly assured in view of the fact that the sporangiophores of a number of Carboniferous representatives were built on the dichotomous plan. As examples there may be noted *Calamostachys* Schimper, which was formerly thought to have peltate sporangiophores (Lacey, 1943), and *Protocalamostachys arranensis* which possessed sporangiophores that branched twice at their apices by cruciate dichotomies to produce four recurved pedicels, each with a single sporangium (Walton, 1949b). It seems very unlikely that forms such as these had been derived from ancestral types with trichotomous sporangiophores.

However, the trichotomous pattern is possibly still traceable in other sphenopsid cones. In *Peltastrobus reedae* (Baxter, 1950), for example, there were six to nine peltate sporangiophores in each whorl. These bore thirty to forty sporangia inserted in two or three concentric circles on the adaxial margin of the disc, the outer ring pedicellate, the inner shortly-stalked or sessile. This arrangement is not so very far removed from that in *Eviostachya*.

It is not difficult to visualize the existence of two parallel trends in the evolution of the sphenopsid sporangiophore, both diverging from a common ancestral type with primitive dichotomous fertile telome trusses. One trend has retained the primitive cruciate dichotomous form of construction, while the other quickly acquired a fixed trichotomous pattern as a result of a very early, unequal condensation of the fertile telome trusses of the ancestral type. *Protohyenia* is of considerable interest in that it exhibits a stage at which the sporangia had not yet assumed an anatropous position.

The other feature of interest arising from recent studies is that in the *Sphenopsida* it is at once evident that the bractless cone, composed exclusively of fertile appendages, preceded cones with sterile members intercalated between the whorls of sporangiophores as was the rule in most examples from younger Palaeozoic strata. It is not intended to pursue the subject of the evolution of these latter forms, about which much has already been written. It should be noted that at least two bractless cone-types are known from Carboniferous rocks (*Protocalamostachys* and *Peltastrobus*), and a similar condition, of course, exists in the modern members of the Equisetales. There now seems less reason, therefore, for attempting to derive the equisetalean cone by a series of reductions from complex earlier types that possessed sterile whorls. It much more likely represents the survival of a primitive condition which has remained virtually unaltered through the course of time.

E. ANCIENT REPRESENTATIVES OF THE *Lycopsida*

One of the more interesting features resulting from researches on Pre-Carboniferous lycopsids has been the demonstration that the possession of forked leaves was more widespread than had generally been suspected. Such an unusual character in the group has long been authenticated for *Protolepidodendron* Krejci, which is best known from the studies of Kräusel and Weyland (1932b, 1940) on *P. scharianum* Krejci.

The genus, which had a widespread distribution and ranges through the Lower and Middle, and probably into the Upper, Devonian, was herbaceous and resembled a modern lycopod in habit and appearance, though it was rather more robust with stems up to 2·0 cm in diameter. The numerous, relatively short leaves, measuring no more than 8·0 mm in length, were arranged in a spiral and forked once. There is little doubt that they were persistent, for leafless axes do not exhibit defined abscission scars as did those of the later arborescent lepidodendra of the Carboniferous. Certain of the leaves, on what may be termed fertile branches, were more widely separated and each bore an elongated sporangium on its upper surface. Internally the axes possessed a solid, triangular protostele which was either exarch or mesarch.

Not many years ago a new genus *Colpodexylon*, based on two species, was described by Banks (1944, 1960) from the Middle and Upper Devonian of North America. This was also herbaceous, with stems somewhat larger and longer than those of *Protolepidodendron*. The leaves, measuring up to 3·0 cm in length, formed a dense clothing to the axes and were inserted in tight spirals or "pseudo-whorls". They were unusual in being

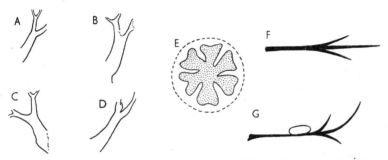

Fig. 4. Early lycopsids. A–D. Dichotomous leaves of *Sugambrophyton pilgeri* Schmidt (traced from photographic plates in Schmidt, 1954). E. *Colpodexylon deatsii* Banks; T.S. stem showing actinoxylic primary strand (redrawn from Banks, 1960). F. *Colpodexylon deatsii*; leaf showing characteristic forking (redrawn from Banks, 1944). G. *Colpodexylon deatsii*; sporophyll bearing an adaxial sporangium (redrawn from Banks, 1960).

three-forked; "the central projection often as long as the lamina up to the point of forking" and the "lateral projections half or less as long as the central" (Fig. 4, F). Some of the leaves bore an elliptical sporangium on the upper surface just below the fork (Fig. 4, G). The vascular structure of the stems was in the form of lobed protostele with exarch or mesarch protoxylem (Fig. 4, E).

More recently Schmidt (1954) has reported another type, *Sugambrophyton pilgeri*, with forked leaves. This was a Lower Devonian plant with erect dichotomous axes, as much as 2·6 cm in diameter, which are thought to have arisen from a creeping rhizome. Towards the base of the stem the leaves were thorn-like, but higher up they were bifurcate and still higher up they were forked two or three times (Fig. 4, A–D). Some sporangia-like bodies have been seen and are considered to have been borne individually on the upper surfaces of unmodified leaves. Nothing is known of the vascular organization.

The possession of microphylls and the adaxial position of the sporangia on the sporophylls are the two recurring features which characterize the *Lycopsida* as far back in time as fossil members of the group may be recognized. It has long been customary to consider the microphyll, with its single vein not producing a leaf-gap at its point of departure from the axial stele, as having been derived by the progressive vascularization of an enation. Most readers will be familiar with the suggested evolutionary series which starts with a type such as *Psilophyton*, entirely devoid of leaf-traces, continues through an *Asteroxylon*-type in which cortical traces end singly and abruptly in the bases of the individual leaves, and terminates in a true lycopod form with a vein running the entire length of the leaf. According to this conception the derivation of the microphyll is quite different from that of the megaphyllous leaf which has undoubtedly arisen from the modification of a telome system. The disturbance of the stem-stele by the departing leaf-traces and consequent formation of leaf-gaps exactly similar to those formed when branching of the stems occurs, instantly demonstrates the homology of the leaf and stem in megaphyllous plants.

In seeking to apply his Telome Theory to all vascular plants, Zimmerman has also conceived the origin of the *Lycopsida* from *Rhynia*-type ancestors, and considers the microphyll to represent a reduced telome system. Two important facts seem to point away from this hypothesis: lack of evidence from the fossil record and the absence of leaf-gaps in all members of the group. With the new discoveries already referred to above, it seems that at last the fossils are providing some evidence in favour of Zimmerman's concept. They do not, of course, offer any solution about when, or how, the close adaxial association between sporangium and sporophyll occurred, but they do indicate that the microphyll may have been derived by reduction of a small dichotomous telome system, rather than by vascularization of a small enation. Forked leaves are known in the Carboniferous genus *Eleutherophyllum* Stur (Gothan and Weyland, 1954; Gothan and Zimmerman, 1936; Zimmerman, 1930), which suggests a lingering archaic feature, while the well-known Devonian *Drepanophycus* has been shown by Halle (1936) to have possessed a forked vein in some of its leaves, which favours an interpretation that reduction has occurred. It seems moreover unlikely that the somewhat complicated leaves of *Colpodexylon* could have arisen as enations. Nor is the absence of leaf-gaps particularly difficult to explain if an assumption be made that the telomes destined to evolve into the microphyll had already assumed their new organ status very early in the development of the group, prior to any great modification, such as medullation, of the primary stelar structure of the axes.

In general, studies of early lycopsids indicate that there was a stock of somewhat similar forms, from which there were emerging others with characters that were later to become fixed in the various types of the later Periods. Apart from the adaxial position of the sporangia, they all shared strong basic similarities such as spirally arranged, persistent leaves and the possession of a solid, lobed or ridged, cylinder of xylem with the protoxylem in the exarch, or slightly mesarch, position. They were apparently devoid of secondary wood, and must have been herbaceous in habit, for well-defined leaf-cushions were not yet a feature of defoliated axes. Some of these gave rise to other herbaceous forms, and the habit became stabilized and retained until the present day, but others very soon evolved the arborescent habit which was quickly established as the dominant type during the Carboniferous.

IV. Ontogenetic Studies of Petrified Plant Remains

Over the years a great deal of information about the internal structure of a number of fossil plants has been amassed from the study of petrified remains. Preservation in petrified material is such that an investigator is enabled to prepare thin sections comparable with those made from a living plant. The technique is obviously different, but the end result is much the same and it is often possible to carry out highly detailed and extensive analyses of the anatomy of a fossil plant.

More often than not, however, the material is fragmentary and there is a complete absence of external features, which are obstacles that prevent satisfactory identification and correlation. Sometimes relatively few sections are obtainable, and palaeobotanists reluctantly assign a specific name to the material, though with no more intention than to provide a label and ready means of reference. Thus the literature abounds with a multiplicity of specific forms which, at best, can be regarded as no more than anatomical types.

This aspect of the study of fossil plants has not always been fully appreciated by non-specialists, and there has been a tendency for confusion to arise over the interpretation of a specific name. Quite often the limits of a palaeobotanical "species" have not been accurately maintained and a name has come, through general usage, to mean more than it originally implied, and has even been used for a whole plant. This is unfortunate, for it disregards the possibility of variations in anatomical structure, especially of the stele, which may occur at other levels in the plant and differ quite markedly from that at the level to which the name most strictly applies.

That such variations did not occur in fossil plants seems most improbable, since they commonly occur in living plants. Modern ferns provide very good examples of changing stelar pattern at different levels in the stem, and it seems hardly likely that fossil plants would have behaved otherwise, though they may have possessed other, and perhaps unusual, patterns of internal construction. A study of this so-called "ontogenetic variability" (Morgan, 1959) now forms the basis of a new and very valuable technique in palaeobotany. The method has already been applied with great success to some genera of petrified plants (Delevoryas, 1955; Morgan, 1959; Eggert, 1961), and it will certainly have widespread applications in the future. It seeks to separate the structural change in an individual, at different levels in the plant body, from true specific differences between individuals. The success of any ontogenetic study must be limited, of course, by the availability of material, and it is obvious that the more sections there are to hand for comparison the more nearly correct will be any interpretation that may be made. An important point, which has already emerged from the new approach, is that many previously described anatomical "species" appear to be nothing more than growth stages of a single one.

A. ONTOGENY OF THE CARBONIFEROUS ARBORESCENT Lycopsida

The studies of Eggert in a contribution bearing the title adopted for this section have successfully provided an elucidation of the morphology of the lepidodendra (Eggert, 1961).

In a survey of a large number of stem sections he has found that, in general, small primary xylem cylinders tend to be solid protosteles and large ones tend to be siphonosteles, while mixed protosteles occupy an approximately intermediate size range. Such a finding was not altogether unexpected for it had already been noted in other material (Arnold, 1940a; Delevoryas, 1957; Felix, 1952; Pannell, 1942), and the relationship between stelar medullation and stelar diameter is a well-known phenomenon in ferns. It appears also that any stelar type with almost any diameter of primary cylinder may, or may not, have secondary xylem, and Eggert comes to the conclusion that secondary xylem is of no use as a taxonomic criterion. Instead, it is a measure of the stage of development of any isolated section, broadly reflecting its position in the plant and the relative age of that part of the stem.

Towards the end of last century, Williamson had already realized the relationship between age and level in determining the significance of the extent of secondary vascular tissues, but he failed to provide a correct interpretation regarding the development of the primary body. His schematic representation of the morphology of a lepidodendron correctly shows the maximum development of secondary xylem at the base of the aerial stem, with a gradual decrease in thickness towards the apices of the branches. It errs, however, in representing the primary stele as steadily changing from a large siphonostele in the older, basal part of the stem, where pith and primary xylem have their maximum dimensions, to a slender solid protostele in the relatively immature terminal branches, with accompanying decrease in diameter and in the amount of component tissues. This interpretation of the primary structure was based on the postulate that as the axes aged there had been a "de-differentiation" of tissues to produce the siphonostelic condition, followed by divisions in the medulla and xylem which gave rise to greater amounts of these tissues in the larger primary cylinders. Such a condition is wholly unsupported by living plants, and there is no evidence from the fossils that it may have occurred in the extinct genera. A combination of stelar characters both primary and secondary, as suggested by Williamson would surely have produced a plant with a conical stem, much elongated it is true, but still not the cylindrical stem which is so marked a feature of the lepidodendra. Only if there had been an excessive production of secondary cortical tissues in the more distal parts of the trunk would it have maintained a virtually unchanged diameter. Granted there were large amounts of secondary cortical tissues, but they certainly did not reach their maximum development high in the trunk. Had they done so, compression fossils would show the leaf-cushions to be much distorted or even absent. This is not so. Instead, conditions are the reverse, which points to there having been increasing development of the secondary tissues towards the base.

Eggert's study has very satisfactorily demonstrated that, within the trunk up to the point at which the crown was produced, the primary vascular morphology changes from a small solid protostele at the base, to a large siphonostele, with maximum dimensions of pith and primary xylem, at the highest point. Superimposed on this acropetally expanding primary body were the secondary tissues which produced a columnar, or perhaps slightly tapering, trunk (Fig. 5, A). Therefore, large sections with small solid protosteles surrounded by a great thickness of secondary xylem come from the lowest part of the trunk, and sections with the larger primary siphonosteles showing a maximum of pith and primary xylem, but surrounded by relatively less secondary xylem, come from higher levels (Fig. 5, B and C). It appears that this interpretation may be applied to all arborescent lycopods, but it is most clearly demonstrated from material of *Lepidophloios wuenschianus* Carruthers studied by Walton (1935) some few years ago. Large trunk bases were found containing several fragments of stele in a vertical position. These must have dropped from higher levels of the trunk, and reference to the angle of departure of the leaf traces allowed correct alignment of the fragments with respect to one another. One protostelic fragment usually occurred with several siphonostelic pieces, which could be placed in a

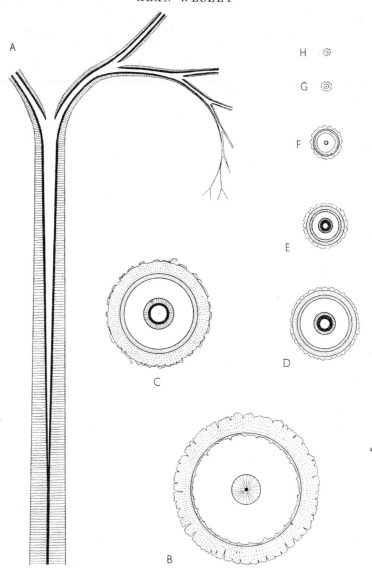

Fig. 5. A. Partial and generalized diagrammatic reconstruction of the xylem system of the aerial part of an arborescent lycopod in median L.S.; primary xylem (solid black). secondary xylem (horizontal lines) and medulla (unshaded) alone are represented. B–H. Diagrammatic transverse sections of various levels of the aerial part of an arborescent lycopod; primary xylem (solid black), secondary xylem (radiating lines), periderm (broken radiating lines). B. Base of trunk. C. Level of maximum primary body dimensions in upper part of trunk. D–F. Siphonostelic branches of the crown at successively higher levels. G and H. Protostelic twigs from progressively more distal parts of the crown. (All figures redrawn from Eggert, 1961.)

sequence showing increasing diameter of primary xylem and medulla towards the apex. On the other hand, the thickness of the secondary xylem was inversely correlated with the diameter of the primary stele.

This interpretation of the stelar morphology in the trunk allows for the explanation of the development of the plant from a small protostelic sporeling, in a similar manner to that occurring in many modern vascular plants. For such development, resulting in a progressively larger and generally more complex primary body, Eggert has instituted the term "epidogenesis".

The expansive phase of primary development is considered to have ceased with the onset of branching, from which point the primary stelar system becomes progressively reduced in size, and medullation less pronounced (Fig. 5, A). In the ultimate branches the primary stele is composed of a slender rod of primary xylem without a pith (Fig. 5, G and H). The amount of secondary xylem decreases gradually with branching until there is none in the more distal twigs (Fig. 5, D–F). For this type of continued development, characterized by distal reduction of the primary body both in size and complexity, Eggert has applied the term "apoxogenesis", and he interprets it as most likely indicating an ultimate cessation of the vegetative growth of the plant. He suggests that it also occurred in the negatively geotropic stigmarian axes of the lepidodendra.

While there can be no proof that the arborescent lycopods had a determinate plant body, Eggert considers it to have been a possibility. Numerous examples may be cited from living plants in which there is a consistent correlation between apoxogenetic growth and determinate growth, whereby the phase of elongation is terminated by maturation of the apical meristem and differentiation of its cells into some mature structure. There is also the other possibility that the apical meristem of the trunk may have produced a number of branch meristems, which themselves produced meristems for the higher orders of branching prior to the development into shoots. Rapid elongation of these branches, to produce the crown, may have then occurred before the differentiation of the stelar system, leaf-bases and leaves. Very appropriate to this suggestion of Eggert is his point that "the structure of the mature branches does not favour the presence of any extensive elongation of the branch after the differentiation of the protoxylem and leaf bases".

B. STELAR MORPHOLOGY IN THE MEDULLOSAE

Ever since the turn of the century when it was demonstrated that they were ovule-bearing plants, the Palaeozoic pteridosperms have featured prominently in discussions of the phylogeny of the seed plants. This applies especially to the two better known families, the Medullosaceae and the Lyginopteridaceae, each of which in turn has been claimed as the group ancestral to the living cycads.

Their morphology and internal structure are very fully known in some cases, since their abundant remains occur both as compressions and casts and in the petrified state. Certain features of the stelar morphology, particularly of the Medullosaceae, have long perplexed palaeobotanists, but new and extended investigations embodying the ontogenetic approach have now provided clarification of some points (Delevoryas, 1955; Stewart and Delevoryas, 1952, 1956). These studies have been based mainly on an examination of every available American specimen, and have revealed that, as might have been forecast, many of the so-called "species", including even some of recent institution (Andrews and Mamay, 1953; Baxter, 1949; Stewart, 1951; Stewart and Delevoryas, 1952), represent nothing more than different levels in the stems of relatively few distinct types. In addition, it appears that "certain of the stages of development in *Medullosa* stems indicate that it is not always possible to interpret these plants . . . on the same bases used for extant floras" (Stewart and Delevoryas, 1956).

The three stem genera of the Medullosaceae (*Medullosa* Cotta, *Colpoxylon* Brongn. and *Sutcliffia* Scott) are characteristically polystelic with the individual steles consisting of a central mass of primary xylem surrounded by secondary wood and phloem. The species vary tremendously in size and in complexity of their stelar systems. The steles may be distributed at random or they may be arranged in a polycyclic pattern. In some species, they may be elongated tangentially, while in others there may be a complete "bicollateral" cylinder enclosing other smaller steles. It appears, however, that in all medullosan stems the steles tend to divide and anastomose to a varying degree at different levels in the plant (Fig. 6), so that the exact number is not a reliable taxonomic character.

The unreliability of stelar number as a taxonomic character has been very clearly demonstrated by Delevoryas (1955) in *Medullosa noei* (Steidtmann, 1937, 1944). He has been able to observe the sequence of development of stems from a level just below the apex to an old stage typified by an abundance of secondary vascular tissue, secondary parenchyma, and an absence of leaf-bases that had been sloughed off as the stem expanded in diameter. It is essentially a three-steled species, and now includes specimens formerly called *M. distelica* (Schopf, 1939) and *M. pandurata* (Stewart, 1951), both of which represent nothing other than earlier stages of growth. In a young region, the steles are small and there is little development of primary and secondary tissues.

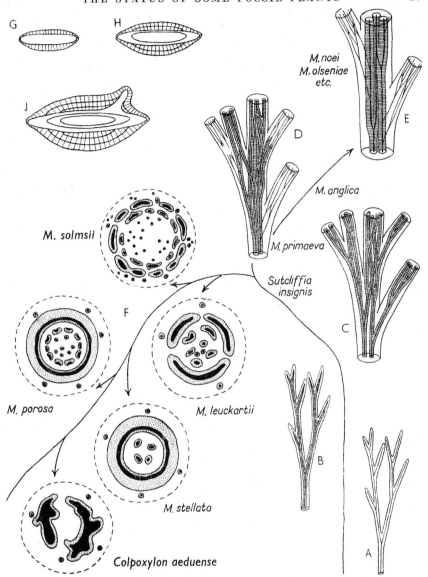

FIG. 6. Stelar morphology in the *Medullosae*. A–F. Evolutionary developments. A. Dichotomous branching system. B. Acquisition of secondary vascular tissue. C. syngenesis of telomes—retention of stelar dichotomy. D. Stelar dichotomy and fusion—some reduction in leaf-traces. E. Phyletic fusion of steles—leaf-traces primary. F. Tangential expansion of peripheral steles, followed by phetic fusion (primary xylem solid black, secondary xylem dotted). G–J. Sequence of enlargement of primary and secondary xylem of the steles of *Medullosa noei* Steidtmann with increasing age; primary xylem unshaded, secondary xylem indicated by radiating lines, former extents of primary and secondary xylem indicated by concentric lines. (All figures redrawn from Delevoryas, 1955.)

As the stem became older there was an increase in thickness of the secondary tissues as might be expected, as well as an increase in the amount of primary xylem. There seems to be a very definite relationship between size of primary xylem and development of secondary wood, both increasing with age (Fig. 6, G–J). This is a unique feature amongst vascular plants, and Delevoryas has suggested that the primary xylem retained some potentiality for increase in volume, even after secondary wood had started to develop. He considers that the rather large amount of parenchyma-like tissue in the primary xylem represents a persistent pro-cambium which divided and produced further xylem elements after elongation had ceased.

Medullosa primaeva (Baxter, 1949) offers a striking example of change in stelar number, which may be quite rapid over a short distance. In a specimen of this species, to which the name *M. heterostelica* was originally given (Stewart and Delevoryas, 1952), it was shown that there were two steles in the part immediately beneath the node. In the nodal region, there was a prolific ramification to a total of twenty-three steles, some of which functioned as leaf-traces, while others fused to reconstitute a smaller number, probably two, at a higher level. It appears that there was also some enlargement of the primary steles during ontogeny, but not so much as in *M. noei*. Delevoryas has shown, too, that other conspecific forms of *M. primaeva* are *M. elongata* (Baxter, 1949), *M. grandis* (Andrews and Mamay, 1953) and *M. anglica* var. *ioensis* (Andrews and Kernen, 1946), as well as some of the specimens which Andrews and Mamay (1953) identified as *M. thompsonii* (Andrews, 1945) and *M. distelica*.

The vascular supply to a medullosan leaf consists of several strands, which arise at differing levels as branches from the main axial steles and then pass upwards and outwards to the leaf-bases and petioles where they divide and fuse many times. Within the leaf there is a repetition of this sequence of events, and the traces to the pinnae are formed simply as branches from the vascular strands of the petiole. There is thus a striking similarity to stem branches. A big difference, however, is that stem steles are encircled by secondary wood, while leaf-traces are usually wholly primary at their point of origin. This difference is less apparent in a type such as *M. primaeva* where certain of the leaf-traces are nothing more than branch steles invested with secondary xylem which is not lost until higher levels. Even in *M. anglica* Scott, in which the leaf-traces are somewhat differentiated from stem steles, there is a proximal investment of secondary wood. These points have led Delevoryas to claim that the leaf-trace with secondary vascular tissue, especially when it most resembles a stem stele, represents the primitive condition in the Medullosaceae. The primary trace in other species, where it is least like a stem stele, is to be considered as the derived condition.

The similarity of the vascular architecture in both the stems and the leaves of certain forms lends great support to an interpretation that they are homologous structures, and that the leaves are modified branches or telome systems. An immediate ancestral type to the known members of the Medullosaceae may be conceived as a form "with many steles, some of which branched and were incorporated into lateral axes, and there continued to divide" (Delevoryas, 1955). All the steles of this hypothetical type would have been surrounded by a cylinder of second-ary xylem (Fig. 6 C), since the more primitive form of leaf is thought to have been the one in which there is little difference between the stem steles and the bundles entering a leaf. Specialization of the lateral branches, which were destined to become modified as leaves, involved a progressive loss of secondary vascular tissues as the leaf-traces became differentiated. As additional evidence for this proposed line of develop-ment, Delevoryas cites an example from *M. primaeva* which "appears to show a stem system 'caught in the act' of becoming differentiated into a leaf". In this particular case, the vascular structure of the branch showed no recognizable difference from that occasionally seen in some of the leaves of the same species.

Paralleling this differentiation of the leaf-traces there is an obvious trend towards reduction in the number of stem steles. Delevoryas considers that this is the result of gradual phyletic fusion, and a com-bination of the two processes forms the basis of one of the two distinct patterns of evolutionary development which he recognizes in the Medullosaceae. All the Carboniferous members of the genus *Medullosa*, and one from the Permian, are examples of this line of development which has its origins in a primitive type such as *M. primaeva*, passes through an intermediate stage as represented by *M. anglica*, and ter-minates in a group of advanced forms which includes such species as *M. noei* and *M. olseniae* (Roberts and Barghoorn, 1952) characterized by wholly primary leaf-traces (Fig. 6, D and E).

The other pattern of development involves a tangential expansion and lateral fusion of the peripheral stem steles to form ultimately a hollow stele, together with retention of secondary elements around the leaf-traces (Fig. 6, F). This line includes all the European Permian forms and *Colpoxylon aeduense* Brongn. The basal member of the line is repre-sented by a type such as *M. leuckartii* Goepp. and Stenz., which had several tangentially expanded peripheral steles surrounding a number of smaller, nearly cylindrical or only slightly expanded, steles. From this, by lateral fusion of the outer ring of steles and some reduction of the enclosed central strands, *M. stellata* Cotta may have been derived, and ultimately some form in which the central members had disappeared. *Colpoxylon* comes nearest to this condition, but its large, irregularly outlined tangentially expanded steles had never reached the stage at which lateral fusion had occurred. The other two species, *M. solmsii* Schenk, in which there are usually two peripheral cycles of slightly expanded steles surrounding a number of smaller cylindrical strands, and *M. porosa* Cotta, which had a peripheral cylinder enclosing one or more rings of slightly expanded and often strongly endocentric bundles as well as small central strands, fit very well in intermediate positions in this scheme of development.

These two lines of evolution are considered by Delevoryas to have diverged from his hypothetical ancestral type, already referred to, with many steles surrounded by secondary wood in both the main axes and lateral branches. This type, in its turn, is derived from the Psilophytales through a line which is characterized by the acquisition of secondary vascular tissues followed by a fusion of some of the axes, but with a retention of dichotomizing stelar systems (Fig. 6, A–C). An anastomosis of the vascular system then followed. One of the elementary processes of the Telome Theory is syngenesis of telomes resulting in several dichotomizing stelar cylinders within a single axis. It very appropriately applies here. Further, *Sutcliffia insignis* Scott possesses a stelar system which retains many characteristics of a form representing dichotomizing vascular cylinders within fused axes, and it fits suitably into the scheme somewhere at this point.

Though the arguments in favour of the polystelic types as the more primitive members of the family seem most convincing, it must be admitted that there is a complete lack of information about forms presumed to be intermediate between these and the Psilophytales. It is possible that there was a connection as visualized by Delevoryas, but it should be remembered that other polystelic types, such as *Xenocladia* (Arnold, 1940b, 1952), *Steloxylon* Solms-Laubach and *Cladoxylon* Unger (summarily surveyed and illustrated recently by Andrews and Mamay, 1955, and Andrews, 1961), were in existence during the late Devonian and early Carboniferous, and may have included the ancestral forms of the Medullosaceae.

At this point it becomes necessary to consider whether these latest notions about the evolution of the Medullosaceae significantly alter ideas of the origin of the cycads. Arnold's very useful discussion of the relative merits of the two rival theories gives a reasonably clear indication of current attitudes (1953), so that only a brief summary of their main points is needed here.

Worsdell (1906) postulated a medullosan origin and considered that the vascular system of the living group consisted of one-sided remnants of a number of steles which, during the passage of time, had lost the cambial activity on their inner sides. He based his theory largely on the structure of the cotyledonary node, where there are concentric stele-like bundles arranged around a larger central one, and interpreted the bundle complex of the cone stalk of *Stangeria paradoxa* and the occasional, small and anomalous, internal bundles of certain genera with reverse orientation of xylem and phloem, as the relics of concentric cylinders normal in the ancestral type. He also believed that the lyginopterid-type had developed by reduction from a medullosan ancestor.

Scott (1909), on the other hand, favoured a lyginopterid origin for the cycads from a type similar to *Lyginopteris*, in which there was a ring of collateral bundles in the stem. He considered that the vascular system of *Lyginopteris* was a medullated monostele which had developed from a protostele, and that the centripetal metaxylem was its last remnant. *Heterangium* Corda, another member of the Lyginopteridaceae, had a mixed protostele, as did the lately discovered *Microspermopteris* (Baxter, 1949). These together surely make Scott's theory more attractive than that of Worsdell. Two strong objections may also be made against Worsdell's ideas. In the first place, it is by no means certain that anatomical characters exhibited in seedlings are ancestral characters: it is possible that they may be the result of response to some physiological stimulus. Secondly, there is no proof that the collateral stem bundle of *Lyginopteris*, or the bundle in the stalk of a cone of a cycad, is the one-sided remnant of a single concentric stele.

Delevoryas and Stewart incline to the view that it is no great step from a form such as *Medullosa stellata* to the cycadean anatomy. They envisage a loss of the small internal bundles and of the inner cambial ring. Indeed, *Colpoxylon* has no inner bundles. Yet there is no later stage leading to the cycads. It seems very unlikely that these authors would attempt to derive *Lyginopteris*, as did Worsdell, from the medullosans by a further process of reduction. Why, therefore, is not similar reasoning applied to the origin of the cycads, a group which, after all, shares many anatomical characters in common with the Lyginopteridaceae? Apart from anatomy, the reproductive organs of the two groups, especially the microsporangiate organs, are sufficiently different as to render the origin of the cycads from the Medullosaceae, at least the known members, as highly improbable.

V. NOEGGERATHIALES

The exact taxonomic position of a small group of late Palaeozoic remains, comprising heterosporous cone-like organs and associated

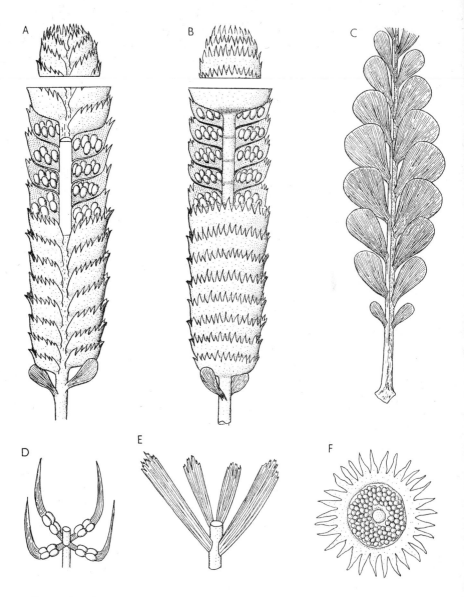

FIG. 7. A and B. Restoration of *Noeggerathiostrobus bohemicus* O. Feist. in two views at right angles to each other; in A, part of the cone is shown in median L.S.; in B, some of the sporophylls have been removed (redrawn from Halle, 1954). C. *Noeggerathia foliosa* Sternb. D. *Tingiostachya tetralocularis* Kon'no; whorl of four sporophylls with sporangia (redrawn from Zimmerman, 1959). E. *Tingia* sp.; part of a vegetative axis (redrawn from Zimmerman, 1959). F. Saucer-like sporophyll of *Discinites jongmansi* Hirmer (reconstructed from a photograph in Zimmerman, 1959).

foliage for which the name Noeggerathiales was instituted some years ago (Němejc, 1931), is still problematical. Petrifactions which would provide valuable clues have yet to be discovered, but details of the organization of the fructifications have been reasonably demonstrated from studies of compression material.

Halle (1954) has recently confirmed that the fructification known as *Noeggerathiostrobus bohemicus* O. Feist. had a fairly stout central axis which bore two opposite rows of sporophylls, that were shaped like half-saucers and fringed along their rounded outer edges (Fig. 7, A and B). On the adaxial surfaces of the sporophylls were numerous sporangia containing sixteen megaspores. Sporangia with microspores are also known, but the relative distribution of the two types of sporangia within a fructification remains ill-defined.

The foliage has long been known as *Noeggerathia foliosa* Sternb. (Fig. 7, C) but opinion is somewhat divided on the question whether the fossils represent individual pinnate fronds, extremely cycad-like in appearance, or shoot systems composed of axes with entire leaves in distichous arrangement. Fructifications and foliage organs are not known in complete organic connection, but their close association in the same plant bed and the occasional presence of laminate structures at the base of the cone, similar to those of the foliage, have generally been held as sufficiently conclusive confirmation that they belonged to the same type of plant. Zimmerman has lately given a reconstruction showing connection, but this is not based upon an actual specimen or new evidence (Zimmerman, 1959, Fig. 122A).

Three other types of cone have been included in the Noeggerathiales. *Tingiostachya* Kon'no consists of tetramerous whorls of free sporophylls, each upturned at its end and with a large tetralocular synangium on its adaxial surface (Fig. 7, D). The cones were borne in pairs on bifurcated branches of flattened shoots, called *Tingia* Halle, which carried four rows of more or less linear leaves (Fig. 7, E). *Discinites* K. Feistm. is very different, with a number of sporophyllar discs (possibly formed by a lateral fusion of several sporophylls) borne in verticils and covered on their adaxial surfaces with numerous sessile micro- and megasporangia (Fig. 7, F), the latter containing from one to sixteen megaspores according to the species. It has sometimes been thought that the *Archaeopteris*-like fronds known as *Palaeopteridium* Kidst. may have been the sterile foliage of the plant which bore *Discinites*, but much doubt has now been cast on this idea as a result of the discovery of specimens of *Discinites* in the Permian, at horizons younger than those in which *Palaeopteridium* is at present known to have occurred (Mamay, 1954). *Saarodiscites* Hirmer and its associated foliage, *Saaropteris* Hirmer are essentially similar.

3

The unusual combination of characters shown by these fossils defies their satisfactory classification within any existing taxonomic group. Yet there has been a tendency to regard them as having relationships with the *Sphenopsida*, or even as ferns. The latter, very startling, conclusion was arrived at by Hirmer 20 years ago (1940, 1941), but it is doubtful whether any botanist would be in agreement, since it presents a radical departure from the accepted principles of fern morphology and classification. Thus, according to Hirmer's arguments, the fructifications represent fertile fern fronds, but their morphology is explained by reference to features of the sterile frond, a dangerous and often misleading form of deduction (Halle, 1954). Of the reproductive organs, only *Noeggerathiostrobus* is bilateral and perhaps, therefore, had claim to remote affinities with the ferns, but all the other known forms were radially constructed and could hardly be considered as other than strobili consisting of an aggregation of sporophylls. To regard them as individual sporophylls seems quite revolutionary.

Noeggerathia may certainly have been a pinnate leaf, rather than a shoot of limited growth with distichously inserted leaves (a rare type of phyllotaxis amongst the more primitive vascular plants), but the evidence that *Noeggerathiostrobus* was borne terminally on *Noeggerathia* is not at all convincing. Some bennettitalean "flowers" had simple bracts at their bases similar to the individual pinnae of the corresponding sterile fronds, but the "flowers" were certainly not borne at the tips of the leaves. It is possible, therefore, that a similar relationship existed between *Noeggerathiostrobus* and *Noeggerathia*, and too much importance should not be attached to the similarity between the leaflets at the base of the cone and those of the sterile parts. The acceptance of Hirmer's arguments would demand that *Tingia* also be regarded as a fern frond. This cannot be so since there is no doubt that Tingia was a shoot.

Relationships with the *Sphenopsida* seem equally remote. There is a complete absence of articulation from both the strobili and the sterile parts. There is no suggestion of such a character in *Tingia*, or in *Noeggerathia* even if one favours regarding it as a shoot. Certainly the fern-like foliage of *Palaeopteridium* and *Saaropteris* can hardly have belonged to articulate plants. The main argument for an alliance with the *Sphenopsida* has been based on the supposed stalk beneath the sporangia of *Tingiostachya* but, as noted above, the sterile parts are not articulate. The sessile sporangia of the other three cone-genera also seem to invalidate this argument.

Halle came to the conclusion that these fossils should be regarded "as an isolated group of *pteridophyta incertae sedis*", a view which has recently been reiterated by Andrews (1961). Zimmerman (1959) even places them in a separate Division, the Noeggerathiophyta, with the genera distributed amongst three Orders: Noeggerathiales (*Noeggerathiostrobus* and *Noeggerathia*), Discinitales (*Discinites*, *Palaeopteridium*, *Saarodiscites* and *Saaropteris*), and Tingiales (*Tingiostachya* and *Tingia*). Whether there is sufficient evidence to warrant this is debatable, but the separation of *Tingiostachya* and *Tingia* is plausible for they are much different from the other forms.

VI. TECHNIQUE IN MESOZOIC PALAEOBOTANY

By far the greater number of Mesozoic plant remains occur as compressions. These often appear at a first glance to be no more than unpromising-looking material from which only external form may be deduced. However, in a large number of cases, the fossils still retain some traces of the original plant material from which the cuticle or cutinized membranes, themselves very resistant to decay and virtually unaltered after many millions of years, can be separated for examination under a microscope.

The cuticle not only forms an intimate investing layer closely following the surface contours of a plant organ, but also permeates the walls of the adjacent cells to varying depths. Hence it is neither flat nor of constant thickness, and when all the non-cutinized plant substance has been removed the resulting membrane presents an exact negative replica of the features of the surface at which it was secreted. It can then give remarkably clear and accurate information about the epidermal and stomatal structure, in which "it is often possible to recognize family characters, as in the arrangement of the cells round the stomata; generic characters, as in the grouping of the stomata; and specific characters, as in certain details of cell shape" (Harris, 1956).

That the cuticle may often still be present on the surface of a compression has been known for more than a century, but its value as an aid to the identification of fossil plants, particularly of Mesozoic age, has only really become evident and fully exploited over the last 30 years. This is readily apparent in the numerous and masterful writings of Thomas, Harris and Florin, who are foremost amongst a group of workers that includes Kendall (1947, 1948, 1949a, 1949b, 1949c, 1952), Lundblad (1950, 1959), Wesley (1956, 1958), and Jacob and Jacob (1954).

In the light of the many successful investigations on Mesozoic plants, it seems that much useful information might be gained from plant remains of earlier Periods by an application of similar techniques, for a study of the cuticle allows the separation of morphologically similar organs as well as enabling isolated organs to be identified as parts of a single species. This latter aspect is of special importance, since fossil plants are rarely preserved in their entirety with all their parts attached. Though repeated association of separated organs may point to their having belonged to a single species, it is unsafe alone as evidence on which to base final conclusions, and it merely indicates that any set of organs has come from plants that grew in the same kind of vegetation.

The uses of the cuticle are not confined to gathering information about the epidermis and stomata alone. The vast subject of palynology is entirely dependent upon the fact that the walls of fossil spores and pollen grains are cutinized. It should not be forgotten, moreover, that occasionally, as in the case of seeds, a certain amount can be learnt about internal organization from a study of cutinized membranes (Harris, 1954, 1958).

A. FOSSIL CYCADS

With the discovery of abundant remains of cycad-like fronds and trunks in strata of Mesozoic age during the early and middle parts of last century, it was quickly assumed that the Cycadales were the dominant group of plants during the Period. Some years were to elapse before it was to be demonstrated that many of these fronds and trunks belonged to another Order of gymnospermous plants, the now extinct Bennettitales, which differed markedly from the cycads in the construction of their reproductive organs.

Much confusion arose and not until some character, more reliable and diagnostic than gross morphology, had been recognized, could the problem of the satisfactory identification of isolated organs, especially leaves, be resolved. It soon became clear from the intensive investigations of cuticles by Thomas and Bancroft (1913), Florin (1920, 1931, 1933) and Harris (1932) that the distinguishing feature was to be found in the structure of the stomatal apparatus. In the Bennettitales, and alone amongst living plants in the genera *Welwitschia* and *Gnetum*, the stomatal apparatus is of the syndetocheilic type and consists of two guard cells bordered laterally by two subsidiary cells, all four cells having been derived by parallel divisions of a single initial cell (Fig. 8, A). On the other hand, all other gymnosperms, both fossil and living, are characterized by the haplocheilic type of stomatal apparatus, in which the two guard cells originate from a common mother cell and the subsidiary cells become modified from certain of the neighbouring epidermal cells (Fig. 8, B).

The result has been a demonstration that a large proportion of the cycad-like fossil leaves belonged to the Bennettitales and that the previously supposed dominance of the cycads during the Mesozoic was more apparent than real. Indeed, the amount of material conclusively proved to belong to the Cycadales is very small (Harris, 1961b). It includes leaves and reproductive structures alone, for the discovery of genuine cycadean trunks has yet to be made.

The most informative fossil cycad material consists of three series of related organs from the Jurassic of Yorkshire. The first of these was assembled by Harris in 1941 and includes leaves (*Nilssonia compta* (Phillips) Bronn), microsporangiate cones (*Androstrobus manis* Harris) and ovulate cones (*Beania gracilis* Carruthers), to which a scale leaf (*Deltolepis crepidota*) was added a year later (Harris, 1942b). More recently, Thomas and Harris (1960) have assembled the other two series which comprise, on the one hand, leaves (*Nilssonia tenuinervis* Nath.), microsporangiate cones (*Androstrobus wonnacotti* Harris), ovulate cones (*Beania mamayi* Thomas and Harris) and an undescribed species of scale leaf and, on the other hand, leaves (*Pseudoctenis lanei*

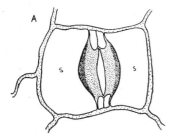 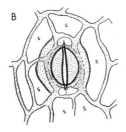

Fig. 8. Stomatal structure in *Bennettitales* and *Cycadales*. A. Generalized bennettitalean stomatal apparatus of the syndetocheilic type with two subsidiary cells (s) flanking the guard cells. B. Typical haplocheilic stomatal apparatus of a living cycad with several subsidiary cells (s).

Thomas) and microsporangiate cones (*Androstrobus prisma* Thomas and Harris).

All three forms of *Androstrobus* consist of spirally inserted microsporophylls, which broaden distally and bear numerous crowded and sessile sporangia on their undersides (Fig. 9, B–D). The type (*A. prisma*) related to the leaves called *Pseudoctenis* was fairly massive and compact, and very much like the microsporangiate cone of a modern cycad, but the other two species, which belonged to plants bearing *Nilssonia*-type leaves, were smaller and less substantial, and only measured up to 7·5 cm in length and 2·0 cm in diameter. These latter were stalkless and, since they were somewhat loosely constructed, were believed to have been pendulous like catkins (Harris, 1961b).

Pseudoctenis lanei is a pinnate leaf looking very much like one from a plant of *Zamia* or *Encephalartos*, with which it agrees in "size, form, venation, general topography of the cuticle and in the fine details of

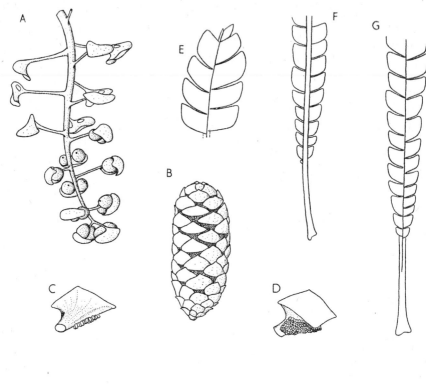

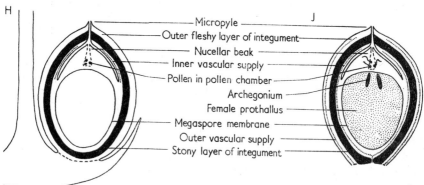

FIG. 9. Fossil cycads. A. Ovulate cone, *Beania gracilis* Carruthers, of which the upper, more mature, megasporophylls have shed their seeds (redrawn from Harris, 1941). B. Microsporangiate cone, *Androstrobus manis* Harris (redrawn from Harris, 1941). C and D. Two views of a microsporophyll of *Androstrobus manis* (redrawn from Harris, 1941). E–G. *Nilssonia compta* (Phillips) Bronn; parts of three leaves viewed from the upper surface (E and G) and below (F). H. Part of a megasporophyll of *Beania gracilis* with an attached seed showing parts so far demonstrated by maceration (based on Harris, 1941 and 1961b). J. Generalized diagram of the seed of a living cycad (redrawn from Harris, 1961b).

the epidermal cells and stomata" (Thomas and Harris, 1960). These facts, coupled with quite close agreement between various features of *Androstrobus prisma* and those of the microsporangiate cones of some living genera, are very suggestive that genuine members of the Zamioideae were already in existence by middle Mesozoic time. Until ovulate cones have been found, however, this must remain as pure speculation.

The foliage genus *Nilssonia* Brongn. comprises a large number of species, and specimens of 40·0 cm length are not uncommon. It is unlike a modern cycad leaf in gross morphology, for the lamina is divided into a number of truncate segments which vary a good deal in size and are often quite irregularly formed, so that the segmentation to one side of the stout rachis is unrelated to that on the other. It also differs in that the lamina arises from the top of the rachis instead of towards the sides, and in that there is a clean abscission from the stem at the leaf base (Fig. 9, E–G). Yet in other characters, such as venation and structure of the vascular bundles and of the cuticle, the leaf is very cycadean.

The ovulate strobili, with their attached seeds, have been known for a good number of years, but it is only during the past 20 years that really full knowledge of the construction of *Beania* and its seeds has become available (Harris, 1941). *Beania* was a loosely constructed strobilus consisting of a number of megasporophylls inserted in a spiral on the central axis (Fig. 9, A). The larger of the two Yorkshire species (*B. mamayi*) measured about 18·0 cm in length, with an axis about 10·0 mm wide at its base. *B. gracilis* reached at least 10·0 cm in length with an axis about 4·0–5·0 mm wide. The sporophylls were distantly spaced and arose at right angles from the axis. No subtending bracts were present. Each sporophyll consisted of a slender stalk which widened distally into a flattened head that bore two seeds, one on either side of the stalk, on its inwardly facing surface (Fig. 9, A). Both young and old strobili are known, and it appears that at all stages of development the strobili were always loosely constructed and open (Thomas and Harris, 1960). *Beania* has been likened to "a loose bunch of dates on their stalk" (Harris, 1961b).

Mature seeds of *B. gracilis* are oval and measure, on an average, about 1·6 cm by 1·3 cm, but those of *B. mamayi* are somewhat smaller. Harris (1941) has made a noteworthy investigation of the seed of *B. gracilis*, which has enabled him to show clearly how much it resembles the seed of a living cycad (Fig. 9, H and J). In the integument, which is only free from the nucellus at the micropylar end, two distinct layers have been recognized—an outer fleshy, possibly fibrous, zone and an inner stony one. The first macerations of seeds revealed the imprint of tracheids on the adherent cuticles of the surface of the nucellus and the inside of the integument (Harris, 1941). This led Harris to suggest that the nucellus might be vascularized, but additional material conclusively showed that the vascular strands were actually in a fleshy layer of the integument internal to the stone (Thomas and Harris, 1960). Vascular tissue has so far not been seen in any other part of the seed.

The entire outside surface of the seed, the micropylar canal, and the inner surface of the integument were cutinized. The free surface of the nucellus was also cutinized, except for a small area at its apex which was probably associated with the formation of a pollen chamber. No pollen grains were found in either the micropyles or the pollen chambers of the first seeds that Harris examined (1941), but two good examples have recently been recorded and figured which have pollen grains inside the seeds (Thomas and Harris, 1960). One of these showed numerous pollen grains near the top of the nucellus, while the other, quite re- markably, showed two germinated grains enclosed within the nucellus. The megaspore membrane is represented by a sac of cuticle extending from within the free tip of the nucellus to near the base of the seed, but nothing is known of its cellular structure. Within the limits of the information obtained from the seed of *Beania*, it is clear how fully it may be matched with the seed of a recent cycad (Fig. 9, H and J). Though some important features still remain unknown, there is no hint of a difference.

From what has been written above there should arise little doubt that there is overwhelming evidence for including all these fossils within the Cycadales. They exhibit purely cycadean characters, so that their general classification can be accepted without reservation. But within the Order the position is more difficult. The remarkable uni- formity of the microsporangiate organs of the living cycads is almost exactly repeated in the known fossil specimens. Yet the difference between the cone associated with *Pseudoctenis* and those associated with *Nilssonia* is suggestive that here are representatives of two separate groups. This idea is enhanced by the fact that the ovulate strobilus of the *Nilssonia*-bearing plant was a loosely constructed organ.

Harris (1941) originally gave a reconstruction of *Beania* as standing upright following an analogy with living cycads, but he later came to the conclusion that it more likely hung downwards like the micro- sporangiate cones. These suggestions, together with the fact that both the microsporangiate and ovulate cones are more loose and open, indeed the ovulate cone is always open, than those of the Zamioideae (to which *Androstrobus-Pseudoctenis* appears to show greater resemblance), have led Thomas and Harris to prefer a reference of *Beania* and its associated organs to a separate subfamily, the Nilssonieae. This seems a good idea, especially as the leaves show some differences in their gross morphology and insertion (clean abscission). The separation of the Nilssonieae from the Zamioideae would be very apt if the suggested tree-like habit proves to be correct; as indeed it might, since loose pendulous cones would be poorly placed on a relatively low growing trunk amongst a cluster of leaves as in most modern cycads.

VII. The Reproductive Organs of Coniferophytes

The exhaustive studies of Florin on the female reproductive organs of the cordaites and conifers (including the taxads) have resulted in the most completely worked-out evolutionary sequences in the plant kingdom. His original, and quite monumental, account appeared in a series of eight magnificent parts in Germany over the period 1938–45, and was followed by an equally important and condensed account in English (1951). Other condensed versions of selected parts of the research have also been published (Florin, 1939, 1950a, 1950b, 1954), and the outcome has been a universal acceptance of his findings.

A. CORDAITALES

Florin's observations have given a clarified picture of both the male and female reproductive organs of this group. These are generally called *Cordaianthus* Grand'eury, but Fry (1955) has recently suggested that the name ought, more correctly, to be *Cordaitanthus* O. Feist.

Both the male and female organs are now known to be built on the same plan, and consist of lax inflorescences composed of a central axis with two opposite rows of bracts in the axil of each of which was a single bud-like body (Fig. 10, A).

Each bud-like structure of the male inflorescence had a central axis with a large number of spirally arranged, uni-nerved scales; the basal ones sterile, the terminal ones (microsporophylls) fertile and with an apical cluster of elongated microsporangia arranged in two pairs or two triads (Fig. 10, B). The vascular strand was mesarch and forked dichotomously at the apex of the sporophyll to send a branch to the base of each sporangium (Fig. 10, C–H). Apart from minor differences, this basic structure is repeated in newly described American species (Delevoryas, 1953; Fry, 1956).

Two types of female inflorescences are known; an earlier primitive type of Westphalian age (*Cordaianthus pseudofluitans* Kidst.), in which the axis of the bud carried simple, sterile scales at its base, and elongate, branched and projecting, megasporophylls at its apex, each with two or more terminal ovules that were more or less pendulous and flattened (Fig. 10, J); a later, reduced type of Stephanian age (*C. zeilleri* Renault), in which short, unbranched, uni-ovulate megasporophylls were con-

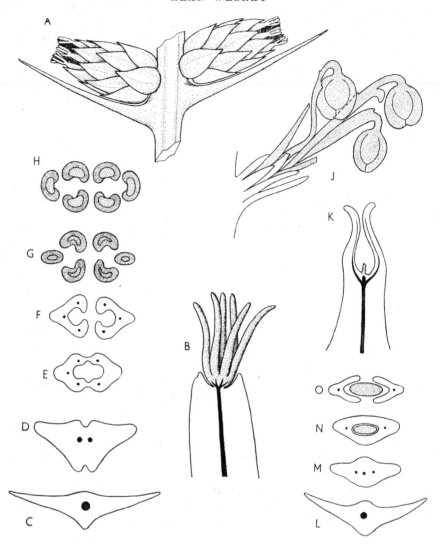

Fig. 10. Reproductive organs of the *Cordaitales*. A. Reconstruction of part of the inflorescence of *Cordaianthus concinnus* Delevoryas showing two axillary fertile shoots. The bracts of this species are unusually long (redrawn from Delevoryas, 1953). B. Distal part of a single microsporophyll of *Cordaianthus penjoni* Renault with six erect terminal sporangia (redrawn from Florin, 1938–1945, Pt. 6). C–H. Diagrammatic transverse sections through successive levels of a microsporophyll and microsporangia of *Cordaianthus penjoni* (redrawn from Florin, 1938–1945, Pt. 6). J. Single fertile shoot of *Cordaianthus pseudofluitans* Kidst.; megasporophylls and ovules dotted (redrawn from Florin, 1938–1945, P. 7). K. Apex of a megasporophyll of *Cordaianthus williamsoni* Renault with an aborting ovule (redrawn from Florin, 1938–1945, Pt. 7). L–O. Diagrammatic transverse sections through successive levels of a fertile megasporophyll and ovule of *Cordaianthus williamsoni* (redrawn from Florin, 1938–1945, Pt. 7).

cealed among sterile scales. The integuments of the ovules were single and consisted of two halves united in a median line on both sides of the nucellus. All scales had a single median bundle, but in the megasporophyll it divided into three just below an ovule. From this point the centre strand continued to the base of the sporangium and the two laterals ran into the fusing lobes of the integument. In the older forms the ovules gave rise to the platyspermic *Samaropsis*-type of seed, and in the younger to the *Cordaicarpus*-type with narrow wings.

Florin has demonstrated quite clearly that, in both the male and female inflorescences, the sporophylls and sterile scales are borne in the same genetic spiral and are therefore homologous structures and that, contrary to Renault's original interpretation, the sporophylls are never axillary to sterile bracts. Thus each bud is quite definitely a microstrobilus (or "simple flower") rather than a partial inflorescence as was thought by Renault.

The microsporophylls are clearly primitive organs, in which traces of primary radial symmetry still exist in the cruciate dichotomy of their apices and in the presence of mesarch xylem in the vascular strands. It seems more than likely that they have developed, by condensation and flattening, from a primitive radial and dichotomizing truss of fertile telomes, whose ultimate branches formed erect and elongate, cylindrical sporangia. The sporangia have retained their primitive form, but are placed terminally on a uni-nerved, leaf-like sporophyll. The microsporophylls, therefore, cannot be interpreted as metamorphozed foliage leaves, but appear to have evolved along an independent line from morphologically little differentiated telome systems. A similar interpretation is also applicable to the megasporophylls, since the fertile members of the early type forked dichotomously and cruciately as in primitive symmetrical axes. A particularly significant discovery that some of the megasporophylls of one species (*C. williamsoni* Renault) were non-functional and had aborting ovules at their apices, yet fully formed integumental lobes, indicates that the integument may be interpreted as consisting of two sterile telomes (Fig. 10, K).

B. CONIFERALES

Florin's splendid work on the fossil conifers has conclusively solved a problem that has given rise to much controversy for more than a century. This was the problem of the phylogeny of the female cone of that group, and there was much speculation about the morphology of the ovuliferous scale and its subtending bract in modern conifers such as *Pinus*. This was inevitable until the structure of the fossil representatives had been elucidated, though a majority of botanists had tended to favour the Brachyblast Theory, according to which the ovuliferous scale is a modified and reduced axillary shoot of the ovuliferous bract.

Sterile shoots of fossil conifers, of Upper Carboniferous (Stephanian) and Lower Permian age, had long been known under the general name of *Walchia* Sternb., but detailed studies of the epidermal structure and

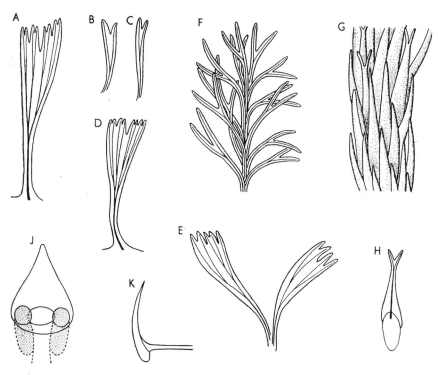

Fig. 11. Fossil conifers. A–E. Leaves on shoots of penultimate and ultimate orders of *Buriadia heterophylla* Seward and Sahni. F. Leaves on a shoot of ultimate order of *Carpentieria frondosa* (Goepp.) Florin. G. Entire and bifurcated leaves on the base of a shoot bearing an ovulate cone of *Lebachia piniformis* (Schloth. *pars*) Florin. H. Bifurcated leaf on a shoot of penultimate order of *Lebachia laxifolia* Florin. J and K. Microsporophyll of *Lebachia hypnoides* (Brongn.) Florin in surface and side views ; sporangia dotted. (All figures redrawn from Florin, 1938–1945, Pt. 6).

the discovery of female cones enabled Florin to recognize two distinct forms, for which he introduced the new genera *Lebachia* and *Ernestiodendron*. Both forms were exclusive to the northern hemisphere, and *Lebachia* was the dominant genus of early conifers with a wide distribution in Lower Permian time.

The walchias were probably very much like the living *Araucaria excelsa* in habit with apparent whorls of lateral spreading shoot systems.

These lateral shoot systems consisted of a shoot of first or penultimate order with small radially arranged leaves, together with two rows of lateral branchlets flattened into one plane. Two types of foliage leaf were present; those on the stem and lateral axes of penultimate and antepenultimate order, which were nearly always bifurcate, with a single median bundle that also forked at the apex (Fig. 11, H); and entire leaves on lateral axes of ultimate, or sometimes penultimate, order, which were uni-nerved. It seems probable that the biforked leaves represent an intermediate stage between juvenile and adult forms, and most likely gave rise to the simple condition by reduction. It would seem also that the dichotomized leaf is more primitive than the entire leaf which it has preceded in the course of evolution. This seems plausible in the light of evidence from two other contemporaneous genera, *Carpentieria* Němejc and Augusta and *Buriadia* Seward and Sahni. In *Carpentieria* all the leaves were forked once (Fig. 11, F), while in *Buriadia* those on ultimate order branchlets were usually forked once, but those on branches of penultimate order dichotomized several times (Fig. 11, A–E).

The ovulate cones of both *Lebachia* and *Ernestiodendron* were built on a common plan, and consisted of a central axis with spirally inserted, bifid bracts, each with a fertile dwarf shoot in its axil. A fertile shoot of *Lebachia* was radially symmetrical or, exceptionally, slightly flattened, and possessed a short axis with spirally inserted scales, of which all but one were usually sterile (Fig. 12, A–E). The fertile scale (seed-scale, megasporophyll) was placed in a posterior position, i.e. on the side facing the cone-axis, and carried a single, erect ovule at its tip. The ovule was flattened and had a single integument completely enclosing the nucellus and formed as a direct continuation of the megasporophyll. On the other hand, *Ernestiodendron* had more or less flattened fertile shoots. The reduced axis mostly bore only a few sterile scales, sometimes none at all, and from three to seven megasporophylls in the middle and distal region. Each megasporophyll bore a single ovule at its tip, which was either erect or inverted (Fig. 12, F and G). *Lebachia* represents the more primitive of the two types, except in the number of its seed-scales. *Ernestiodendron* is more advanced in several respects; with regard to the asymmetry of the dwarf shoot, reduction of the shoot axis and number of sterile scales, and the recurving of the sporophyll tip resulting in the inversion of the ovule in certain species. The variability of the fertile shoots, from species to species, contrasts with the uniform appearance of the bracts, but it is evident that they are homologous structures in all cases.

The elucidation of the organization of the ovulate cones of these early conifers clearly infers that the compound strobilus, or inflorescence, is

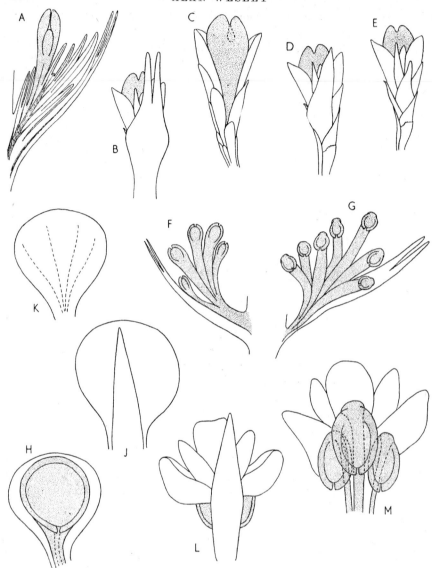

Fig. 12. Reproductive organs of fossil conifers. A–E. Single seed-scale complex of *Lebachia piniformia* (Schloth. *pars*) Florin (redrawn from Florin, 1938–1945, Pt. 6). A. in median L.S. B. viewed from cone axis (posterior side). C–E. viewed from outside (anterior side); in C the bract has been removed, and in D the larger sterile scale also. F. Single seed-scale complex of *Ernestiodendron filiciforme* (Schloth. *pars*) Florin, a species with erect ovules (redrawn from Florin, 1938–1945, Pt. 6). G. *Walchia* (*Ernestiodendron?*) *germanica* Florin, a species with anatropous ovules (redrawn from Florin, 1938–1945, Pt. 6). H–K. Seed-scale complex of *Ullmannia bronnii* Goepp. (redrawn from Florin, 1938–1945, Pt. 7). H. viewed from cone axis (posterior side). J. viewed from outside (anterior side). K. sterile part of five fused scales. L and M. Seed-scale complex of *Pseudovoltzia liebeana* (H. B. Gein.) Florin (redrawn from Florin, 1938–1945, Pt. 7). L. viewed from outside (anterior side). M. viewed from cone axis (posterior side). (In all figures the megasporophylls and ovules are dotted.)

the primary form of the female reproductive organ in the group, and that the bracts and their axillary "seed-scale complexes" had very early become united into a more or less compact cone. It would seem that the ovulate organs of their precursors were constructed very much to the plan of those of the Cordaitales, for there is no doubt whatsoever that the structures of the two groups are homologous. Yet a derivation from the Cordaitales must not be assumed, for the plan of the male organ is different.

It will be remembered, from the previous section, that the male organs of the Cordaitales were compound strobili and homologous with the females. In the walchias, on the other hand, the microsporangiate cones were simple and each represented the exact equivalent of a single seed-scale complex of their own group, or one of the fertile, axillary dwarf-shoots of both male and female *Cordaianthus*. They occurred laterally on leafy shoots, with the vegetative leaves gradually becoming replaced by microsporophylls. The individual cone was radially symmetrical and each sporophyll bore two free microsporangia (Fig. 11, J and K). There were no sterile scales. Clearly these cones have not resulted from the reduction of a compound strobilus, and it is obvious that they could not have been derived from a cordaitean type. While *Cordaianthus* gives a good clue to the basic organization of the seed-scale complex, it is certain that the Cordaitales cannot be considered ancestral to the conifers. Finally, it may be noted that Florin's researches have produced no evidence in support of the postulation of Chamberlain (1935) that the unisexual cone had been derived by reduction from a heterosporangiate ancestral type.

Evidence from later genera has revealed a gradual modification and differentiation of the seed-scale complex into a fertile part facing the cone-axis and a sterile part, the ovuliferous scale, in an anterior position. This has been accompanied by flattening of the shoot, and a change from spiral to decussate arrangement of the parts, together with a later fusion in varying degree between the fertile part, the ovuliferous scale and the subtending bract. Essentially the trend has been towards an elimination of all but a few sterile scales which became fused together as the ovuliferous scale, and the ultimate suppression of the megasporophylls which eventually became incorporated in the proximal part of the ovuliferous scale.

In the Upper Permian *Pseudovoltzia liebeana* (H. B. Gein.) Florin, the flattened seed-scale complex consisted of a short axis bearing five unequal sterile scales on its anterior side, and three fertile stalk-like megasporophylls on the posterior side, each with a single inverted ovule (Fig. 12, L and M), while in the contemporaneous *Ullmannia bronnii* Goepp. reduction and fusion had been carried further to the extent that the five sterile scales were fused in the form of a single disc-shaped structure, and the fertile part was represented by only one sporophyll with a large inverted ovule (Fig. 12, H–K). In both genera, the bract was entire, but free from its axillary complex.

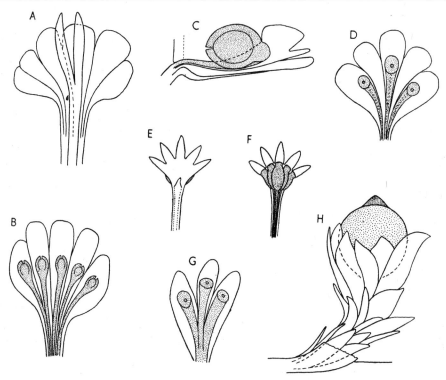

Fig. 13. Reproductive organs of fossil conifers and a taxad. A and B. Seed-scale complex of *Voltziopsis africana* Seward viewed from outside (anterior side) in A, and from cone axis (posterior side) in B. C and D. Seed-scale complex of *Voltzia* sp. C. side view showing one ovule, D. viewed from cone axis (posterior side) showing scars of three ovules. E and F. *Swedenborgia cryptomerioides* Nath.; seed-scale complex viewed from outside (anterior side) (E) and from the cone axis (posterior side) (F). G. Seed-scale complex of *Schizolepis hörensis* Antevs viewed from posterior side and showing scars of three ovules. H. *Palaeotaxus rediviva* Nath.; ovuliferous shoot. (All figures redrawn from Florin, 1938–1945, Pt. 7; megasporophylls and ovules dotted).

Several other forms are known, though they do not fit into a perfect sequence. The Triassic *Voltziopsis africana* Seward had five equal sterile scales partly fused at their bases, but it still retained five free fertile members, with probably erect ovules, and its free bract was bifurcate as in the primitive types (Fig. 13, A and B). On the other hand, *Voltzia* Brongn., with five equal sterile scales fused together for the greater part, showed an important advance in the incorporation of the stalks of the three sporophylls into the fused sterile scales, so that only the recurved ovules were free (Fig. 13, C and D). In this genus, the bract was entire, but still free. *Schizolepis* C. F. W. Braun, from the Upper Permian and Lower Jurassic, was similar to *Voltzia*, but there were only three sterile scales (Fig. 13, G). Finally, there is the Lower Jurassic *Swedenborgia cryptomerioides* Nath. which is chiefly of interest in that, though there were still five megasporophylls, the sterile part was fused not only to them but also to the bract for the greater part of its length (Fig. 13, E and F).

In conclusion it may be said that the most interesting and remarkable thing about conifers is that the evolution of the seed-scale complex had been very nearly completed by the end of the Palaeozoic, some 150 million years ago, and that relatively little evolution has occurred within the group since that time.

C. TAXALES

Although it has been the custom to classify members of this group with the conifers, Florin's studies seem to point conclusively to their having constituted a distinct taxonomic unit, at least as far as can be inferred from the fossil record (Florin, 1951, 1954).

In the modern *Taxus*, the ovule is terminal on a short lateral shoot and partly enclosed by a fleshy aril. Such a structure is recorded with virtually no change as far back as the Jurassic, and *Taxus jurassica* Florin differs little from the living species. Even in the Upper Triassic *Palaeotaxus rediviva* Nath., there were solitary ovules (Fig. 13, H), so that there seems little likelihood that the ovuliferous shoot of the taxads has been derived by reduction of the compound type of ovulate strobilus that characterizes the conifers. Yet it must not be forgotten that some of the members of the Podocarpaceae are reduced to such an extent that there are single fertile shoots. Perhaps such reduction also occurred in the taxads but, if so, then it must have occurred in very remote geological time and certainly before the Trias.

VIII. Pentoxyleae—A Unique Group of Gymnosperms

Based mainly on well-preserved and silicified material from the Jurassic of India, the Pentoxyleae are a group of gymnosperms covering a unique assemblage of fossil plants. The group was instituted in 1948 by Sahni, though detailed studies of the various organs had already been made by Srivastava (1935, 1937, 1946), Sahni (1932, 1938) and Rao (1943). More recently, Vishnu-Mittre (1953), has described what he believes to be the male organs, and on the basis of new specimens has suggested a modification of Sahni's interpretation of how the female organs were borne on the parent plant (Sahni, 1948).

The stems (*Pentoxylon sahnii* Sriv.) may reach several centimetres in diameter, and characteristically contain a ring of five, or occasionally six, large vascular bundles lying close together and occupying most of the cross-sectional area. Each xylem bundle consists of a tangentially elongated mesarch primary strand enclosed in its own zone of secondary wood. The secondary wood is remarkably endocentric, and the activity of the cambium apparently soon ceased on the cortical side (Fig. 14, C). As a result, in older stems, the bundles tend to meet in the centre and laterally, so that the pith and rays are crushed. The secondary wood, in general, is typically coniferous and has well-marked growth rings. It is compact and includes uniseriate rays. The radial walls of the tracheids show bordered pits in either uniseriate or multiseriate distribution, but nothing is known of the softer tissues which are poorly preserved or have been crushed by the secondary growth. Alternating with the main bundles were five much smaller ones consisting mainly of secondary wood, but they are not thought to have been leaf-traces. On these stems were smaller "short shoots", 5·0–7·5 mm thick, covered with an armour of elongated leaf-cushions that were rhomboidal in shape and arranged in a dense spiral (Fig. 14, B).

The leaves (*Nipaniophyllum raoi* Sahni) were coriaceous and strap-shaped, up to 7·0 cm in length but usually less than 1·0 cm broad. The lamina tapered at the base to form a narrow wing along the sides of the petiole, and the venation was of the *Taeniopteris*-type, i.e., there was a marked midrib reaching to the apex and lateral veins, unbranched or forking once, diverging almost at a right angle (Fig. 14, B). There were five to nine bundles in the petiole and midrib which were continuous with a similar number of leaf-traces that may be seen as scars on the leaf-cushions of the short shoots. The bundles are arranged in a very shallow arc and are of the diploxylic type (i.e. mesarch protoxylem separates a large mass of centripetal xylem from a smaller mass of centrifugal xylem), though the more distal and lateral veins are exarch. The traces are composed exclusively of scalariform tracheids, and the vascular anatomy of the leaf is very cycadean. Yet the architecture of the leaf-trace system differs from that of a modern cycad. It does not show the deeply Ω-shaped alignment of the bundles, nor do the bundles encircle the stem. Instead, they pursue a straight course to a leaf on the same side. The stomata, on the other hand, are syndetocheilic (see p. 36) and essentially bennettitalean.

The ovulate organ consists of a peduncle bearing a number of spirally inserted pedicels, each terminating in a cone (Fig. 14, D). The pedicels are shortly decurrent and diverge at a fairly abrupt angle. It was originally thought that they were bifurcated and are shown so in

Sahni's reconstruction (1948, Fig. 46), but Vishnu-Mittre has shown that they were unbranched (1953, Fig. 12).

Two types of cone are known, the one spherical to oblong measuring from $10{\cdot}0 \times 7{\cdot}0$ mm to as much as $20{\cdot}0 \times 10{\cdot}0$ mm (*Carnoconites compactum* Sriv.), the other cylindrical, longer, more slender and with more ovules (*C. laxum* Sriv.). There were neither megasporophylls nor interseminal scales, and the ovules were sessile and inserted directly

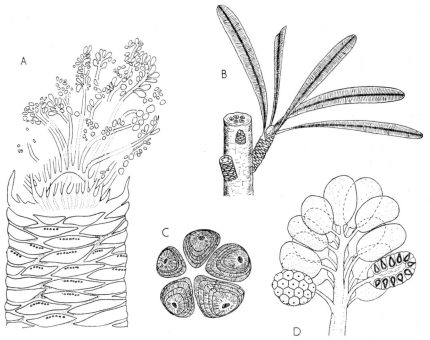

Fig. 14. The *Pentoxyleae*. A. "Male flower", *Sahnia nipaniensis* Vishnu–Mittre (re-drawn from Vishnu-Mittre, 1953). B. Reconstruction of a leafy shoot (redrawn from Sahni, 1948). C. Cross-section of the five bundles in the stem, *Pentoxylon sahnii* Sriv. (traced from photograph, Fig. 9, in Sahni, 1948). D. Reconstruction of the ovulate "infructescence", *Carnoconites compactum* Sriv. (based on Sahni, 1948, and Vishnu-Mittre, 1953).

on the cone axis. The micropyles pointed outwards, and the very thick integument of each ovule consisted of a well-developed stony layer and a very thick outer fleshy layer. The fleshy layers of the adjacent ovules were closely concrescent, so that the whole cone must have been very much like a fleshy mulberry (Fig. 14, D). The nucellus was free to the base, and a single vascular strand supplied the chalazal region and penetrated into the stony layer of the integument. Nothing is known about the mode of pollination, or of the structure of either the mega-prothallus or the embryo. Clearly the stachyosporous ovules may be likened to those of the Bennettitales, but there resemblances cease for *Carnoconites* was a compound "infructescence" and its individual cones lacked both interseminal scales and a perianth.

The "male flowers" (*Sahnia nipaniensis* Vishnu-Mittre) were borne terminally on dwarf shoots resembling those of *Pentoxylon sahnii*. The "flower" consisted of a ring of filiform, spirally branched microsporophylls which were fused at the base to form a disc that surrounded a broad, conical receptacle (Fig. 14, A). Unilocular sporangia were borne at the ends of the ultimate branchlets, and contained monocolpate pollen grains with smooth walls. Here, as in the leaves, there is an unexpected combination of characters. Though of the general bennettitalean pattern, the microsporophylls are erect and never circinate, and they are not bilateral pinnate organs, but branch in a spiral fashion. Nor are the unilocular sporangia in any sense aggregated into synangia. The sporangia alone, and their pollen grains are more cycadean.

It is clear that the Pentoxyleae must stand as an individual group, exhibiting as they do a synthesis of several features characteristic of other groups of gymnosperms. Sahni was well aware of this, and pointed out the difficulty of ascertaining the place of the fossils in a broad classification. The uniqueness of the group and its rather isolated position became even more apparent with the discovery of the microsporangiate organs. Though the fossils combine one or more characteristics of the conifers, Medullosaceae, Cycadales and Bennettitales, the differences are so great as to render their classification with any one of these groups as quite out of the question.

IX. The Fructifications of *Glossopteris*

Glossopteris, first used by Brongniart more than a hundred years ago, has been a favourite name for a large assembly of leaves of Gondwana age from the southern hemisphere. The ubiquity of the fronds and their common occurrence is recorded in the name of the flora of which they form a large constituent—the *Glossopteris* Flora. Until recently, nothing had been known of the reproductive organs of the plants that produced these leaves. But a few years ago, a considerable collection of material was described, in which several sorts of "fructification" had been discovered attached to *Glossopteris*-type leaves (Plumstead, 1952, 1956a, 1956b, 1958a, 1958b). This alone was an exciting contribution to palaeobotany, but it must be admitted that its effectiveness has been much reduced as a result of the highly provocative interpretation that was applied to the material.

The name *Glossopteris* has been used for entire, spathulate leaves with reticulate or open dichotomous venation. It includes forms either with or without a defined midrib, and characters based on venation have served for the institution of two other genera apart from *Glossopteris*. These are *Gangamopteris* McCoy and *Palaeovittaria* O. Feist. However, many of the characters of one genus merge into those of another in certain species, and it then becomes difficult to refer specimens to one or other of the genera. Much confusion still surrounds the use of the names, and the limits of many species are not at all well defined. Added confusion also arises from the fact that a number of old illustrations are not always accurate, since many of the early authors were not well served by their artists.

It seems that no real distinguishing characters can be sharply defined for the three leaf-genera, and that criteria, like the presence or absence of a midrib and the extent of anastomosis between secondary veins, are too inconsistent to be relied upon as of generic value. The conclusion is almost forced upon one, therefore, that the *Glossopteris*-type leaf represents a generalized morphological pattern that was common to a diversity of plants. This is borne out, in part by the differing types of reproductive organs, in part by cuticular studies. Studies of the cuticles of many leaves have shown that considerable differences exist between them, and it is now possible to recognize six groups, possibly of generic rank, but all agree in showing characteristics of seed plants (Surange and Srivastava, 1956; Pant, 1958). At the moment, however, there is no obvious correlation between a grouping based on epidermal characters alone and that based on morphology of the reproductive organs.

Most of our knowledge of the "fructifications" has resulted from the studies of Plumstead in South Africa, who grouped these peculiar southern hemisphere fossils in a new Class, Glossopteridae. Other studies have also been carried out in India by Sen (1954, 1955a, 1955b, 1955c, 1956), and by Thomas (1958) on younger South African material, while more recently "fructifications" have been discovered in Central Africa (Lacey, 1959, 1961) and New South Wales (Rigby, 1961).

"*Scutum* is a bilaterally symmetrical, two-sided cupule usually borne on its own short pedicel which grows, according to the species, from the midrib or from the top of the petiole of the leaf which in every other respect resembles the vegetative leaf of a species of *Glossopteris*. The fructification is believed to be axillary but the pedicel is adnate for the greater part of its length. The cupule resembles a purse" and "is supported on its edge by the pedicel". It is ovate, or rounded, in outline, measuring from 2·0 to 6·0 cm in diameter, and is thought to have been open during one stage of its development (Fig. 15, A), but usually the two halves fitted closely together. "On the outside each half consists of a central part, or head, which is thick, or raised and has fan-

FIG. 15. The fructifications of *Glossopteris*. A–C. *Scutum* sp. (redrawn from Plumstead, 1956a). A. two-sided cupule in position of growth at an immature stage. B. inwardly facing surface of "fertile" half. C. inwardly facing surface of outer "male" half with four presumed staminate organs attached. D–F. *Hirsutum* sp. (redrawn from Plumstead, 1956a). D. in position of growth showing inner "fertile" half alone. E. interior surface of empty outer half. F. interior surface of outer half at an early stage with presumed staminate organs still attached. G and H. *Ottokaria* sp.; inside views of the two halves (redrawn from Plumstead, 1956b). J and K. *Lanceolatus* sp. (redrawn from Plumstead, 1956a). J. immature stage showing cupule and swelling on midrib. K. mature stage with cupule removed. L. "*Cistella*" sp. (redrawn from Plumstead, 1958a). M and N. *Pluma* sp. (redrawn from Plumstead, 1958a). M. "Mature male fructification". N. "Mature female fructification". P. Group of fertile leaves of *Lidgettonia africana* Thomas showing stalked cupules (redrawn from Thomas, 1958).

shaped venation, and a surrounding wing which is often fluted, striated and has a dentate margin. The pedicel is attached to the base of the head and not to the edge of the wing". On the inside of the half nearest the leaf, "called the fertile half" (Fig. 15, B), there is a number of "small oval sacs embedded in the tissue of the central head, whilst in the vast majority of cases the opposite half is empty and apparently only protective. Its head area is concave . . ." (Plumstead, 1956a). The so-called sacs usually measure about 1·5–2·0 mm in length, and "do not separate when ripe but burst with a crater-like opening, and contain a round hard core . . ." (Plumstead, 1952). In an extremely small number of specimens "the inside of the fertile half exhibits sacs which are immature and ring-shaped instead of swollen and oval, in addition there is a small mark or dot in the centre of each". The other half "instead of being empty, bears a varying number of long, broad projections . . . . These grow from the head area and extend well beyond the margin of the wing" (Fig. 15, C), and are regarded by Plumstead as being "extremely short lived and the purpose they served, very temporary" since only eight, out of more than two hundred, specimens showed such structures (1956a).

The closely similar *Hirsutum* was originally described as a species of *Scutum*, but the discovery of numerous, thin filaments, instead of the larger broad projections, on the inner face of the outer half of the cupule during its open stage suggested to Plumstead its separation as a distinct type (1958a) (Fig. 15, D–F).

New specimens of *Ottokaria*, a genus described many years previously by Zeiller (1902) and Seward and Sahni (1920), were also shown by Plumstead to be built to a similar pattern. The nature of this fossil genus had been problematical and remained a mystery, until the South African material revealed that it, too, was basically a cupulate structure of *Scutum*-type (Plumstead, 1956b). It consists of a round head, several centimetres in diameter, supported on a long, strong pedicel which is possibly axillary in origin. The head consists of "two round disc-like halves", of which the half nearer the associated leaf consists on its inside of "a central depressed area covered with small oval objects and surrounded by a raised rim of short thick and fleshy individual *bracts* which differ in size (Fig. 15, H). These so-called "bracts" correspond in position with the continuous wing on the fertile half of the cupule of *Scutum*. The other, protective half of the cupule has a strong fan-shaped venation radiating from its base and a dentate margin, but there are no "bracts" (Fig. 15, G). Both halves of the cupule are fused, but there is some suggestion that they may have been open at some stage. The Indian type is associated with a species of *Glossopteris*, but the South African specimens are in association with leaves of *Gangamopteris*.

Cistella, like the three previous genera, also has two halves, but neither half has any wing. Its pedicel appears to have been adnate to the midrib of the leaf for its entire length (Fig. 15, L). The name *Cistella* has no validity since it has already been used for both an orchid (*Cistella* Blume, 1825) and a discomycetous fungus (*Cistella* Quélet, 1886) (Cash and Stevenson, 1961). As the whole problem of the nomenclature of the fossils will eventually have to be clarified, there is, however, no value in making a change of name at this point.

The most intriguing reproductive organ is that named *Lanceolatus*. Two South African species are known, of which one is attached to a species of *Glossopteris*, and the other to a species of *Palaeovittaria*. It characteristically occurs on the surface of the leaf, about one-third from the base of the blade, and the whole pedicel is adnate as well as the fertile half. The lanceolate protective "cupule" is free, except at its base, during the earlier stages of development (Fig. 15, J), but in mature stages it becomes fused to the leaf and covers the fertile area. This fertile area exhibits a unique feature in that "the greater part of its development is believed to have occurred below the epidermis for, except in the most mature stages, the venation of the leaf stretches over it" (Plumstead, 1956a). When mature, the raised surface of the fertile area is described as "covered by the veined epidermis of the leaf but on top of this an additional layer or membrane which carries small oval sacs. Finally, when the head is ripe, no venation is visible; the surface is dented and the sacs have become flattened discs" (Fig. 15, K). These features are representative of *Lanceolatus lerouxides*, the type-species occurring on a *Glossopteris* leaf, but in the less clear *L. palaeovittarius*, borne on a leaf of *Palaeovittaria*-type, the veins part around and do not cover the fertile area on which sacs have yet to be recorded.

The specimens, to which the name *Pluma* has been given, are completely unlike the previous five types, and their inclusion in the same group is based on their supposed attachment to leaves called *Glossopteris*. The better, but nevertheless incompletely, known species, *Pluma longicaulis*, occurs in two forms, each of which is said to be attached to separate leaves of a single species of *Glossopteris*. The one, regarded as equivalent to the protective half of the other genera, consists of an axis which is attached to the petiole of the leaf. It arches outwards, and enlarges to a head area from which droops a fringe of elongated structures which are interpreted as equivalent to the "long, broad projections" on the protective half of *Scutum*, or the numerous thin filaments in a similar position on *Hirsutum* (Plumstead, 1958a) (Fig. 15, M). The other, regarded by Plumstead as the female fructification, is equivalent to the fertile half of the other types, but lacks an enveloping cupule. It consists of a group of rounded objects which are thought to be similar to the sacs on the fertile half of a cupulate type (Fig. 15, N).

Though the preservation of the South African material leaves much to be desired, there seems little doubt, considering the number of specimens which have been examined, that the reproductive organs were borne either upon, or in very intimate association with, *Glossopteris*-type leaves, and that they show varying degrees of adnation to the leaf. In certain cases, it is true, there is no conclusive evidence of attachment, but in others it is so strong that it can hardly be doubted. The bilateral construction of all the forms, except *Pluma*, seems sufficiently convincing. Yet it has been suggested that the fertile half, nearer to the leaf, might represent a greatly compressed strobilus, and that the cupule, with wing, represents the flattened, laminate extensions of the distal ends of sac-like bodies on the strobilar axis (Walton, in Plumstead, 1952). In *Ottokaria*, there is a possibility that the wing of the cupule consists of a number of individual segments, but in *Scutum* and *Hirsutum* there is little to suggest that the wing was other than entire, while in *Cistella* there is a complete lack of a wing around the cupule. It seems, therefore, that the cupule was a discrete structure and had not been formed by the flattening together, during fossilization, of a mass of separate parts.

In her first publication Plumstead (1952) advanced the notion that the fertile half of *Scutum* was a megasporophyll, and that the sacs contained tiny seeds. She tells how originally the sacs were believed to have been of sporangial nature, but all efforts to obtain conclusive proof of the presence of spores had produced negative results. Additional, closed specimens then revealed sacs in every stage of development, including several in which the sacs had burst. Some contained a small central core, while others were empty. Each sac was regarded as containing a seed, rather than being of sporangial nature, as she thought it unlikely that sporangia would become enclosed in a sealed cover. The outer half of the cupule was regarded as entirely protective in function, particularly as it became sealed with the sac-bearing, or fertile, half. It was suggested, therefore, that the two halves must have been partly open until fertilization had occurred, and that they then became fused, or in very intimate contact, and the closed organ "served as a carpel in which the seeds developed". A similar interpretation was applied to *Lanceolatus*, and both types of structure were regarded as female organs, and the plants which bore them as pteridosperms.

With the discovery of a few open specimens of *Scutum* bearing "long, broad projections" on the protective half, and swollen sacs with "a small mark or dot in the centre of each" (1956a), the claim was made that they were bisexual fructifications and that these particular specimens represented the early pollination, or "floral", stage. Maceration of the little remaining carbonaceous matter on the "projection" revealed grains which, since they lacked tri-radiate markings and wings, were considered as pollen grains rather than microspores, and the "projections" were accordingly interpreted as staminate organs. Hence, the outer half, though protective throughout most of its existence, also functioned as the male for a short period. The small mark in the centre of each of the sacs was considered as being possibly connected with fertilization, and representing an opening or tube through which the pollen could enter; i.e. that it "may be analogous with a stigma or possibly the remains or base of a style". Thus Plumstead believed that the whole reproductive organ was a hermaphrodite flower, of which one side bears the gynoecium on its inner surface, and the other the androecium, showing a combination of characters more of the angiosperms than any known gymnosperm.

From what has been summarized in the few preceding paragraphs, it must be at once obvious why the effectiveness of Plumstead's contributions has been much reduced. For there is not only an unfortunate misuse of certain descriptive terms of standard botanical terminology, but also a sad misconception of what constitutes an angiospermous flower. Nevertheless, the factual information which has been presented stands as a very important and valuable advance in palaeobotany.

It has already been pointed out that five of the reproductive genera were, more than likely, built to a bilateral plan. There seems little doubt, too, that the fertile half was the side of the fructification nearer the leaf, but beyond this little more is really known. There is no evidence that they were seed-producing, for though the sac-like structures on the fertile halves may have contained seeds, it is equally possible that the sacs were true sporangia. Nor is there any evidence that they were bisexual. Much doubt exists about the supposed staminate organs, since there is no proof whatsoever that the pollen found on those of *Scutum* had originated from inside.

Even if it is assumed for the moment that these fossils are the remains of some novel type of ancient angiosperm, it must not be forgotten that they were of a very specialized nature, and must have evolved far beyond a "plastic" stage from which other forms might have arisen. They clearly represent the end of an evolutionary line, but should it eventually be shown otherwise and that they were indeed true angiosperms, then modern ideas about the origin of the living angiosperms will need very serious reconsideration.

Nevertheless, Melville has accepted the evidence of *Scutum* and incorporated it as part of one of the lineages in his "gonophyll theory" of the angiosperm flower (1960). This theory is based on the conviction that the basic structure of the angiosperm ovary was a leaf bearing a dichotomous fertile branch on its midrib or petiole. If this so-called gonophyll were bisexual (i.e. an "androgynophyll"), condensation from a branch system with sporangia epiphyllous on leaflets could take the form of a leaf bearing a pair of leaf-like structures, one male, the other female. These would be opposed on the primary dichotomy of the epiphyllous branch, and their ovules and microsporangia would be attached to the opposing surfaces exactly as in *Scutum*. While the new theory is to be welcomed, too much weight should not be attached to *Scutum* since so little is known about it.

This section on the glossopterid fructifications would not be complete without reference to the studies of Sen and Thomas.

Sen's material includes some specimens which are essentially similar to some of the South African forms, but he has also described some which are quite different and which he believes to be microsporangia (Sen, 1956). These are small cutinized sporangia and associated pollen grains of *Pityosporites*-type which have been isolated from coal. For a long time, Indian palaeobotanists have assumed that the very common *Pityosporites*-type of pollen belonged to *Glossopteris*, though the assumption has only been based on association. Now Sen believes that his sporangia, from their similarity in size and shape, could have come from the "sori" that have been recorded on many leaves of *Glossopteris*, and that they are the male reproductive organs (Sen, 1955c). However, direct evidence is not available and the conclusions remain unproved. Plumstead suggests that the "sori" might be fungal associations.

The appreciably younger *Lidgettonia* is a somewhat different reproductive organ from South Africa (Thomas, 1958). The fertile leaves are quite small (about one-third the length of the associated sterile *Glossopteris* leaves), and bear a number of slenderly stalked campanulate, or peltate disc-like, bodies, 5·0–6·0 mm in diameter, on one of the surfaces (Fig. 15, P). In all probability these are reproductive structures, but there is no direct evidence as to the nature of the bodies borne in the cupules. However, small elongated sporangia and seeds occur in considerable abundance in the matrix around the fertile leaves, and there is therefore likelihood that the cupules may have contained one or other, or even both, of these structures. Clearly these are generically different from the older types, and it may be that they do not belong to the same group of plants. The only feature in common is that they were all produced on foliar organs.

X. Distribution of Floras Prior to the Tertiary Period

A brief survey of some current ideas about the distribution of early fossil floras is not without place, for certain of the fairly recent summaries (e.g. Just, 1952; Krishnan, 1954) may not easily be available, and the general reader may not be fully acquainted with the essential problems involved.

It is well to remember the words of Sir Joseph Hooker that "we have not in the fossilized condition a fraction of the plants that have existed, and not a fraction of those we have are recognizable specifically". Also, since the accumulation of sediments is most permanent at low altitudes, where erosion is least active, the fossil record must be limited largely to lowland habitats. This means that the overall picture of the vegetation at any one moment in time must necessarily always be incomplete. To these difficulties must be added the great problem of correlating floras of roughly comparable ages, but of distant geographical locations. Not only must their position in the vertical rock series be considered, but also the factor of latitude (and hence climatic conditions). The time/space relationship is very important, since similar floras occurring in different areas may just as likely be homotaxial as synchronous, and any discrepancy in age as determined by other data will then represent the time required for migration, or expansion, from the older to the younger locality. This is of particular importance if two similar floras are separated by wide latitudinal range. For while it is doubtlessly true that ancient climatic zones may have been less distinct than at present, there is no basis for supposing that at any time in the past has the influence of latitude on land climate and vegetation been negligible. It also seems unlikely that conditions responsible for the growth of terrestrial plants could ever have been uniform over a very wide range of latitude.

A. PALAEOZOIC VEGETATION

Available evidence suggests that the earliest floras were cosmopolitan, since no regional differences occurred in such widely separated stations as South America, U.S.A., Europe, Spitzbergen, Africa and Australasia. The main records are of a particular type of environment, however, and there is reason to believe that the uniformity of the floras may be more apparent than real. Not until Carboniferous time is there a manifest differentiation into more restricted local groups, and from then onwards world-wide uniformity of vegetation was never again achieved.

1. The earlier floras

The earliest authenticated vascular land flora is still that containing *Baragwanathia* in Australian rocks of late Silurian age. Claims that *Aldanophyton*, from the Cambrian of Siberia, may have been a vascular plant have not been substantiated, and the presence of resistant spores in rocks of a similar age does not prove the existence of a vascular flora (see p. 8 *et seq.*). There can be no certainty as regards the relationships of *sporae dispersae* unless they are in organic union with recognizable parent material. Nevertheless, the discoveries of fragments of tracheids, coupled with the fact that by Silurian times there were well-developed vascular plants, must suggest that tracheophytes were already being evolved in the Cambrian.

There is a clear distinction between the Lower and Middle Devonian floras on the one hand, and those of the Upper stage of the Period on the other. At the debut of the Devonian, only the *Psilopsida* and *Lycopsida* were quite distinct. But by the close of the Period all basic types of vascular organization had been differentiated, and all potential morphological variations had been released, so that the major lines of evolution were already well indicated, except perhaps that the ovule was not yet in existence.

The earlier Devonian floras were predominantly psilophytic, but there were many associated genera of which the systematic positions are still the source of much controversy. The flora of the Upper Devonian is in distinct contrast with those of the earlier part of the Period, for the world-wide uniformity has now shifted to the pteridophytes. Axelrod (1959) has questioned the absence of regional differentiation in the Devonian, on the ground that the extant records are those of a particular ecological environment in which stability over widespread regions may be expected. He suggests that evidence from logs and spores points to the diversity of the vegetation in adjacent uplands which included plants morphologically more advanced than the psilophytes. He also comes to the conclusion that the successive phases of the Devonian lowland vegetation, as represented by the *Psilophyton* (Lower), *Hyenia* (Middle) and *Archaeopteris* (Upper) floras, represent no more than replacement of different ecologic units migrating from upland into lowland areas, and not evolutionary change as such.

Floristically the top of the Upper Devonian is marked by the disappearance of *Archaeopteris*, and its replacement by a transition flora characterized by *Lepidodendropsis*, *Rhacopteris* and *Triphyllopteris*, with which *Cyclostigma* continues as an associate from the Upper Devonian. There are distinct relationships between the very oldest Carboniferous flora and that of the Upper Devonian, and both agree on the whole in their relative uniformity throughout the world. In Australia, there has been distinguished a "*Lepidodendron veltheimianum* flora" and a slightly younger *Rhacopteris* flora, both of Lower Carboniferous age, of which the latter is interesting in that it lacks some important northern genera such as *Neuropteris* and *Alethopteris*.

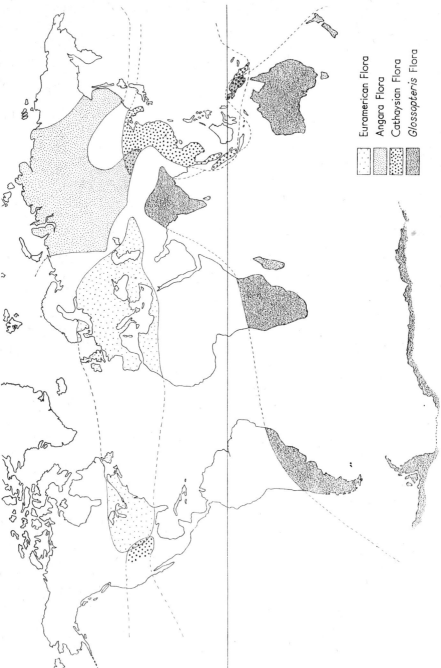

Euramerican Flora

Angara Flora

Cathaysian Flora

Glossopteris Flora

Fig. 16. Distribution of floras during late Palaeozoic time. The overlap of the Angara flora into the Cathaysian province in Kansu and that of the Cathaysian flora into the Euramerican province in North America represent time lapses. The overlap of the Cathaysian and *Glossopteris* floral provinces in N. e Guinea represents a real overlap.

2. Establishment of floral provinces during Permo-Carboniferous time

The floras of Namurian and Westphalian (A–D) time were in general uniform all over the world, though minor local differences did occur. During the early part of the Westphalian they began to become more restricted in their distribution, and by the middle of this period (Westphalian C) differences were beginning to become marked between the floras of the paralic and limnic basins (see later) of Europe (Jongmans, 1950, 1951, 1952, 1954).

From the beginning of Stephanian time, four distinct world floral provinces were in existence (Fig. 16), due to climatic changes associated with the Permo-Carboniferous glaciation of the southern hemisphere. These comprise three Arcto-Carboniferous provinces with a northern range, and a single, Antarcto-Carboniferous, province confined to former southern lands, and all four can be traced to the beginning of the Triassic Period.

i. Arcto-Carboniferous provinces. a. The Euramerican (European Permo-Carboniferous) flora extended from eastern U.S.A., throughout Europe, to the Urals, Iran, Turkestan and Morocco. This is the flora with which Western readers will be most familiar, for it includes the well-known assemblage of Coal-Measure plants. It is remarkable for the prevalence of arborescent genera, and the abundance of pteridosperms which are represented by numerous fern-like fronds. By the latest stages of the Carboniferous, the first conifers (*Lebachia*) had appeared.

Within the province two distinct types are recognizable—the one (Stephanian) confined to Central France and Central Europe, the other (Westphalian E) characterizing the rest of the flora and with a considerable geographic distribution.

The Stephanian-type is confined to the interior limnic basins, which were independent areas in lakes and hilly parts of the continent and isolated from the rest of the province by some geographical barrier. They may be likened to islands, for they feature a restricted number of species of which many are endemic.

The Westphalian E-type, on the other hand, is representative of the exterior paralic basins which formed in big plains along, or near, the coast on the foreland of the more elevated parts of the continents. In these areas, there was the possibility of vivid floral exchange, and a continuity of types from earlier Westphalian times, which could not occur in the more restricted and isolated limnic basins.

b. The Cathaysian flora extended from Korea and northern China southwards to Sumatra and New Guinea, and in western North America as far as Colorado, Oklahoma and Texas where it met the Euramerican province. In its older parts (e.g. Sumatra) and in the Kaiping basin, it contains a number of Euramerican elements which linger on for a relatively long time. This is to be expected, since the basins are paralic and the flora is thus essentially of the Westphalian E-type. However, it occupies a special place on account of the presence of a number of exclusively eastern types, such as *Lobatannularia* and *Tingia*, and of the early introduction of Mesozoic genera (e.g. *Taeniopteris* and *Clado-phlebis*). The flora also includes some interesting pteridosperms of Permian age which bore their seeds in a superficial position on the fronds. It is also remarkable for the minor lycopod element and the extreme scarcity of conifers, especially Euramerican forms. The characteristic genus *Gigantopteris*, by which the flora is sometimes known, did not appear until the middle of the Permian.

c. The Angara flora, sometimes known as the Kusnezk flora, covered the best defined natural area, and includes the great Siberian coal-fields of Kusnezk and Minussinsk. It was the last of the main province to be formed. It first developed in the interior of Siberia, from where it expanded to reach its full extent, in the Upper Permian, with a range from the Urals to the Pacific coast of Asia, from North Korea to the Bering Strait, and from the North Polar Sea as far south as Mongolia and into Kashmir. At its maximum extent it had replaced the Cathay-sian flora in Kansu, and there is some mixing in Korea. Characteristic members of the flora include *Angaridium*, *Gondwanidium*, *Ginkog-phyllum* and *Vojnovskya*. The conifers were scarce in contrast with their relative abundance in the Euramerican province.

ii. Antarcto-Carboniferous province. d. The *Glossopteris* flora prevailed all over the southern continents, occupying Gondwanaland, Ant-arctica and Peninsular India. As with the other provinces, it is important to note that the older Carboniferous rocks in these areas contain floras with pronounced Euramerican elements (*Lepidodendropsis*/*Rhacopteris*-type), and that the *Glossopteris* flora itself is separated from the basal floras by a stratigraphical break.

There seems little doubt that the origin of the *Glossopteris* flora was connected with the cooling of climate towards the end of the Carboni-ferous, when refrigeration of Gondwanaland resulted in the large-scale extinction of the pre-*Glossopteris* cosmopolitan elements. This is borne out by the comparative purity of the flora and paucity of species, which resulted from evolution under severe temperature conditions and in isolation from the northern provinces. The southern genera are quite distinct and include such forms as *Glossopteris*, *Gangamopteris*, *Schizo-neura*, *Buriadia* and *Walkomiella*. They appeared very late, and are probably not synchronous with those of the Euramerican province.

The unusual modern distribution astride the equator, of both *Glos-sopteris*-bearing strata and rocks showing indications of a former glaciation, has been one of the most favoured pieces of evidence for Continental Drift, especially with respect to the northward movement of India. It is a fact that there is no mixture of the *Glossopteris* and Cathaysian floras on the eastern borders of India, though Burma and southern China are now close to India. This clearly infers that there must have been an extremely efficient barrier between the two provinces at that time, and that the ancient Tethys Sea, which separated Gondwanaland from the northern continents, must have been exceptionally wide with India occupying a much more southerly location than at present.

The old idea of the presence of *Glossopteris* elements in the Angara province, which had possibly arrived there by migration along a route through Kashmir, is no longer accepted. Records of *Glossopteris* remains in Angaraland have been demonstrated as invalid, and the superficial resemblances between these and genuine *Glossopteris* are now regarded as indicative of parallel evolution under similar climatic, and possibly ecological, conditions. Certainly the Angara flora was of the cool temperate type, for wood specimens are known which show seasonal growth rings. It is important to remember that the Gondwana and Angara provinces were also separated by a west–east belt of uniform flora (represented by the Euramerican and Cathaysian provinces), which flourished under sub-tropical conditions and would prevent interchange of species between the two provinces in a north–south direction.

In the youngest Carboniferous, the *Glossopteris* flora contains certain Euramerican elements of Stephanian-type (e.g. in Rhodesia and Brazil), while in New Guinea there is a mixture with Cathaysian elements. Some writers have regarded these occurrences of northern species as representing a slow southerly migration, but Jongmans prefers to regard them as relics of a former world-wide flora which had become pushed out by the expanding *Glossopteris* flora. This is of special interest, for New Guinea is closely connected with Australia in the concept of Continental Drift and must have been to the south of Tethys and outside the limits of the Cathaysian province.

Though the four provinces remained distinct throughout the Permian, there were changing climatic conditions and topography. The closing days of the Palaeozoic witnessed the disappearance of older plants, and there was an introduction of newer types. Continental uplift was draining swampy lowlands, and climatic changes involving greater extremes of temperature and precipitation were becoming more pronounced. In Europe, coal-forming conditions disappeared, but the Angara flora continued little altered and it now had a new genus *Supaia*. *Supaia* has also been found in the Hermit Shale flora of Arizona.

4

This flora is regarded as a combination of Angara and other Arcto-Carboniferous elements, but it differed from other Permian floras of eastern U.S.A. and Europe by characters suggestive of a much more arid climate. In the Cathaysian province, the distinctive *Gigantopteris* was now present (recorded also from western North America), together with Mesozoic types. On the other hand, in the southern hemisphere the flora of Gondwanaland was still of the typical *Glossopteris* facies.

B. MESOZOIC VEGETATION

At the opening of the Mesozoic Period, the prevalence of desert conditions over a wide area of North America and the Old World was unfavourable to the further development, or continued existence, of members of the rich Permo-Carboniferous floras. Few genera survived, and the plants were now of smaller stature and generally showed xeromorphic features. In Gondwanaland, there was also an impoverishment of the *Glossopteris* flora with its replacement by a new widely spread flora (*Thinnfeldia* flora) which existed through the Trias and Rhaetic.

Yet by late Triassic time conditions were becoming more favourable again, at least in the northern hemisphere, for the remains of the famous great petrified araucarian forest of Arizona, with its associated ferns of a type now living in moist tropical regions and numerous large horsetails, suggest locally humid conditions. Many of the same types of late Triassic plants occur in eastern U.S.A. and Europe, and in the Rhaetic floras of Europe, ferns of now-tropical types, cycadophytes, conifers and ginkgophytes are abundant. Related members occur in such widely separated regions as Greenland and Indochina, though at the latter locality there is indication of intermediate relationship with floras of the southern hemisphere.

It is significant that there is still a marked difference between the floras of the two hemispheres during the Mesozoic. This difference, which first became apparent during the mid-Carboniferous, with its consequent divergent floral evolution, has with little doubt left its imprint on the land vegetation right up to the present day.

As long ago as 1931, Harris pointed out that three distinct floras were in existence in Rhaeto-Liassic times; the *Lepidopteris* flora (followed by the *Thaumatopteris* flora in the Lias) in Greenland and Europe, the flora of Tonkin with many exclusively asiatic species, and the *Thinnfeldia* flora (sometimes known as the *Dicroidium* flora) of the Gondwana province. He also noted that the origin of the European floras of the earlier part of the Mesozoic was probably outside Europe, since it is quite remarkable how many genera of Oolitic, Liassic, Rhaetic and Triassic plants of that region are absent from the preceding stages in Europe, but are present in the preceding stages in other parts of the world.

The northern province apparently had a uniform flora extending through 160° of longitude (25°W–135°E) and 35° of latitude (70°N– 35°N), ranging from east Greenland through Sweden to France, Rumania, Poland, Russia and Siberia, within the same latitudes, to western Japan. The middle province ranged from Tonkin, with localities in China, Pamir, Iran, Armenia, Mexico and British Honduras. The two provinces ran obliquely across the present climatic and latitudinal belts, which offers some support for the idea of a different disposition of the continents during Mesozoic time. The position of these two provinces is in approximate agreement with the land distribution at that time as postulated by adherents of the theory of Continental Drift. The young science of palaeomagnetism may eventually help in such problems and their elucidation. At the moment, the position of the north pole in the Far East during Mesozoic times, as deduced from palaeomagnetic observations, conveniently brings the Greenland flora into temperate latitudes, but unfortunately western Japan assumes a position as near to the fossil pole as Greenland is to the present position of the pole (Opdyke and Runcorn, 1959, Fig. 15).

The Jurassic Period was generally considered to have witnessed the most widespread uniformity of conditions at any time during earth history, but an analysis of the conifers soon revealed that differences existed between the two hemispheres (Florin, 1940). The differences were so great, as indeed they are today and as they were during the Permo-Carboniferous, that one cannot speak of a world-wide uniform flora, and hence widespread uniform conditions. *Araucarites* was the only genus with a distribution to north and south of the equator, but living members of its family, the Araucariaceae, are now confined to the southern hemisphere, as are most of the podocarps. Fossil members of the Podocarpaceae are widely recorded at southern localities during the Mesozoic, but they are only known north of the equator in Peninsular India, which of course was once part of the old Gondwana continent. The provincial character of the Mesozoic floras as a whole is further emphasized by the absence of the northern families, Taxodiaceae, Cupressaceae and Pinaceae, from the southern hemisphere.

In both hemispheres, the cycadophytes formed a dominant part of the vegetation. Members of the Bennettitales were very common, but though their growth-habit was very much like that of present-day cycads, it is by no means certain that they lived under similar conditions of habitat. However, they probably lived in regions where a mild, or warm, temperature prevailed.

It is remarkable that Jurassic plants are recorded as far south as

Graham Land (63°S) and as far north as the New Siberian Islands (75°N). Clearly conditions over such a wide range of latitude could not have been uniform, though the effects of latitude may have been reduced as a result of more general circulation of winds and ocean currents consequent upon a broad submergence of the continents. The abundance of ferns related to genera that now inhabit a frostless environment indicates an equable climate, and the rarity of these in floras at high latitudes is in accord with the suggestion that it was less mild there. Indeed, there are other indications of latitudinal zonation, for the floras of Siberia are particularly characterized by the abundance of ginkgophytes and by the paucity of cycadophytes and conifers. The cycadophytes tended to be restricted to a belt coinciding approximately with the region formerly occupied by the Coal-Measure flora (Edwards, 1955).

REFERENCES

Ananiev, A. R. (1957). *Bot. Zh. S.S.S.R.* **42** (5), 691–702.
Andrews, H. N. (1945). *Ann. Mo. bot. Gdn* **32**, 323–360.
Andrews, H. N. (1959). *Cold Spr. Harb. Symp. quant. Biol.* **24**, 217–234.
Andrews, H. N. (1961). "Studies in Paleobotany", 487 pp. John Wiley & Sons, Inc., New York.
Andrews, H. N. and Alt, K. S. (1956). *Ann. Mo. bot. Gdn* **43**, 355–378.
Andrews, H. N. and Kernen, J. A. (1946). *Ann. Mo. bot. Gdn* **33**, 141–146.
Andrews, H. N. and Mamay, S. H. (1953). *Ann. Mo. bot. Gdn* **40**, 183–209.
Andrews, H. N. and Mamay, S. H. (1955). *Phytomorphology* **5**, 372–393.
Arnold, C. A. (1940a). *Contr. Mus. Geol. Univ. Mich.* **6**, 21–52.
Arnold, C. A. (1940b). *Amer. J. Bot.* **27**, 57–62.
Arnold, C. A. (1952). *Contr. Mus. Geol. Univ. Mich.* **9**, 297–309.
Arnold, C. A. (1953). *Phytomorphology*, **3**, 51–65.
Axelrod, D. I. (1959). *Evolution* **13**, 264–275.
Banks, H. P. (1944). *Amer. J. Bot.* **31**, 649–659.
Banks, H. P. (1960). *Senck. leth.* **41**, 59–88.
Baxter, R. W. (1949). *Ann. Mo. bot. Gdn* **36**, 287–352.
Baxter, R. W. (1950). *Bot. Gaz.* **112**, 174–182.
Cash, E. K. and Stevenson, J. (1961). *Taxon* **10**, 244.
Chaloner, W. G. (1960). *Sci. Progr. Twent. Cent.* **48**, 524–534.
Chamberlain, C. J. (1935). "Gymnosperms: Structure and Evolution", 484 pp. Univ. Chicago Press, Chicago.
Darrah, W. C. (1960). "Principles of Paleobotany", 2nd edn., 295 pp. The Ronald Press Co., New York.
Delevoryas, T. (1953). *Amer. J. Bot.* **40**, 144–150.
Delevoryas, T. (1955). *Palaeontographica* **97B**, 114–167.
Delevoryas, T. (1957). *Amer. J. Bot.* **44**, 654–660.
Edwards, W. N. (1955). *Advanc. Sci., Lond.* **12**, 165–176.
Eggert, D. A. (1961). *Palaeontographica* **108B**, 43–92.
Felix, C. J. (1952). *Ann. Mo. bot. Gdn* **39**, 263–288.
Florin, R. (1920). *Ark. Bot.* **16** (7), 1–10.

Florin, R. (1931). *K. svenska VetenskAkad. Handl.* Ser. 3, **10**, 1–588.
Florin, R. (1933). *K. svenska VetenskAkad. Handl.* Ser. 3, **12**, 1–134.
Florin, R. (1938–45). *Palaeontographica* **85B**, 1–729.
Florin, R. (1939). *Bot. Notiser* **4**, 547–565.
Florin, R. (1940). *K. svenska VetenskAkad Handl.* Ser. 3, **19**, 1–107.
Florin, R. (1950a). *Acta Hort. berg.* **15**, 111–134.
Florin, R. (1950b). *Bot. Rev.* **16**, 258–282.
Florin, R. (1951). *Acta Hort. berg.* **15**, 285–388.
Florin, R. (1954). *Biol. Rev.* **29**, 367–389.
Foster, A. S. and Gifford, E. M. (1959). "Comparative Morphology of Vascular Plants", 555 pp. W. H. Freeman & Co., San Francisco.
Fry, W. L. (1955). *Bull. Torrey bot. Cl.* **82**, 486–490.
Fry, W. L. (1956). *J. Palaeont.* **30**, 35–45.
Ghosh, A. K. and Bose, A. (1947). *Nature, Lond.* **160**, 796.
Ghosh, A. K. and Bose, A. (1952). *Nature, Lond.* **169**, 1056.
Gothan, W. and Weyland, H. (1954). "Lehrbuch der Paläobotanik", 535 pp. Akademie-Verlag, Berlin.
Gothan, W. and Zimmerman, W. (1936). *Jb. preuss. geol. Landesanst.* **56**, 208–210.
Greguss, P. (1959). *Proc. IX Int. Bot. Cong. Montreal* **2**, 142–143.
Greguss, P. (1961). *Acta biol.*, N.S. **7**, 3–30.
Halle, T. G. (1913). *K. svenska VetenskAkad. Handl.* **51** (3), 1–58.
Halle, T. G. (1936). *Palaeont. sinica* Ser. A, **1**, 5–28.
Halle, T. G. (1954). *Svensk bot. Tidskr.* **48** (2), 368–380.
Harris, T. M. (1931). *Biol. Rev.* **6**, 133–162.
Harris, T. M. (1932). *Medd. Grønland* **85** (3), 1–114.
Harris, T. M. (1938). "The British Rhaetic Flora", 84 pp. British Museum (Nat. Hist.), London.
Harris, T. M. (1941). *Phil. Trans.* **231B**, 75–98.
Harris, T. M. (1942a). *Ann. Mag. nat. Hist.*, Ser. 11, **9**, 393–401.
Harris, T. M. (1942b). *Ann. Mag. nat. Hist.*, Ser. 11, **9**, 568–587.
Harris, T. M. (1954). *Svensk bot. Tidskr.* **48** (2), 281–291.
Harris, T. M. (1956). *Endeavour* **15**, 210–214.
Harris, T. M. (1958). *Palaeobotanist* **7**, 93–106.
Harris, T. M. (1961a). "The Yorkshire Jurassic Flora", Vol. I, 212 pp. British Museum (Nat. Hist.), London.
Harris, T. M. (1961b). *Palaeontology* **4**, 313–323.
Hirmer, M. (1940). *Palaeontographica* Suppl. 9, 1–44.
Hirmer, M. (1941). *Biol. gen.* **15**, 134–171.
Høeg, O. A. (1945). *Norsk geol. Tidsskr.* **25**, 183–192.
Jacob, K. and Jacob, C. (1954). *Palaeont. indica*, N.S., **33** (1), 1–34.
Jongmans, W. J. (1950). *Proc. VII Int. Bot. Cong., Stockholm*, 591–594.
Jongmans, W. J. (1951). *C.R. 3e Congr. Av. Étud. Strat. carb*, 295–306.
Jongmans, W. J. (1952). *Second Conf. Origin and Const. Coal, Crystal Cliffs, Nova Scotia*, 3–28.
Jongmans, W. J. (1954). *C.R. VIII Congr. Int. Bot., Paris, Sect.* 5, 135–138.
Just, T. (1952). *Bull. Amer. Mus. nat. Hist.* **99** (3), 189–203.
Kendall, M. W. (1947). *Ann. Mag. nat. Hist.*, Ser. 11, **14**, 225–251.
Kendall, M. W. (1948). *Ann. Mag. nat. Hist.*, Ser. 12, **1**, 73–108.
Kendall, M. W. (1949a). *Ann. Mag. nat. Hist.*, Ser. 12, **2**, 299–307.
Kendall, M. W. (1949b). *Ann. Mag. nat. Hist.*, Ser. 12, **2**, 308–320.
Kendall, M. W. (1949c). *Ann. Bot.*, N.S., **13**, 151–161.

Kendall, M. W. (1952). *Ann. Mag. nat. Hist.*, Ser. 12, 5, 583–594.

Kidston, R. and Lang, W. H. (1917–21). *Trans. roy. Soc. Edinb.* 51–52, 761–784; 603–627; 643–680; 831–854; 855–902.

Knox, E. (1939). *Trans. bot. Soc. Edinb.* 32, 477–487.

Kozlowski, R. and Greguss, P. (1959). *Acta Palaeont. polon.* 4, 1.

Kräusel, R. (1950). "Versunkene Floren", 152 pp. Verlag Waldemar Kramer, Frankfurt am Main.

Kräusel, R. and Weyland, H. (1926). *Abh. senckenb. naturf. Ges.* 40, 115–155.

Kräusel, R. and Weyland, H. (1929). *Abh. senckenb. naturf. Ges.* 41, 315–360.

Kräusel, R. and Weyland, H. (1932a). *Senckenbergiana* 14, 274–280.

Kräusel, R. and Weyland, H. (1932b). *Senckenbergiana* 14, 391–403.

Kräusel, R. and Weyland, H. (1935). *Palaeontographica* 80B, 171–190.

Kräusel, R. and Weyland, H. (1940). *Senckenbergiana* 22, 6–16.

Krishnan, M. S. (1954). "History of the Gondwana Era in Relation to the Distribution and Development of Flora", 15 pp. Birbal Sahni Institute of Palaeobotany, Lucknow.

Kryshtofovich, A. N. (1953). *Dokl. Akad. Nauk. S.S.S.R.* 91, 1377–1379.

Lacey, W. S. (1943). *New Phytol.* 42, 1–4.

Lacey, W. S. (1959). *Nature, Lond.* 184, 1592–1593.

Lacey, W. S. (1961). *Proc. Rhod. sci. Ass.* 49, 26–53.

Lang, W. H. and Cookson, I. C. (1935). *Phil. Trans.* 224B, 421–449.

Leclercq, S. (1940). *Mém. Acad. R. Belg.* in-4°, Sér. 2, 12 (3), 1–67.

Leclercq, S. (1954). *Svensk bot. Tidskr.* 48 (2), 301–315.

Leclercq, S. (1956). *Evolution* 10, 109–114.

Leclercq, S. (1957). *Mém. Acad. R. Belg.* in-4°, Sér. 2, 14 (3), 1–39.

Leclercq, S. (1959). *Proc. IX Int. Bot. Cong., Montreal* 2, 219.

Leclercq, S. (1961). *In* "Recent Advances in Botany", II, pp. 968–971. Univ. Toronto Press, Toronto.

Leclercq, S. and Andrews, H. N. (1959). *Proc. IX Int. Bot. Congress, Montreal* 2, 219.

Leclercq, S. and Andrews, H. N. (1960). *Ann. Mo. bot. Gdn* 47, 1–23.

Lundblad, B. (1950). *K. svenska VetenskAkad. Handl.* Ser. 4, 1 (8), 1–82.

Lundblad, B. (1954). *Svensk Bot. Tidskr.* 48 (2), 381–417.

Lundblad, B. (1955). *Bot. Notiser* 108 (1), 22–39.

Lundblad, B. (1959). *K. svenska VetenskAkad. Handl.* Ser. 4, 6 (2), 1–38.

Mägdefrau, K. (1942). "Paläobiologie der Pflanzen", 2e auf., 438 pp. Verlag Gustav Fischer, Jena.

Mamay, S. H. (1954). *J. Wash. Acad. Sci.* 44 (1), 7–11.

Medwell, L. M. (1954). *Proc. roy. Soc. Vict.* 65, 63–111.

Melville, R. (1960). *Nature, Lond.* 188, 14–18.

Merker, H. (1958). *Bot. Notiser* 111 (4), 608–618.

Merker, H. (1959). *Bot. Notiser* 112 (4), 441–452.

Merker, H. (1961). *Bot. Notiser* 114 (1), 88–102.

Morgan, J. (1959). *Illinois biol. Monogr.* 27, 1–108.

Nêmejc, F. (1931). *Preslia* 10, 111–114.

Obrhel, J. (1958). *Geologie* 7, 969–983.

Opdyke, N. D. and Runcorn, S. K. (1959). *Endeavour* 18, 26–34.

Pannell, E. (1942). *Ann. Mo. bot. Gdn* 29, 245–260.

Pant, D. D. (1958). *Bull. Brit. Mus. (nat. hist.)* A, 3 (4), 127–175.

Plumstead, E. P. (1952). *Trans. geol. Soc. S. Afr.* 55, 281–328.

Plumstead, E. P. (1956a). *Palaeontographica* 100B, 1–25.

Plumstead, E. P. (1956b). *Trans. geol. Soc. S. Afr.* **59**, 211–236.

Plumstead, E. P. (1958a). *Trans. geol. Soc. S. Afr.* **61**, 51–79.

Plumstead, E. P. (1958b). *Trans. geol. Soc. S. Afr.* **61**, 81–96.

Radforth, N. W. and McGregor, D. C. (1954). *Canad. J. Bot.* **32**, 601–621.

Radforth, N. W. and McGregor, D. C. (1956). *Trans. roy. Soc. Can.* Ser. 3, **50**, 27–33.

Rao, A. R. (1943). *Proc. nat. Acad. Sci., India* **13**, 333–355.

Rigby, J. F. (1961). *Aust. J. Sci.* **23**, 230.

Roberts, D. C. and Barghoorn, E. S. (1952). *Bot. Mus. Leafl. Harv.* **15**, 191–200.

Sahni, B. (1932). *Proc. XIX Ind. Sci. Cong. Bangalore,* 322.

Sahni, B. (1938). *Proc. XXV Ind. Sci. Cong., Calcutta,* 133–176.

Sahni, B. (1948). *Bot. Gaz.* **110**, 47–80.

Schmidt, W. (1954). *Palaeontographica* **97**B, 1–22.

Schopf, J. M. (1939). *Amer. J. Bot.* **26**, 196–207.

Scott, D. H. (1909). "Studies in Fossil Botany", Vol. II, 2nd edn., 322 pp. A. & C. Black Ltd., London.

Sen, J. (1954). *Sci. & Cult.* **20**, 202–203.

Sen, J. (1955a). *Proc. nat. Inst. Sci., India* **21**B, 48–52.

Sen, J. (1955b). *Bot. Notiser* **108** (2), 244–252.

Sen, J. (1955c). *Nature, Lond.* **176**, 742–743.

Sen, J. (1956). *Nature, Lond.* **177**, 337–338.

Seward, A. C. and Sahni, B. (1920). *Palaeont. indica,* N.S., **7** (1), 1–55.

Sitholey, R. V., Varma, C. P. and Srivastava, P. N. (1953). *J. sci. industr. Res.* **12**B (12), 645–647.

Srivastava, B. P. (1935). *Proc. XXII Ind. Sci. Cong. Calcutta,* 285.

Srivastava, B. P. (1937). *Proc. XXIV Ind. Sci. Cong., Hyderabad-Deccan,* 273–274.

Srivastava, B. P. (1946). *Proc. nat. Acad. Sci., India* **15**, 185–211.

Steere, W. C. (1946). *Amer. Midl. Nat.* **36**, 298–324.

Steidtmann, W. E. (1937). *Amer. J. Bot.* **24**, 124–125.

Steidtmann, W. E. (1944). *Contr. Mus. Geol. Univ. Mich.* **6**, 131–166.

Stewart, W. N. (1951). *Amer. J. Bot.* **38**, 709–717.

Stewart, W. N. (1959). *Proc. IX Int. Bot. Cong., Montreal* **2**, 382–383.

Stewart, W. N. (1960). *Plant Sci. Bull.* **6** (5), 1–5.

Stewart, W. N. (1961). *In* "Recent Advances in Botany", II, 960–963. Univ. Toronto Press, Toronto.

Stewart, W. N. and Delevoryas, T. (1952). *Amer. J. Bot.* **39**, 505–516.

Stewart, W. N. and Delevoryas, T. (1956). *Bot. Rev.* **22**, 45–80.

Stockmans, F. W. (1948). *Mém. Mus. Hist. nat. Belg.* **110**, 1–85.

Surange, K. R. and Srivastava, P. N. (1956). *Palaeobotanist* **5**, 46–49.

Thomas, H. H. (1958). *Bull. Brit. Mus. (nat. Hist.),* A **3** (5), 179–189.

Thomas, H. H. and Bancroft, N. (1913). *Trans. Linn. Soc. Lond. (Bot.)* **8**, 155–204.

Thomas, H. H. and Harris, T. M. (1960). *Senck. leth.* **41**, 139–161.

Vishnu-Mittre. (1953). *Palaeobotanist* **2**, 75–84.

Walton, J. (1925). *Ann. Bot.* **39**, 563–572.

Walton, J. (1928). *Ann. Bot.* **42**, 707–716.

Walton, J. (1935). *Trans. roy. Soc. Edinb.* **58**, 313–337.

Walton, J. (1949a). *Trans. geol. Soc. Glasg.* **21** (2), 278–280.

Walton, J. (1949b). *Trans. roy. Soc. Edinb.* **61**, 729–736.

Wesley, A. (1956). *Mem. Ist. geol. Univ. Padova* **19**, 1–69.

Wesley, A. (1958). *Mem. Ist. geol. Univ. Padova* **21**, 1–57.

Worsdell, W. C. (1906). *Ann. Bot.* **20**, 129–155.

Zeiller, R. (1902). *Palaeont. indica.*, N.S., **2** (1), 1–40.

Zimmerman, F. (1930). *Arb. Inst. Paläobot. Petr. Brennsteine* **2**, 83–102.

Zimmerman, W. (1959). "Die Phylogenie der Pflanzen", 2e auf., 777 pp. Gustav Fischer Verlag, Stuttgart.

Growth Regulation by Metals and Chelates
HANS BURSTRÖM

Institute of Plant Physiology, Lund University, Lund, Sweden

I. INTRODUCTION

In a monograph by Martell and Calvin (1952) a chelate is described as a compound with a metal atom joined to two or more electron-donor groups of a single organic molecule. The donors are usually O, N, or S atoms. This gives a ring configuration, in which the metal is firmly held as if between claws, and the complex has charge and solubility properties alien to the metal itself. As an example may be cited ethylenediamine-tetraacetic acid $(HOOC—CH_2)_2—N—CH_2—CH_2—N—(CH_2—COOH)_2$ binding Fe^{2+}, thus:

$$
\begin{array}{ccc}
HOOC . CH_2 & & CH_2 . COOH \\
\diagdown & & \diagup \\
& N—CH_2—CH_2—N & \\
H_2C & \diagdown \quad \diagup & CH_2 \\
| & Fe & | \\
OC & \diagup \quad \diagdown & CO \\
& O \qquad O &
\end{array}
$$

The stability of the chelate depends upon both metal and ligand, but ought to follow the same order of metal ions for all ligands (Mellor and Maley, 1947, 1948). Data given by Martell and Calvin (loc. cit.) indicate for the physiologically interesting ions the order:

$$Fe^{3+} > Cu > Ni > Co > Zn > Fe^{2+} > Mn > Ca > Mg.$$

4§

The ability of organic compounds to form metal-chelates is wide-spread; an extensive survey of the chemical properties of chelates and their occurrence is found in the quoted book by Martell and Calvin and in Chaberek and Martell (1959). Chelation is encountered with such common cell constituents as peptides, the acids of the Krebs cycle, porphyrines, and organic phosphates. It can safely be assumed that di- and tri-valent metals within the plant normally occur to a large extent as chelates. The chelation of compounds is of paramount importance for cell structures and for the function of enzymes, the metals thereby serving as activators or as mediators in electron transfer. On the other hand, the addition of a chelate-forming compound to a nutrient medium may affect plant growth, which tacitly must be assumed to depend upon the chelation of metals. This involves some problems of current interest, with regard to both growth mechanisms and heavy metal functions, which will be dealt with in the present paper.

The topic has to be limited to some specific problems concerned with growth. Severe deficiency of an indispensable metal will lead to cessation of growth, irrespective of the mode of action of the metal in question; the metal may be involved in the basal metabolism or in the growth process in a restricted sense. If a growth response to a heavy metal or a chelate ligand is very rapid it is likely that such substances affect the specific growth mechanism directly; when the response comes to the fore slowly then the substance probably acts through disturbance of the basal syntheses and ensuing starvation.

The mode of action of a chelating agent is illustrated by the fact that chelates were originally used in plant physiology for two different pur-poses. In the 1920's citric acid was already employed in order to keep Fe dissolved in a nutrient solution and subsequently it has been re-placed in routine work by ethylenediaminetetraacetic acid (EDTA) or some of its derivatives. On the other hand, strong chelate-forming agents like diethyldithiocarbamate (DIECA) were used in order to purify solutions by removing heavy metal traces, and others such as 8-oxyquinoline (oxin) (also referred to as 8-hydroxyquinoline) as fungi-cide or bactericide by depriving the cells of essential metals. Depending upon the relative stabilities of the complexes formed by a metal with an externally applied compound and with the native chelating consti-tuents of the cells, a chelating agent may either carry a metal to the plant or remove one. There is a competition between internal and external chelating agents for available metals (Brown et al., 1960). An externally added chelating agent can supply or remove the same metal depending upon the ratio between metal and ligand concentrations and the presence of other metals competing for the compound. The sodium salt of EDTA may remove an essential metal from the plant, but the presence of another metal may release the essential one again to the benefit of the plant. This duplicity of the chelating agents has to be considered in interpretations of chelate actions.

Chelating compounds have been used extensively for curing mineral nutrient deficiencies, for example iron chlorosis, both in field and laboratory work (cf. Weinstein *et al.*, 1954). These authors also quote examples, to which many could be added, of inhibitions of enzyme reactions by chelation, owing to removal of metals. Such disturbances of the basal metabolism are only mentioned exceptionally in the following discussion, which will be directed to instances of chelate actions studied in relation to growth response.

Gäumann *et al.* (1957) made use of the duplicity of the chelates in order to explain the actions of the toxins lycomarasmin and fusaric acid in the plant. As a matter of fact, the studies by Gäumann's school on these toxins produced by fungi seem to have been the first ones in which the physiological activity of biogenic compounds in the plant was explained by their chelating ability. Fusaric acid, analogous in appearance to gibberellic acid, was assumed to act by chelating and redistributing Fe inside the plant, thus causing deficiency or excess in diverse organs.

The mode of action of chelating agents on growth has been approached from two different directions. Attempts have been made to trace a similarity between chelating agents and natural auxins, thus either lifting the chelate problem into the auxin research or explaining auxin actions by a chelation of metals. The other approach has been to identify the metals responsible for the growth actions by added chelating compounds, particularly EDTA. The short history of the problems involved is of some importance for an understanding of the present situation and will be briefly outlined.

II. SIMILARITY BETWEEN AUXINS AND CHELATING AGENTS

A. LITERATURE REVIEW

The history in this field of research does not begin until in 1956. In a lecture held in 1955 Bennet-Clark (1956) mentioned that in a coleoptile section test for auxin an addition of EDTA 10^{-5}–10^{-3} mole/litre caused a growth promotion resembling that of an auxin. In simultaneously performed experiments Heath and Clark (1956a) corroborated this statement. They found in wheat coleoptile straight-growth tests that EDTA increased growth in the concentrations 10^{-9}–10^{-4} mole/litre. These experiments were carried out with the sodium salt of EDTA in unbuffered solution. The same effect was found with seven other chelating agents, amongst them DIECA and oxin. Analogues of oxin, for example α-naphthol and quinoline, devoid of chelating ability, proved to cause only inhibitions. Indole-3-acetic acid exerted the same type of activity in the same concentrations, and auxin and chelating compounds could replace each other indiscriminately. This strongly indicated that IAA also acts by virtue of a capacity to form chelates, thus regulating the condition of some metal in the plant. A still more striking support of this view was found in a mutual antagonism between the active compounds (Heath and Clark, 1956b). In a wheat root inhibition test each one of the compounds in concentrations of 10^{-11} mole/litre reversed the action of any one of the others in 10^{-5} mole/litre concentration (Table I). The same was found with supra-optimal additions to coleoptiles: a mutual antagonism between concentrations of 10^{-9} and 10^{-3} mole/litre. IAA reacted in every respect as a

TABLE I

Interaction between IAA and EDTA on Root Growth of Wheat var. Eclipse
(from Heath and Clark, 1956a)

Concentrations in mole/litre; values denote root length cm per plant.

IAA	EDTA				
	0	10^{-11}	10^{-9}	10^{-7}	10^{-5}
0	25	26	21	16	9
10^{-11}	26	26	24	23	23
10^{-9}	25	25	25	22	21
10^{-7}	17	21	23	20	15
10^{-5}	7	24	23	16	11

chelating agent as far as the influence on the over-all growth was concerned. The concentration ratio between the two mutual antagonists, up to 1 000 000 : 1 is remarkable.

The problem was promptly taken up in other quarters. Weinstein *et al.* (1956) reported that they had already made the same discovery with EDTA added to *Lupinus* hypocotyls in 1953. Fawcett *et al.* (1956) were declared opponents. They tried the herbicidal carbamates and xanthates, with structures indicative of chelate formation, in cylinder and other tests along with IAA and the compounds of Heath and Clark, finding much less growth activity of chelating compounds than of auxins, hardly more than of ethyl alcohol. They assumed the action of the non-auxin compounds to be pre-mortal effects and rejected the opinion that auxins are chelate-forming.

In a study of the effect of EDTA on roots of intact plants in water culture, Burström and Tullin (1957) stated that cell divisions in the apical meristem were inhibited, whereas auxins were known primarily to inhibit cell elongation. EDTA-inhibition gave roots with few, long cells, auxin-inhibition roots with many, short cells. The similarity in over-all growth would be superficial only, and the compounds can hardly substitute each other physiologically. This does not exclude the possibility that the auxins, nevertheless, are chelating compounds. Moreover, this objection is not applicable to coleoptile growth, which depends entirely upon cell elongation. The growth activity of chelating agents was again dealt with by Fawcett *et al.* (1959). About forty chelating compounds tested with regard tó pea curvatures, tomato epinasty, and coleoptile cylinder growth gave very irregular responses, and no consistent connection was found with chelating ability. Finally, Perrin (1961) has pointed out that IAA and 2,4-D possess a chelating ability of the same order of magnitude as acetic acid. The concentrations of Zn and Cu complexed with auxins inside the plant must be negligible.

A full report on the experiments by Heath and Clark (1960) has settled some of the discrepancies between their results and those of Wain's group. The main differences lie in technique and in the sensitivity of the wheat strains employed. More examples are given of the unusual mutual antagonism in root growth between auxin and chelating agents, although all responses are smaller in these experiments. The

TABLE II

Interaction between IAA and EDTA on Root Growth of Wheat var. Holdfast
(from Heath and Clark, 1960)

Concentrations in mole/litre; values denote root length cm per plant.

IAA	EDTA			
	0	10^{-9}	10^{-7}	10^{-5}
0	23	23	21	3
10^{-9}	22	25	19	6
10^{-7}	18	10	19	22
10^{-5}	1	7	22	5

antagonism is most evident between EDTA and IAA (see Table II) with a full reciprocal reversal of the inhibition by 10^{-7} mole/litre of one compound through 10^{-5} of another. With coleoptiles very small growth responses were obtained; significant inhibitions by supraoptimal concentrations of IAA and EDTA are antagonized by the other compound, but no distinct reciprocal antagonism is recorded for combinations of two chelating agents. The results are much less striking than in the preliminary experiments.

Experience so far indicates that chelating agents are growth-active, and that EDTA affects growth of both roots and coleoptiles in concentrations corresponding to those of auxins. It is also clearly evident that auxins and non-auxin chelating compounds appear as antagonists. This does not mean that the actions must be physiologically similar. The analysis by Burström and Tullin (1957) shows that they are not, in the case of root growth. What is involved in this antagonism is, on the contrary, an intricate question. The analysis does not necessarily imply that auxins chelate metals but does not exclude this possibility either.

Heath and Clark (1960) emphasize that the actions of EDTA and IAA cannot be identical. The ability of one compound to reverse the action of another in the ratio of 1 : 100 or 1 : 10 000 excludes the possibility that they bind the same metal. By the same token the three mutually antagonistic chelating agents EDTA, oxin, and DIECA must inactivate three different metals, but four metals can hardly produce even superficially similar growth responses. Some kind of interaction seems to occur, but a further insight into the relations is required. Thimann and Takahashi (1961) suggest that EDTA inactivates some metal system destroying IAA. There is much in favour of this view as regards shoot parts, but in roots, as mentioned, EDTA and auxin inhibitions are of different nature.

B. SUPPLEMENTARY EXPERIMENTS

For this reason the experiments of Heath and Clark have been repeated and imitated as exactly as possible with the generous cooperation of Dr. Heath. We are much obliged to him for furnishing us with a copy of the original schedule of the methods used by Dr. Clark and for supplementary information about the procedure employed by these two workers.

A few comments on the technique are necessary. Tests were made by Heath and Clark with wheat plants grown for 8–9 days in test tubes of carefully purified water with no additions other than the compounds under investigation. Root growth was recorded as total lengths per plant. The solutions were regularly renewed and the experiments run in darkness. Some occasional weak light could not be avoided, but this could certainly be neglected. We used the same quality and size of glassware, the same temperature—kept constant—and duration of experiment. Our water was doubly quartz distilled, and the solutions renewed every second day. The plants were mounted according to the prescriptions. We used another variety of wheat, Weibull's *Eroica*, of a sensitivity resembling that of var. *Holdfast* in the 1956 experiments of Heath and Clark (cf. Table II) and a green safe light was employed for unavoidable temporary illumination. The concentration of EDTA was chosen so as to give about the same response as 10^{-7} mole/litre of IAA. Control plants exhibited very nearly the same growth as those of Heath and Clark.

Their results could be partly verified. An interaction between EDTA and IAA exists, although less regular than in the experiments of Heath and Clark. We ascribe this irregularity to the unfavourable, unbalanced growth medium and the small number of replicate plants (the same as with Heath and Clark). EDTA in a concentration of 10^{-7} mole/litre antagonizes an inhibition of root growth by 10^{-5} mole/litre of IAA. On the other hand, we were unable to verify clearly and reproducibly the reciprocal antagonism of a low concentration of IAA against EDTA. A ratio of 1 : 1 was sometimes favourable.

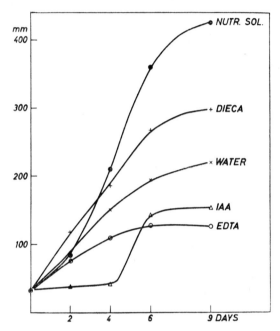

Fig. 1. Total root length mm per plant in distilled water, a complete nutrient solution, IAA 10^{-5} mole/litre, and EDTA and DIECA 10^{-4} mole/litre. Methods according to Heath and Clark (1960).

In order to explain the interaction it is necessary to analyse the growth inhibitions of the two compounds, because an inhibition of total root length produced may be brought about in different ways. The time curves of the actions of EDTA, IAA, and DIECA in distilled water and the growth in a complete nutrient solution are compared in Fig. 1. This shows that in an experiment of 8–9 days' duration the total growth ceases. This is due partly to mineral nutrient deficiency as revealed by the greater growth in the nutrient solution, partly to a detrimental effect of the unbalanced medium. However, the main cause is the well-known fact that wheat seeds are depleted of nutrients after about 8 days (Bosemark, 1954). This is not at variance with the greater growth

obtained in the nutrient solution, because the shoots consume the
bulk of the available nutrients. With mineral salts added the roots
manage to appropriate more of the organic food before the reserves
are exhausted. Under the laboratory conditions 8 days is the time limit
for the shift from heterotrophous to autotrophous nutrition. The growth
recorded after 8–9 days is thus no good measure of the growth rate but
more an expression of a nutrient level.

EDTA causes a regular progressive decrease in the actual growth
rate and a cessation of growth after about 6 days, but the IAA-treated

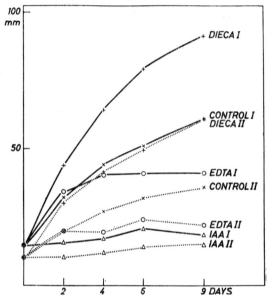

FIG. 2. Growth of individual seminal roots of wheat plants: I, median radicle; II, first
pair of adventitious roots. The same material as in Fig. 1.

roots are at first totally inhibited; they seem to recover after the 4th
day and then grow rapidly for the next 2 days. In spite of the similar
growth values recorded after 9 days the actions of EDTA and IAA
are fundamentally different. What is behind the growth responses will
be shown in Figs. 2–4.

The wheat seedlings have three seminal roots: the true radicle and
one pair of adventitious roots. After a lag period of a few days there
emerges a varying but for each growing condition rather constant
number of additional adventitious roots. All these roots do not react
in the same way to the treatments. The total root growth recorded
depends upon the growth exhibited by each root and the number of
adventitious roots formed.

The three seminal roots usually grow very much alike (Fig. 2). The notable result is that the seminal roots are totally inhibited in this concentration of IAA, whereas with EDTA the growth rate is reduced only. The growth recorded in IAA is due entirely to the larger number of later adventitious roots. The increased initiation of adventitious roots by auxin has repeatedly been recorded in the literature. With chelating agents the number of such roots is normally three, or at the most four, and their elongation rate is similar to that of the seminal roots. This explains the fundamental differences between the growth responses of IAA and EDTA (Fig. 1).

EDTA added to IAA does not reverse the inhibition of the seminal roots but affects the adventitious roots only (Figs. 3 and 4). Figure 3

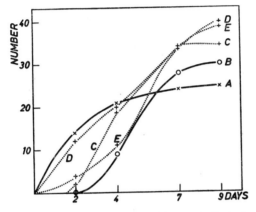

FIG. 3. Methods as in Fig. 1. Number of adventitious roots formed per seven plants in: A, distilled water; B, IAA 10^{-5}; C, IAA 10^{-5} + EDTA 10^{-9}; D, IAA 10^{-5} + EDTA 10^{-7}; and E, IAA 10^{-5} + EDTA 10^{-5} mole/litre.

depicts the number of adventitious roots formed in IAA with low additions of EDTA. Root production is increased by 10^{-7} mole/litre of EDTA because the lag-period of initiation of adventitious roots disappears and their number increases. The growth of the individual roots is unchanged (Fig. 4). The rate of emergence of adventitious roots is likewise unaffected. The main meaning of the superficial "reversal" of the IAA inhibition is, under the present circumstances, only an earlier start of the emergence of adventitious roots compensating the absent growth of the seminal roots. But since the rate of formation of adventitious roots is not hampered by IAA there is no question of an antagonism between the compounds. They act independently. This results in a root growth per plant with IAA + EDTA equalling that of the control, because this is the limit set to root growth by the nutrient conditions.

The results can be summarized as follows. IAA prevents growth of the seminal roots but increases the initiation of adventitious roots, which grow rapidly in spite of repeated IAA additions. EDTA uniformly reduces the longitudinal growth of all roots. There is no resemblance between the two agents. The total growth made in both instances is limited by nutrient conditions.

This extensive treatment of the range of importance of the results of Heath and Clark has been necessary in order to make clear both the significance of the similarity between auxins and chelating agents and

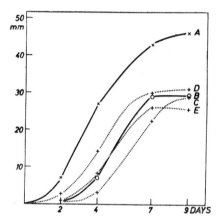

FIG. 4. As in Fig. 3. Growth in length of adventitious roots formed up to the 4th day. Later roots disregarded.

because an interpretation of a growth action must rest upon a study of what growth implies in each instance.

III. DESCRIPTION OF THE GROWTH ACTION BY EDTA

Growth inhibitions by EDTA and homologous compounds are assumed to depend upon a successful competition with natural ligands in the cell for an essential metal. *Mutatis mutandis*, growth promotions should depend upon the removal of a toxic metal. This must not be taken literally but may imply the lowering of a supra-optimal concentration of an otherwise indispensable metal. It is, of course, of importance methodologically that a promotion should not depend upon the detoxication of an accidental contamination of the nutrient medium by a metal alien to the plant, or the prevention of an external precipitation of a nutrient metal. This is what EDTA does to Fe in an ordinary nutrient solution, but it is physiologically of minor interest.

The true competition ought to occur inside the tissue by EDTA taken up by the plant (Burström, 1961). It is known to be taken up by cells, as was clearly demonstrated by Weinstein et al. (1954) and Weinstein et al. (1956 a, b). Opinions differ, however, whether ligand and metal are taken up in combination or separately (Wallace and North, 1953; Brown and Tiffin, 1960; Tiffin et al., 1960). Tiffin and Brown (1961) have demonstrated that Fe is taken up independently of the ligand and in relation to the initial level of Fe. EDTA and homologues were practically not taken up at all. Circumstantial evidence points to an uptake of the chelate (Burström, 1961) as does also the high stability of the complexes. According to Hill-Cottingham and Lloyd-Jones (1961) EDTA is actually metabolized in the plant, which is supposed to be the way in which strongly chelated metals like Fe^{3+} are released and utilized. An exchange would then hardly occur inside the plant. It must be left open to what extent this generally holds true. It is obvious that the results of different authors are controversial, and the conditions for the uptake of chelating ligands of this class are not well defined. There may also be quantitative differences between the recorded results, which cannot easily be evaluated from the data presented.

Most actions described have been growth inhibitions; promotions have been recorded in auxin tests on shoots (see Section II, A), and in two instances on root growth as well. Weinstein et al. (1956) found maximal growth of soybeans at 5–10 ppm (about $2 . 10^{-6}$ mole/litre) of EDTA and Majumber and Dunn (1960) with corn at $2 \cdot 5$–$5 . 10^{-6}$ mole/litre. These results were obtained in long-time water culture experiments with entire plants and it was not possible to specify the mode of action of EDTA.

In 1954 Weinstein et al. had already stated that inhibitory effects of EDTA ought to be due to removal from enzymes of activating metals, to which could be added metals entering into prosthetic groups. Only few attempts have been made to follow up this line on plants. Weinstein et al. (1956 a, b) studied the composition and activities of soybean plants grown to maturity with EDTA. In concentrations promoting growth they noticed a decrease in nitrogen content, increases in polyphenol oxidase and catalase activities, and decreases in cytochrome oxidase and peroxidase activities. Higher, growth-inhibiting concentrations generally increased enzyme activities. They conclude that the aberrant pattern cannot be associated with definite metal requirements. This may depend upon the complicated conditions in old plants suffering from diverse disturbances. Honda (1957) has reported reduction of root respiration by 10^{-3}–10^{-2} mole/litre of DIECA and oxin, partly in a 10^{-2} mole/litre EDTA-buffer, which seems to have been without effect itself. Decreased respiration has also been recorded by Hanson (1960) in roots together with a destruction of RNA, but at concentrations of EDTA exceeding 10^{-3} mole/litre, which are 1000 times higher than those causing a 50% reduction of root growth in darkness (Burström, 1961). Higuchi and Uemura (1959) obtained similar results with high concentrations added to yeast. These metabolic and enzymatic inhibitions can hardly be advanced as explanations of growth effects by very low concentrations of the chelating agents.

An attempt to specify the action of chelates on cell elongation was made by Ochs and Pohl (1959) by studying the effect of EDTA and of the likewise chelating kojic acid on sugar uptake and synthesis of cellulose in *Avena* coleoptile sections at supra-optimal additions. In spite of a complete cessation of growth neither part of the carbohydrate transformations was affected. The result is in accord with the finding that strong growth inhibitions do not hamper the cell wall synthesis either (Burström, 1958), but gives no clue to the point of action of the chelating agents in the growth system.

TABLE III

The Action of EDTA on Root Growth in Light and Darkness
(from Burström, 1960)

Basal nutrient solution without heavy metals but with an excess of Ca and Mg.
Concentration of EDTA 10^{-6} mole/litre.

Treatment	Root length (mm)	Cell length (μ)	Cell number
Light, control	31·8	152	209
+ EDTA	19·0	119	160
Darkness, control	52·2	157	339
+EDTA	19·7	154	128

Another approach has been an attempt to elucidate the part of the growth process affected by EDTA and other chelating compounds. Information on this point is furnished in the auxin tests on coleoptiles by Bennet-Clark (1956) and Heath and Clark (1956a), because these tissues are said to grow only by cell elongation. Growth inhibitions, mainly studied on roots, are of different nature, according to the prevailing conditions and concentrations of the chelating agents. Detailed studies have been performed on roots only. Burström and Tullin (1957) demonstrated that in an unbalanced solution with intact plants, EDTA in low concentrations inhibits only the cell multiplication but not the real cell elongation. Experiments with intact plants are difficult to interpret (cf. Section II) and no distinction was made between rate and duration of the cell elongation.

This was done in more extensive studies with isolated roots (Burström, 1960), which also revealed that the action of EDTA is light-dependent (Table III). The cell length is reduced by such a low addition of EDTA as 10^{-6} mole/litre, but only in light. This was shown to depend upon a decrease in rate of elongation proper, whereas the meristem activity is reduced in darkness only. This latter effect is the predominating growth inhibition by EDTA in low concentrations.

This is also clearly borne out by Fig. 5, depicting experiments with a constant addition of Fe and varied EDTA. The cell multiplication in the apex is strongly affected by EDTA in the dark but not in the light, when it becomes practically insensitive to EDTA. It has been

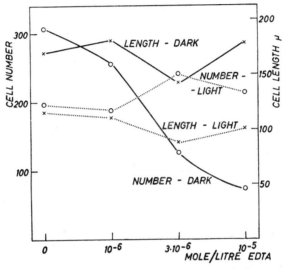

FIG. 5. The influence of Na_2 EDTA on the growth of isolated wheat roots in light and darkness. Growth computed as cell division rate (cell number) and cell elongation (cell length). Basal solution with Fe 10^{-6} mole/litre.

estimated that the roots are at least twenty-five times more sensitive to EDTA in darkness than in light. The normal light inhibition of root growth resembling a prevented etiolation and due to a reduced meristem activity is converted by EDTA into a light promotion. This clear-cut interaction between EDTA-inhibition of root meristem activity and illumination must be accounted for in an explanation of the chelate action on growth.

Higher concentrations of EDTA than 10^{-5} mole/litre will virtually kill isolated wheat roots in a weak nutrient solution (Burström, 1960) even with a forty-fold excess of Ca and Mg. They also cause deformations of *Brassica* roots resembling those induced by oxalate (Cormack, 1959) or certain auxinic compounds (Hansen, 1954). It should be observed that these concentrations are still very low compared to those found to inhibit metabolism in general and special enzyme reactions. Growth seems to contain some process very sensitive to chelated metals.

IV. IDENTIFICATION OF THE ACTIVE METALS

A. THE CA- AND CU-THEORIES

Granted that chelates increase shoot growth and decrease root growth by removing or adding metals, the next question is what metals these may be. In the earlier papers dealing with this problem the conclusions were based mainly on deductions. Interest was focused on Ca, because it was known to be especially implicated in the growth mechanism. Ca more than other cations promotes root growth (Burström, 1952); 10^{-5} mole/litre added to wheat roots will yield maximal growth. It also inhibits coleoptile elongation (Cooil and Bonner, 1957), but this effect was found only at higher concentrations of Ca. About 1/300 mole/litre was required for 50% inhibition.

Bennet-Clark (1956) then launched the idea that EDTA chelates Ca from the cell wall. Growth should depend upon the plasticity of the wall pectins; this increases by methylation whereas Ca-bridges between carboxyls lower the plasticity and thus growth. This principle has attracted considerable attention as one mechanism of regulation of cell elongation. It is beyond the scope of the present review to decide whether the principle is acceptable or not. The current problem is whether it can explain the action of externally added chelating compounds. Even if this does not hold good the theory may, nevertheless, be correct.

Heath and Clark (1956a) supported this idea but only with some reservations in 1960, and without direct evidence of their own. In a lecture Thimann and Takahashi (1958) claimed to have found a removal of Ca from the cell walls by EDTA and IAA, a statement which later could not be verified (Thimann and Takahashi, 1961). In an incomplete nutrient solution the EDTA-inhibition of cell multiplication in roots can be reversed by Ca and Mn (Burström and Tullin, 1957). However, it is emphasized that neither metal need be the active one, even if Ca-deficiency was likely to occur under those circumstances.

Objections to the Ca-chelate hypothesis were raised by Carr and Ng (1959a). They pointed out that a classical agent chelating Ca is citric acid. From cell wall preparations at pH 5·5 citric acid removes about 80% of the bound Ca, but EDTA none whatsoever. They further confirm (Ng and Carr, 1959) that EDTA yields some stimulation of coleoptile growth at pH 4·5–6·5, which they say cannot depend upon a removal of Ca. However, their claim that Ca is not chelated by EDTA under these circumstances is open to serious doubt; a correct estimation will show a chelation of 50% at pH 4 and such a small excess of EDTA as 10%, and at pH 4·5 90% chelation. With an EDTA : Ca ratio of 2 : 1 the chelation at pH 4·5 exceeds 95% (cf. Chaberek and Martell, 1959, Appendix 4). Finally Carr and Ng (1959b) show that the EDTA-action on growth requires aerobiosis and is most likely connected with some kind of metabolism, not simply an exchange of Ca in the cell wall. This connection of the ultimate growth response with metabolism does not in itself exclude the possibility that the primary action of the chelating agent is to bind Ca, but it is inconsistent with the simple plasticity concept.

Against the theory could also be mentioned that in cultures of excised roots under certain conditions EDTA considerably inhibits growth even with a 300-fold excess of Ca (Burström, 1960, 1961). However, it could be that EDTA may remove some other metal as well as, under different circumstances, Ca.

Masuda (1959, 1961) has explained the similarity between auxin and chelating agents, avoiding the inability of auxins to form metal-chelates, in the following way. He adheres to the Ca-bridge theory of growth as formulated by Bennet-Clark. Auxins increase the binding of Ca to RNA in the cytoplasm causing a shift of Ca from cell wall to cytoplasm, thereby increasing the growth of coleoptiles. Chelating compounds directly remove Ca from the cell wall with the same result as regards growth. Hanson (1960), on the contrary, states that EDTA causes loss of nucleotides and decreased respiration in roots, which is partially reversed by Ca in the high concentration of 10^{-3} mole/litre; Ca stabilizes the nucleotides in the cytoplasm. Both Masuda and Hanson assume that EDTA removes Ca but from different parts of the cell, cell wall and cytoplasma respectively, but only Masuda has traced a connection with the growth response. Hanson's experiments have been extended over a long time, which may explain some difference.

It is attractive to assume a cell elongation action of chelating compounds through a binding of Ca, but there is at present only circumstantial evidence for a participation of Ca in the chelating system affecting growth.

Cohen et al. (1958) introduced Cu into the picture. They found by means of spectral analyses indications of a complex formation between Cu and auxins with fairly regular ratios of Cu : auxin of 1 : 1 or 1 : 2. A binding of Fe was also mentioned. This was partly confirmed by Recaldin and Heath (1958), but they were unable to find the regular Cu : auxin ratios. Fe catalysed the breakdown of IAA. Again Fawcett (1959) and Fawcett et al. firmly dissented (1959). They explained the spectral changes as due to ordinary dissociations and associations, denying all chelation by auxin. Finally Perrin (1961) as already mentioned has arrived at the same conclusion.

The Cu-hypothesis was launched under the influence of the assumption by Heath and Clark of an identity between auxins and chelating compounds. The possibility that Cu is chelated by auxins can obviously be rejected. There are, furthermore, no reports in the literature of growth actions of Cu which could account for the responses to established chelating compounds either.

B. METHOD OF IDENTIFYING A GROWTH ACTIVE METAL

A reliable method must be founded on the principles of exchange of metals on a chelating agent.

Assuming that K_{en} stands for an endogenous chelating compound present in a cell and activated by a metal M to a complex $K_{en}M$ of a certain physiological activity, and another, exogenous chelating agent K_{ex} is added to the nutrient solution, an equilibrium

$$K_{en}M + K_{ex} \rightleftharpoons K_{en} + K_{ex}M \tag{1}$$

is obtained. If the complex constant of $K_{ex}M$ is considerably higher than the constant of $K_{en}M$, the equilibrium will be shifted more or less towards the right-hand side, and $K_{en}M$ inactivated accordingly. It is unimportant whether the exchange takes place inside the cell by K_{ex}, or is mediated by an exudation of free metal M, provided that an equilibrium will be attained or approached.

To this system an exogenous metal M^1 is added. Assuming that $K_{ex}M^1$ is sufficiently more stable than $K_{ex}M$, the metal M will be liberated

$$K_{ex}M + M^1 \rightleftharpoons K_{ex}M^1 + M, \tag{2}$$

and, since it was assumed in (1) that K_{ex} forms more stable complexes than K_{en}

$$K_{en} + K_{ex}M + M^1 \rightleftharpoons K_{ex}M^1 + K_{en}M. \tag{3}$$

With the equilibrium shifted towards the right-hand side $K_{en}M$ is restored, and M^1 has reversed the inactivation by K_{ex}. Experimentally K_{ex} and M^1 are added simultaneously and an inactivation of $K_{en}M$ by K_{ex} is prevented. To what degree the physiological action of $K_{en}M$ is restored depends upon the complex constants and the concentrations. In order to facilitate an interpretation it is convenient to assume that K_{ex} and M^1 are added in equivalent amounts and that the concentrations of both exceed that of $K_{en}M$. This means experimentally that K_{ex} added alone causes maximal inactivation of $K_{en}M$. The complex constant for $K_{ex}M^1$ should be known. By determining the reversal of the inhibitions by a series of exogenous metals M^1, M^2 ... of known complex constants with K_{ex}, it is possible to estimate $K_{ex}M$ in the following way.

1. In equation (2) the concentrations of K_{ex} and M^1 are known as well as their stability constant. The theoretical liberation of another metal M can be determined for a series of added metals M^1, M^2 .... This has been computed, as an example, assuming $M = Fe^{2+}$, with a Fe-EDTA of $10^{14.5}$ (Martell and Calvin, 1952) and K_{ex} as EDTA. As metals M^1, M^2 ... were chosen Fe^{3+}, Cu, Ni, Co, Zn, Fe^{2+}, Mn, and Ca. The result has been depicted in Fig. 6, with free Fe^{2+} plotted against

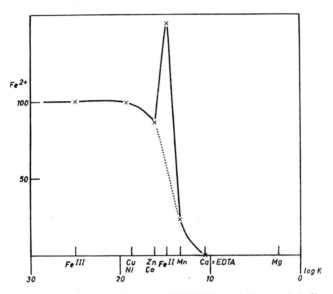

Fig. 6. The theoretical exchange between Fe^{II} EDTA and the metals indicated on the abscissa in inorganic salts. All compounds in concentrations of 10^{-6} mole/litre. On the abscissa log K of the metals with EDTA, on the ordinate relative amount of free Fe^{2+} liberated by the other metal added. Full-drawn curve with Fe^{II} among the salts added broken part without additional Fe^{II}.

the complex constants for the different metals. The graph has the expected shape of a dissociation curve, with a rapid decrease at Fe^{2+} EDTA, or since the series of metals happens to contain Fe^{2+} itself, with a peak at its constant, followed by a sharp drop. This is the theoretical shape of any such curve, and if M had another complex constant the curve would only move along the abscissa. It is in this way possible to identify the liberated metal M.

2. Experimentally it has to be assumed that M is bound to K_{en} according to equation (3) and that the activity of the restored complex $K_{en}M$ can be measured, and stands in some simple relation to the amount of $K_{en}M$.

This system contains two chelating agents K_{en} in the cell and K_{ex} added externally, and two metals M, endogenous, the identity of which is to be established, and one exogenous metal M^1, which is known. The chelating agents compete for the two metals. The principle can be put to experimental tests, by determining the physiological activity ascribed to the complex $K_{en}M$ (a) in the normal condition of the plant, (b) with the addition of an inactivating amount of K_{ex}, and (c) with the same plus a series of competing metals M^1.

3. This has been done with a well-known physiological system: the action of EDTA on the cell multiplication in wheat roots in darkness

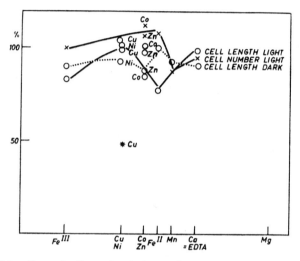

Fig. 7. Cell lengths and cell number in light of wheat roots grown with Na_2 EDTA and additions of the metals indicated on the abscissa as inorganic salts. No basal addition of Fe. All concentrations 10^{-6} mole/litre. Abscissa as in Fig. 6, ordinate in per cent of values with neither EDTA nor additional metals.

(Burström, 1961), which is the predominating root growth inhibition by EDTA. To isolated roots was added Na_2EDTA alone or together with the mentioned series of metals, all 10^{-6} mole/litre. This very low concentration was chosen, since it had been found that maximal inhibition without deterioration of the roots was obtained with $1.1 . 10^{-6}$ mole/litre (Burström, 1961). Series were performed with and without a constant basal addition of Fe^{3+} EDTA in order to avoid a general Fe-deficiency, and experiments were run in light and darkness.

With regard to cell elongation in light and darkness and cell multiplication in light no significant response to EDTA or to the metal additions was obtained, save for the cell length reduction by both Fe^{3+} and Fe^{2+} in light and a probably specific and strong toxicity of Cu

(Fig. 7). In darkness, however, reversal curves were obtained for the cell multiplication (Fig. 8) strikingly resembling that in Fig. 6, leaving no doubt but that an endogenous metal removed by EDTA and responsible for the growth inhibitions is Fe^{2+}. There is an indication of a rise in activity for Zn as well; it may contribute to the growth effect, but Fe^{2+} is mainly responsible for the cell division activity in the root meristem.

It must be especially emphasized that an unknown endogenous metal can be identified only by a comparison between a series of externally added metals. The EDTA-inhibition is reversed also by, for example, Ni and Co; if they were tested alone it could be inferred that

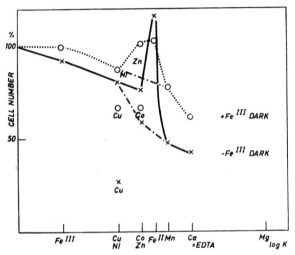

FIG. 8. As Fig. 7 but cell multiplication in darkness. Experiments with (— — — —) and without (————) a constant addition of Fe^{III} EDTA. (·—·—·—·—) indicates probable course if Zn and Fe^{II} had not been added.

they were the endogenous metals and possessed physiological activities of their own. There is no reason for such an assumption; metals with a lower constant such as Zn and Fe^{2+} would then have been unable to reverse the EDTA-inhibition. The differences between the complex constants are so great that Zn cannot compete with Ni, nor Fe^{2+} with Co. It is thus easy to draw erroneous conclusions from tests with single external metals added.

It should be pointed out that of the two closely similar graphs of Figs. 6 and 8, Fig. 6 is based on theoretical computations from complex constants, but Fig. 8 on determinations of rates of cell multiplication. These are not even directly measured but computed from two independent measurements (Burström, 1961) one of increase in root length, another of final cell lengths attained after finished cell elongation. The agreement between Fig. 6 and Fig. 8 must be regarded as most satisfactory.

It is assumed in equations (1)–(3) that the order of complex binding is the same for the endogenous chelating agents and EDTA. This may hold true as a principle, but in physiological tests the order may be disturbed. DeKock (1956) obtained the order of

$$Cu > Ni > Co > Zn > Cr > Mn\dagger$$

for the toxicity in *Sinapis*, which is the theoretical order if it were due to a chelation, but Hunter and Vergnano (1953) record exceptions for metals producing chlorosis in oats; $Ni > Cu > Co > Cr > Zn > Mn$. Since Ni and Cu, and Co and Zn respectively may fall very close to each other, as they do in binding to EDTA, the deviations are of little importance. Decidedly specific effects falling entirely outside the unspecific chelation principle occur, however. This is the case in the mentioned wheat root studies for Cu on cell multiplication and for Ni on cell elongation under certain conditions (Burström, 1961).

V. Position of the Metals in the Growth Mechanism

Exogenous chelating agents are in themselves of little physiological interest as growth regulators; they are only tools, by which it is possible to study requirements and functions of di- and tri-valent metals. There are at present three metals which are suspected of being particularly involved in growth in a rather wide sense: calcium, iron, and cobalt. A summary of their position in the growth mechanism as far as it can be surveyed is warranted.

A. CALCIUM

The literature on calcium relevant to the chelate action has already been mentioned, but some remarks should be added.

The recent literature is dominated by the Ca-bridge theory of Bennet-Clark (1956) and its application. However, it has caused some confusion that roots and shoots react differently to Ca and to chelating agents, which has, for example, led Cleland (1960) to the conclusion that the principle could not be applied to roots. This is, in point of fact, an open question. It is necessary to recall the ancient finding that Ca is an indispensable nutrient for probably all plants down to fungi and bacteria.

---

† This series is similar to that given by Irving and Williams (*J. Chem. Soc.* 3192, 1953) for the stability of nearly all complexes irrespective of the nature of the ligand, in so far as the metals investigated by the two groups of workers overlap.—Ed.

It is not known that coleoptiles and hypocotyles, employed in routine growth tests, should form exceptions, but their Ca-requirements are virtually unknown. However, they are amply supplied with Ca from their seeds. The requirement of wheat roots has been estimated to be an external supply of 10^{-5} mole/litre for the cell elongation and 10^{-6} mole/litre for the cell multiplication in the apical meristem under certain defined conditions (Burström, 1952). Two points should be emphasized: the very low concentrations and the smaller requirement for the cell multiplication.

Two attempts have been made to explain the Ca-action, both of them with regard to the regulation of cell elongation. Based on observations of the relation between Ca and auxin and on determinations of cell wall tensilities, the following picture was outlined for the *suboptimal Ca-effect* studied on roots (Burström, 1952, 1954). Ca appears as an antagonist of auxin; this must not necessarily mean that they react directly with each other, but that they regulate the same part of the cell elongation mechanism. This involves a deposition of new cell-wall matter and the role of Ca should be to stabilize the wall. This could be compared with the well-known stabilizing effect on cytoplasmic colloids, which is regarded as one of the main physiological functions of Ca. With Ca-deficiency elongation stops under a loss of all elasticity, which was regarded as a deterioration of the cell wall.

The theory of Bennet-Clark, subsequently called the Ca-bridge theory, implies a formation of Ca-salt bridges, hardening the wall to the point that elongation is prevented. It is based on results with *supraoptimal Ca-effects* studied on shoot parts. It is probably untenable, implying a chelation of Ca by auxin, even if Cleland's criticism is not entirely conclusive. He can account for about 10% of the cell wall Ca in his analyses, but the state and behaviour of the remaining bulk of Ca, derived from the seed, are unknown. Nevertheless, the theory may hold good for explaining the regulation of cell elongation by Ca in the given instances, even if auxins cannot also be involved in the picture.

The difference between the reactions studied in root and shoot experiments is also illustrated by the external Ca-concentrations employed: with roots concentrations around 10^{-6}–10^{-5} mole/litre, for shoot parts 10^{-3} mole/litre and higher. The difference ought to be significant even considering the facilitated Ca-uptake in roots. The low-Ca stabilization principle at Ca-deficiency and the high-Ca bridge principle should certainly be kept apart, notwithstanding the fact that the two theories are rather similar. It is assumed that Ca in both instances increases the rigidity of the cell walls, from a disorganization not allowing a regular growth to a structure optimal for elongation, and from that stage further to a stiffening preventing further elongation.

It may be pertinent to recall that Letham (1960) has used EDTA for a mild maceration of tissues, explained by a chelation of Ca from pectins. It is true that physiologically very high concentrations were employed, but maceration is also a drastic derangement; the technique rests on the assumption that a certain Ca content of pectins is necessary for normal structure.

Without stretching the comparison too far, a reference to the action of boron on cell walls is revelant. Based on growth studies combined with plasmometric and mechanical measurements of wall tensility Odhnoff (1961) has outlined a picture of the action of boron on cell elongation. Boron may form diols with carbohydrates or *bis*diols constituting bridges between carbohydrate chains. Under conditions con-

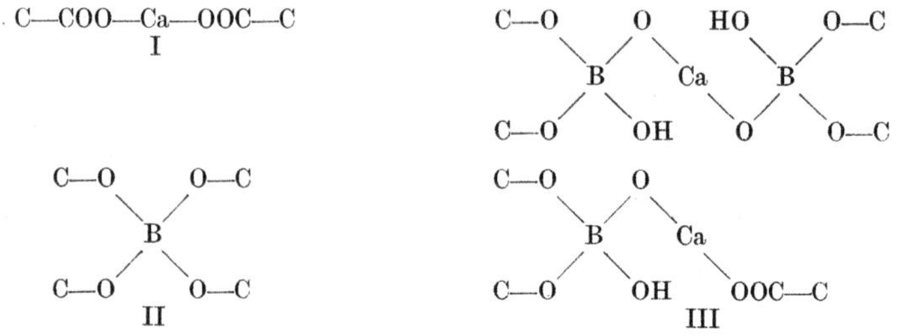

FIG. 9. Some assumed types of bridges formed by Ca and B between carbohydrate chains. I. Ca-bridge between carboxyls according to Bennet-Clark. II. B-*bis*diol-bridge between hydroxyls according to Odhnoff. III. Tentative combinations of Ca and B-bridges.

ducive to *bis*diol formation the wall tensilities decrease as expected, and also the cell elongation. Boron is as indispensable as calcium and the requirements of roots are of the same order of magnitude. The similarity goes so far that *bis*diol formation structurally ought to correspond to a Ca-bridge, even if, as pointed out by Odhnoff, the B-bridges might join microfibrils, not pectins. A possibility worthy of consideration is, of course, a direct reaction between Ca and the partly complexed boric acid (Fig. 9). A physiological interrelation between these two nutrients has repeatedly been emphasized in the literature (see Odhnoff). Ca-bridges could easily be formed to diol-bound boric acids. It cannot be inferred from the available literature (see Chaberek and Martell, 1959) whether also indole-3-acetic acid could join such a system as a ligand.

The action of external chelating agents like EDTA follows from these considerations. A chelation of Ca from shoots is likely to occur: it is borne out by experiments of Cleland (1960, Table 1) but must always depend upon the relative stability of the endogenous complexes. Carr and Ng (1959a found that EDTA prevented a removal of Ca by citric

acid, which is against all expectations. The reactivity of roots illustrates the higher Ca-requirement of cell elongation (Burström and
Tullin, 1957). Without additions of Ca or heavy metals EDTA does
not affect cell elongation, which is already minimal, evidently owing
to lack of Ca. It reduces cell multiplication, but with Ca supplied this is
regulated by iron.

B. IRON

Two different actions of iron on growth can be distinguished (Burström, 1960, 1961). An iron addition to roots in a Fe-free nutrient solution results in a reduced rate of elongation in light and an enhancement
in darkness. In spite of the low concentrations employed, around 10^{-6}
mole/litre this may be called a *high-Fe* effect. Additions of EDTA
strongly inhibits cell multiplication in darkness by chelating primarily
Fe. This is a *low-Fe* effect produced by amounts lower than those carried
to the plant with the seed or as an external contamination.

That Fe is necessary for meristematic activity is hardly surprising.
What needs to be explained is that EDTA chelates Fe^{2+} from roots
only in darkness and that the roots are insensitive to chelation in light.
They are also light-insensitive in the absence of iron. It was advocated
that this leads to two probable assumptions, namely that Fe functions
in a system which inhibits cell divisions in light, and that Fe in light
cannot be removed by EDTA.

The solution may lie in the fact that reduced Fe forms less stable
complexes than the oxidized Fe (Martell and Calvin, 1952). It is also
known that Fe can be oxidized in light (literature see Burström, 1961)
and it will then be more firmly bound. Of course, this holds true of both
a chelation by EDTA and the endogenous agent. Fe^{3+} EDTA is in itself
light-sensitive and undergoes photo-reduction in UV-light (Hill-
Cottingham, 1955) under partial destruction of the ligand. The inability
of EDTA to inhibit cell divisions in the light can hardly depend upon a
light-induced decomposition by this mechanism for the following
reasons. The effect of added EDTA is lacking in light even in the
absence of external Fe and with an excess of EDTA, and other salts of
EDTA than with Fe should be light-insensitive. It might be possible
for a photo-reduction to occur, catalysed by the endogenous Fe, inside
the tissues, but it is not very likely with a sensitivity maximum at
260 mμ. Furthermore, this oxidation should lead to a formation of
Fe(OH) EDTA and subsequent precipitation of ferric hydroxide, which
means a removal of Fe in light, whereas the experiments indicate that
Fe is not removed under these conditions. This possibility seems to be
excluded, unless some different light-destruction also exists. It has

also been shown (Burström, 1960) that EDTA has some effect on cell elongation in light although not on cell division.

One possibility of visualizing the relation between light, Fe, and meristem activity is the following one. Provided that Fe in some compound must occur in predominantly reduced form for the normal metabolism in the meristem, cell divisions can be hampered in two ways, either by oxidation of Fe into a more stable and physiologically inactive Fe^{3+}-complex, or simply by removing the less stable Fe^{2+} by means of an external chelating agent. They should lead to the same result: a decrease in the active iron-complex. This is depicted in the following diagram.

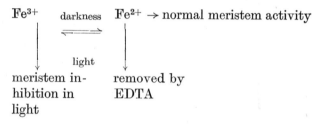

$$Fe^{3+} \quad \text{darkness} \quad Fe^{2+} \rightarrow \text{normal meristem activity}$$

meristem in-   removed by
hibition in   EDTA
light

However, this picture is very tentative since Fe is both promoting and inhibiting growth. It might even be possible that in light Fe is reduced and removed from an indispensable system to an inactive endogenous site. In such a case oxidation might counteract the light inhibition.

Any such system may be independent of the function of Fe in cell elongation, which is light-sensitive only in the presence of an external Fe-supply. In darkness Fe increases the rate of elongation, which might be explained by its function in the basal metabolism. Shibaoka and Yamazaki (1959) found, however, that Fe increased curvatures and cell elongation in Avena coleoptiles in darkness, which could be connected in different ways with the transport and metabolism of auxin.

The light inhibition of root growth resembles somewhat a prevented etiolation in a shoot, especially the Fe-dependent reduction of cell elongation. Nothing seems to be known about a participation of Fe in prevention of etiolation. There is a superficial relation between the growth inhibitions in light and the formation of chlorophyll in the roots (Burström, 1961), but it is not clear whether there is a causal connection or only a common relation to the Fe-status in the tissues. The findings of Shibaoka (1961) of a light production in leaves of a growth inhibitor is of interest in this connection. It is assumed to inhibit SH-groups, because it is reversed by 2,3-dimercaptopropanol (BAL). It tallies in this respect with the Co-inhibition described by Thimann (1956). Since the Fe-effect on roots is likewise influenced by Co it might be possible to co-ordinate these scattered results, although the kind of connection between light, cell elongation, and metals remains obscure. These light reactions are probably entirely different from the promotions studied by Scott *et al.* (1961) likewise on wheat roots.

C. COBALT

Growth actions by cobalt and nickel have been studied in the routine tests for auxins with shoot parts since a positive action was discovered by Miller (1951) (further literature see Burström, 1961). Co promotes growth and, with regard to the nature of the material, it can be assumed that it increases cell elongation (Miller, 1954). Several explanations have been proposed, but only Busse (1959) has tried to outline a detailed mechanism by which Co could increase cell elongation. It involves an inhibition of a reaction which normally limits the elongation. It is connected with a decreased respiration and a decreased incorporation of wall material. Other actions by Co that have been specified are an inhibited flowering and increased critical night length in *Xanthium* (Salisbury, 1959), and inhibitions of bacterial growth, for example in *Proteus*, by blocking the cytochrome system (Petras, 1957). It is difficult to avoid the impression that Co generally acts as an inhibitor of specific reactions, according to Busse (1959) even when increasing cell elongation.

TABLE IV

Interaction between Cobalt and Iron on Cell Elongation of Isolated Wheat Roots
(from Burström, 1961)
All additions 10^{-6} mole/litre. The values denote cell length in μ;
standard error ± 3 μ.

Addition	Light	Darkness
None	157	133
Co	152	119
Fe	103	181
Co + Fe	171	153

In wheat root experiments (Burström, 1960, 1961) Co, like Ni, is able to increase cell elongation but only in the presence of externally added Fe, and when elongation in light is actually reduced by Fe. When in darkness Fe increases elongation, Co reverses this effect as well (Table IV, from Burström, 1961). The Co-action on root elongation obviously depends upon an antagonism against Fe, which must be supposed to be the physiologically active metal. From the principle of chelate exchange it is just as likely for Co to inactivate Ca, and it should be pointed out that the reaction according to Busse inhibited by Co earlier was assumed to be mediated by Ca (Burström, 1952). It is

5

tempting not to ascribe to Co and Ni physiological actions of their own in connection with these growth responses, but to assume that they act as antagonists in exchange for active metals.

VI. SUMMARY

Chelating compounds such as ethylenediaminetetraacetic acid (EDTA) exert growth actions in low concentrations. They may resemble auxins in increasing shoot and decreasing root growth. A detailed study on roots has revealed that the mode of action of the two types of compounds is basically different, and they do not antagonize each other either.

A method has been described for the identification of the physiologically active metals removed from a plant by a chelating agent. It is based on a reversion of the growth action by a series of externally added metals with known ability to enter into chelate bindings. This method has been applied to the dominating action of EDTA on root growth: an inhibition of meristem activity in darkness. This is shown to depend upon a removal of mainly Fe^{II}, probably supplemented by Zn.

The growth actions of the three metals, calcium, iron and cobalt, supposed to be involved in the growth mechanism have been discussed.

Calcium regulates cell elongation; suggested modes of action have been compared and the analogy with a growth action by boron pointed out.

Iron regulates the light sensitivity of roots. Cell elongation is light-sensitive only in the presence of iron, and cell divisions are inhibited by a removal of Fe^{2+} in the dark only. This is discussed in terms of a change in the oxidation of Fe, and the different chelation of Fe^{2+} and Fe^{3+}. Light growth inhibitions require iron.

Cobalt has probably no physiological action of its own in the growth mechanism but exchanges for ions like Fe and possibly Ca in physiologically active complexes in the plant.

ACKNOWLEDGEMENTS

The author is indebted to Dr. O. V. S. Heath, Reading, for valuable information and to Dr. Y. Masuda, Osaka, for reading and criticizing the manuscript. The work has been supported by a grant from the Swedish Natural Science Research Council.

REFERENCES

Bennet-Clark, T. A. (1956). *In* "The Chemistry and Mode of Action of Plant Growth Substances", pp. 284–291.

Bosemark, N.-O. (1954). *Physiol. Plant.* 7, 497–502.

Brown, J. C. and Tiffin, L. O. (1960). *Soil Sci.* 89, 8–15.

Brown, J. C., Tiffin, L. O. and Holmes, R. S. (1960). *Plant Physiol.* 35, 878–886.

Burström, H. (1952). *Physiol. Plant.* 5, 391–402.

Burström, H. (1954). *Physiol. Plant.* 7, 332–342.

Burström, H. (1958). *Fysiogr. Sällsk. Lund Förhandl.* 28, 53–64.

Burström, H. (1960). *Physiol. Plant.* 13, 597–615.

Burström, H. (1961). *Physiol. Plant.* 14, 354–377.

Burström, H. and Tullin, V. (1957). *Physiol. Plant.* 10, 406–417.

Busse, M. (1959). *Planta* 53, 25–44.

Carr, D. J. and Ng, E. K. (1959a). *Physiol. Plant.* 12, 264–274.

Carr, D. J. and Ng, E. K. (1959b). *Austr. J. Biol. Sci.* 12, 373–387.

Chaberek, S. and Martell, A. E. (1959). "Organic Sequestering Agents", 616 pp. John Wiley & Sons, New York.

Cleland, R. (1960). *Plant Physiol.* 35, 581–584.

Cohen, D., Ginzburg, B-Z. and Heiter-Wirguin, C. (1958). *Nature, Lond.* 181, 686–687.

Cooil, B. J. and Bonner, J. (1957). *Planta* 48, 696–723.

Cormack, R. G. H. (1959). *Canad. J. Res. (Botany)* 37, 33–39.

DeKock, P. C. (1956). *Ann. Bot., Lond.* 20, 133–141.

Fawcett, C. H. (1959). *Nature, Lond.* 184, 796–798.

Fawcett, C. H., Wain, R. L. and Wightman, F. (1956). *Nature, Lond.* 178, 972.

Fawcett, C. H., Wain, R. L. and Wightman, F. (1959). *Proc. IX. Int. Bot. Congr.* 2, 113.

Gäumann, E., Bachmann, E. and Hüttner, R. (1957). *Phytopath. Z.* 30, 87–105.

Hansen, B. A. M. (1954). *Bot. Notiser* 107, 230–268.

Hanson, J. B. (1960). *Plant Physiol.* 35, 372–379.

Heath, O. V. S. and Clark, J. E. (1956a). *Nature, Lond.* 177, 1118–1121.

Heath, O. V. S. and Clark, J. E. (1956b). *Nature, Lond.* 178, 600–601.

Heath, O. V. S. and Clark, J. E. (1960). *J. exp. Bot.* 11, 167–186.

Higuchi, M. and Uemura, T. (1959). *Nature, Lond.* 184, 1381–1383.

Hill-Cottingham, D. G. (1955). *Nature, Lond.* 175, 347.

Hill-Cottingham, D. G. and Lloyd-Jones, C. P. (1961). *Nature, Lond.* 189, 312.

Honda, S. (1957). *Plant Physiol.* 32, 23–31.

Hunter, J. G. and Vergnano, O. (1953). *Ann. appl. Biol.* 40, 761–777.

Letham, D. S. (1960). *Exp. Cell. Res.* 21, 353–360.

Majumber, S. K. and Dunn, S. (1960). *Plant Physiol.* 33, 166–169.

Martell, A. E. and Calvin, M. (1952). "Chemistry of the Metal Chelate Compounds", 613 pp. Prentice Hall, New York.

Masuda, Y. (1959). *Physiol. Plant.* 12, 324–335.

Masuda, Y. (1961). *Plant & Cell Physiol.* 2, 129–138.

Mellor, D. P. and Maley, L. (1947). *Nature, Lond.* 159, 370.

Mellor, D. P. and Maley, L. (1948). *Nature, Lond.* 161, 436.

Miller, C. O. (1951). *Arch. Biochem. Biophys.* 32, 216–218.

Miller, C. O. (1954). *Plant Physiol.* 29, 79–82.

Ng, E. K. and Carr, D. J. (1959). *Physiol. Plant.* 12, 275–287.

Ochs, G. and Pohl, R. (1959). *Phyton* 13, 77–87.

Odhnoff, C. (1961). *Physiol. Plant.* 14, 187–220.

100 HANS BURSTRÖM

Perrin, D. D. (1961). *Nature, Lond.* **191**, 213.

Petras, E. (1957). *Archiv. f. Mikrobiol.* **28**, 138–144.

Recaldin, D. A. and Heath, O. V. S. (1958). *Nature, Lond.* **182**, 539–540.

Salisbury, F. B. (1959). *Plant Physiol.* **34**, 598–604.

Scott, E. G., Carter, J. E. and Street, H. E. (1961). *Physiol. Plant.* **14**, 725–733.

Shibaoka, H. (1961). *Plant & Cell Physiol.* **2**, 175–197.

Shibaoka, H. and Yamazaki, T. (1959). *Bot. Mag., Tokyo* **72**, 203–214.

Thimann, K. V. (1956). *Amer. J. Bot.* **43**, 241–250.

Thimann, K. V. and Takahashi, N. (1958). *Plant Physiol.* **33**, xxxiii.

Thimann, K. V. and Takahashi, N. (1961). *In* "Plant Growth Regulation", pp. 363–377. Ames, Iowa.

Tiffin, L. O. and Brown, J. C. (1961). *Plant Physiol.* **36**, 710–714.

Tiffin, L. O., Brown, J. C. and Krauss, R. W. (1960). *Plant Physiol.* **35**, 362–367.

Wallace, A. and North, C. P. (1953). *Calif. Agric.* **7**, 10.

Weinstein, L. H., Meiss, A., Uhler, R. L. and Purvis, E. R. (1956a). *Nature, Lond.* **178**, 1188.

Weinstein, L. H., Meiss, A., Uhler, R. L. and Purvis, E. R. (1956b). *Contrib. Boyce Thompson Inst.* **18**, 357–370.

Weinstein, L. H., Robbins, W. R. and Perkins, H. F. (1954). *Nature, Lond.* **120**, 41–43.

Comparative Anatomy as a Modern Botanical Discipline

With special reference to recent advances in the systematic anatomy of monocotyledons

C. R. METCALFE

The Jodrell Laboratory,
Royal Botanic Gardens, Kew, England

I. Introduction

Three years ago the writer (Metcalfe, 1959) published a historical survey of the progress of plant anatomy from the time of Nehemiah Grew until the present day, and reviewed the status of the subject in its relationship to other branches of botany. The purpose of the present article is to show why it is most desirable that a renewed interest should be taken in the comparative anatomy of the flowering plants and how further neglect of this relatively unfashionable aspect of botany may have serious consequences to the development of plant science as a whole. The writer has met numerous students of botany to whom it is a completely new idea that a knowledge of plant anatomy can be used to solve everyday practical problems, to identify economic plant products, or that it has a place as a serious method of investigation in plant classification. The importance of a knowledge of plant structure as an aid to the more complete understanding of physiology or of the relationship of plants to their environment is more generally realized. Nevertheless there are, even in these fields, surprising gaps in our knowledge. These themes are illustrated by surveying the progress of recent work on the systematic anatomy of the flowering plants or angiosperms, undertaken by the writer and his colleagues and associates, with special reference to the monocotyledons. The very term monocotyledons is so forbidding to those who are not conversant with plant classification that it may not be out of place to remind readers that we shall be considering a group of plants that range from palms, screw-pines and bananas to the cereals, fodder grasses and bamboos as well as to the orchids, daffodils and irises of our gardens to mention but a few of the more familiar monocotyledons. To show how essential it is in these days of specialization to remind ourselves that one cannot take an elementary knowledge of one's subject for granted, it may be worth mentioning that the writer, a few years ago, when discussing the subject of rice with a food chemist, was horrified to find that the chemist did not realize that the rice plant is a grass.

Reverting to the anatomical work with which this article is mainly concerned, it should be noted that the focal centre for these researches is the Jodrell Laboratory at Kew where most of the work has been done and where an extensive programme of research is still in progress. This work on the comparative anatomy of monocotyledons is a natural sequel to the publication in 1950 of two volumes on the anatomy of dicotyledons (Metcalfe and Chalk, 1950).

The corresponding reference book on the monocotyledons has already been initiated by the recent publication of the writer's volume on the anatomy of grasses (Metcalfe, 1960), which represents the outcome of 10 years' work devoted to this family, which includes so many plants of economic importance. The programme of work on monocotyledons is so extensive, however, that the staff of the laboratory have been very glad to receive collaboration from visiting research workers. In particular we are indebted to Dr. P. B. Tomlinson whose work on the palms is quite outstanding. These investigations started at the Jodrell Laboratory and at the University of Leeds, continued at Singapore, in Ghana, and more recently in the U.S.A., have already led to the publication of the second volume of our reference book (Tomlinson, 1960c), which, like the first, is thus devoted to the structure of a group of plants that includes many that are of great economic importance. It is expected that two more volumes will be needed to complete the story of the monocotyledons as a whole, but it will clearly require a further period of years to finish our task. Meanwhile a foretaste of what is to come is provided by a series of articles that have appeared and will continue to be published in scientific journals. For example, besides his work on palms, we are indebted to Tomlinson for a series of articles on the anatomy of members of the families that include ginger (Zingiberaceae), bananas (Musaceae), St. Vincent Arrowroot (Marantaceae), the cannas (Cannaceae) that are so commonly cultivated for ornamental purposes (Tomlinson, 1956, 1959, 1961a, 1961b). Then again Dr. Abraham Fahn (1954) of the Hebrew University of Jerusalem made a special study of the Xanthorrhoeaceae, a fascinating family of Australian plants that are sometimes referred to as Grass Trees. This work was done at Kew during a period of sabbatical leave. The late Miss E. Smithson (1956) of the Northern Polytechnic in London undertook investigations concerning the interesting family Flagellariaceae which seem, in part, to be related to the Gramineae. Dr. L. K. Mann (1952, 1959, 1960) of the University of California who is making a very detailed study of the morphology, anatomy and taxonomy of the genus *Allium*, the genus that includes the onions, has spent two periods at the Jodrell Laboratory in the course of his researches. The writer is also working in close touch with Professor Vernon I. Cheadle (1937–56) of the University of California whose researches on the xylem and phloem of monocotyledons are well known, and with Dr. Sherwin Carlquist of the Rancho Santa Ana Botanic Gardens who is specially interested in the anatomy of Rapateaceae (1961) and Xyridaceae (1960). Indeed it would be true to say that interest in our work has become almost world-wide amongst anatomical workers and the writer is in frequent correspondence on this subject with botanists in many different countries.

Further evidence of the widespread interest in our work is afforded by the helpful way in which botanists in various parts of the world have, often at considerable inconvenience to themselves, taken the trouble to collect specimens of interesting monocotyledons which would not otherwise be available, and have sent them to Kew after preservation in formalin acetic alcohol. It is impossible to mention all of those who have assisted in this way, but some idea of the different parts of the world from which material has been obtained may be given by recording our special indebtedness to R. E. Vaughan in Mauritius, M. R. Levyns in S. Africa, W. M. Curtis in Tasmania, Basset Maguire and J. J. Wurdack in tropical S. America, L. K. Mann and G. L. Stebbins in California, G. Jackson in Nyasaland. V. I. Cheadle has also most generously placed at our disposal material he personally collected for his own researches especially in Australia and S. Africa.

It is always desirable for laboratory workers to collect their own material in the field. Unfortunately it is seldom possible for a botanist engaged in a project based on material from many parts of the world to do his field work in more than a limited number of countries, and the assistance of field workers is therefore very much appreciated. Nevertheless the writer, apart from collecting a great wealth of living material in cultivation at Kew, has studied monocotyledons in the field in various European countries, and amidst the luscious vegetation of the West Indies. What an experience it is for a student of the anatomy of monocotyledons to see bromeliads in the mass; to observe the diverse forms of aroids, to stand in a swampy thicket of the tree-like aroid *Montrichardia* (Pl. I, A), or to see bamboos growing under tropical conditions. Finally to witness the cultivation and harvesting of such crops as sugar cane (Pl. I, F), arrowroot (Pl. I, C, D), coconuts and bananas (Pl. I, E) serves to remind anatomists from temperate regions not only of the diversity of form exhibited by monocotyledons but also that their field of study includes the structure of many important economic crop plants besides the cereals with which we are more familiar in the cooler parts of the world. The sight of these things and the knowledge that only a small fraction of the monocotyledonous species that exist have ever been examined under the microscope provide a stimulating challenge to the plant anatomist.

II. Reasons for Investigating the Anatomy of Monocotyledons

A. HISTORICAL BACKGROUND

The morphologists and anatomists who were active towards the end of the last and during the early part of the present centuries were highly competent observers. From them we have a legacy of information that is for the most part accurate, and provides the sound basis on which our knowledge of the morphology and anatomy of monocotyledons and indeed of all the flowering plants (angiosperms) still rests. The work of the morphologists and anatomists of this period, however, ended when only a small fraction of the flowering plants in the world had been examined. In spite of modern improvements in technique, the anatomical method has always been slow and laborious. Many of the discoveries made by early anatomists seemed to be largely repetitive and for these reasons the illusion grew up that further ana-

PLATE 1. Monocotyledonous plants and crops in the West Indies. A. *Montrichardia* sp. an arboreal member of the Araceae growing in a swamp in Trinidad, West Indies. B. Gigantic inflorescences ("Maypoles") of *Agave* sp. growing in Barbados, West Indies. The apices of the leaves can be seen at the base of the third specimen from the left of the picture. C. A crop of Arrowroot (*Maranta arundinacea* L.) growing in St. Vincent, West Indies. D. A crop of Arrowroot after harvesting. St. Vincent, West Indies. E. Bananas in the foreground; coconut palms in the background. St. Vincent, West Indies. F. Sugarcane arriving at a factory; Barbados, West Indies.

tomical research would not yield data of sufficient importance to justify the necessary expenditure of time. For some years this knowledge of the structure of the flowering plants was passed on from generation to generation, but when other branches of botany subsequently became more fashionable less and less time was devoted to comparative studies in morphology and anatomy. In consequence we now seem to be in danger of raising a generation of botanists with little knowledge of the comparative anatomy of angiosperms and particularly of monocotyledons.

It will at this stage do no harm to remind ourselves that the anatomy of even some monocotyledons that are of well established economic importance is relatively unknown. This was brought home to the writer a few years ago when an agricultural research worker in Ceylon who was interested in the physiology of coconuts asked for sources of information concerning the structure of a whole coconut palm. A search through the literature revealed that the type of information that this agriculturalist needed was not in fact readily available. Similar enquiries were received from a plant pathologist in the West Indies, and again there was no satisfactory source of information. The structure of other crop plants such as pineapples and bananas has been more adequately described, but nevertheless, the information is for the most part in journals where it might be overlooked by research workers unfamiliar with the literature of plant anatomy.

Quite apart from the fact that we ought clearly to learn as much as we can about the anatomical structure of crop plants so as to be able to assist our colleagues who are concerned with the cultivation or physiology of these plants, it must also be remembered that if we cease to teach our students about the histology of economic plant products there will no longer be any botanists who are able to see to it that standards of quality are maintained or that substitutes and adulterants do not reach the markets. It will be still more difficult to find botanists who can write a good description of the diagnostic microscopical characters of an economic plant product or establish the botanical origin of the vegetative parts of plants from their microscopical structure. Some private consultants whose business it is to deal with such problems have told the writer that they prefer to employ technicians who are not graduates to assist with this work because they stick at the task more effectively without becoming bored. So long as the work involved is purely a matter of routine analysis this is understandable. The danger arises, however, that biological material is variable and does not always conform to text-book rules. Furthermore in these days of economic readjustment when new sources of raw materials may replace those that are traditional, or when entirely new economic products may reach the market for the first time, a technician, working by rule of thumb methods, may not have enough knowledge to deal with an anatomical problem which cannot be solved by adopting a standard routine. For example the present renewed interest in native medicinal plants, especially in tropical countries, means that the botanical identity of species of which there are no published histological descriptions has to be established. This is often a matter of extreme difficulty as the writer knows to his cost. Undue reliance on the chemical rather than on the histological assay of vegetable materials can also lead to wrong conclusions. An example of this came to the writer's notice with reference to a sample of poultry food which had caused serious illness to birds to which it had been fed. Chemical tests on this food had been claimed to demonstrate the presence of theobromine, and it was concluded, on this chemical evidence, that toxicity was due to an admixture of cocoa residue. Doubt was thrown on the validity of this conclusion, however, when a microscopical examination of the poultry food, undertaken by the writer, failed to reveal the presence of any cocoa residue. On another occasion the writer was asked to examine an archaeological specimen which, on chemical evidence, was thought to be a portion of human bone. Microscopical examination of this specimen showed quite clearly that the chemical evidence had led to an incorrect conclusion, for the alleged bone was in fact a piece of oak wood. Chemical and histological data should undoubtedly both be used in solving problems of the kinds that have just been mentioned, but the point to be noted is that the solution of all but the most mundane of these problems should be in the hands of botanists with a good basic knowledge of plant histology rather than of technicians performing routine tests by rote.

B. STIMULUS OF MODERN TRAVEL FACILITIES

The increased speed of modern travel has opened up possibilities of which our botanical forefathers would never have dreamed. Air-mail letters can be sent rapidly to botanists in previously inaccessible parts of the world. This makes for effective collaboration with botanists who can supply material of species that have hitherto escaped the notice of anatomists. Even more valuable is the fact that European botanists can work in the field in remote areas, for example during the long vacation.

This increased speed of communication is helping to make us realize that our concepts concerning the classification of plants have, all too often, been at fault because they were formed during a period when the plants of the north temperate region were better known than those from elsewhere. As our knowledge of plants from tropical regions and from the southern hemisphere has increased, it has become more and more apparent that to arrive at a satisfactory taxonomic system it is frequently necessary to start by studying the tropical representatives of a group and then to proceed to those from more northern and more southern climes. Whilst it is becoming generally recognized that this applies to plant classification based on the traditional approach of the herbarium botanists, it is still more true when making use of histological characters for taxonomic purposes because our knowledge of comparative plant structure is so imperfect.

C. USES OF HISTOLOGICAL DATA IN CLASSIFYING PLANTS

Many monocotyledons such as the palms do not lend themselves to study by the traditional methods of herbarium taxonomy. What an aid to the taxonomist it would be if these unwieldy vegetable monsters could be identified, even approximately, by studying the structure of small selected portions of their fronds. And would not great possibilities for the palaeobotanist be opened up if we knew far more than we do at present concerning the comparative structure of the trunks of palms? Then again there are many monocotyledons such as the economically important grass family (Gramineae) which are notoriously difficult to classify on exomorphic characters alone and where the importance of histological data is becoming increasingly recognized. Besides all this there is the well-known fact that there are differences of opinion concerning the delimitation of monocotyledonous families, whilst the evolutionary history and phylogeny of monocotyledons are almost unknown. As we shall see, it is not so much as an aid to the minutiae of plant taxonomy that the study of histology is proving useful, but rather in helping to establish the broad outlines of plant classification.

III. METHODS OF INVESTIGATION

A. NEED FOR SIMPLE TECHNIQUES

In using histological data for taxonomic purposes, the botanist is faced by the difficulties that the preparation of material for microscopical examination takes time, and that only minute portions of plant tissue can be brought into the field of a microscope at any one moment. This means that simple, rapid techniques must be employed to prepare the slides and that attention must be directed to tissues from carefully selected regions of the plant body that are found by experience to yield data of taxonomic value. Here it must be emphasized that freehand sections mounted in dilute glycerine, or stained and mounted in chlorzinc-iodide are often of great service, especially for preliminary observations. Larger sections of more uniform thickness can be taken with the sledge microtome, and, after suitable staining, mounted in Canada balsam and kept for further reference. Sections are best prepared from material that has been fixed in formalin acetic alcohol. With plants such as grasses which contain silica deposits it is usually desirable to remove the silica by treatment with hydrofluoric acid before satisfactory sections of them can be cut. Where one has to deal with flexible material such as the lamina of a leaf this may be held in the microtome clamp between pieces of pith and sections prepared in the same way as if they were being cut freehand with a razor. For these purposes it should be noted that it is quite needless to go to the length of embedding the material in wax or a plastic, and indeed with many monocotyledons superior results are usually obtained by the simple procedure outlined above. When it is necessary to examine an epidermis in surface view, good results are often obtained by using simple methods that involve scraping away the mesophyll with a scalpel and safety razor blade, by a procedure that has been described more fully elsewhere (Metcalfe, 1960, pp. lx–lxi).

B. SELECTION OF MATERIAL

It is impossible here to give precise directions concerning the portions of a monocotyledonous plant that can most usefully be examined for taxonomic purposes. This is because the morphology of monocotyledons is so diverse. Some of them have well-developed bulbs, corms and rhizomes; in others the aerial stems are well developed, or the apparent aerial stem may be no more than a pseudo-stem formed from overlapping leaf bases as in the bananas. When we turn to leaves we have to consider organs ranging in size from the reduced leaves on photosynthetic stems that occur for example in some of the Restionaceae, to the enormous simple leaves of bananas and their allies the heliconias, or the palms with their pinnate or fan-like fronds. In practice some of the most useful diagnostic characters are to be found in leaves, and this is fortunate for they are usually the most readily obtainable part of the plant. Leaves are available even when the plants bearing them are not in flower and this adds considerably to their interest by providing diagnostic characters in vegetative specimens. The structure of leaves can be examined in sections of the blade taken at right angles to or parallel with the veins. Paradermal sections can also be prepared and the epidermis can be examined in surface view. All of these approaches yield information of interest to the taxonomist, but the main difficulty is that local structural differences that occur within a single leaf must put us on our guard against drawing wrong taxonomic conclusions through failing to restrict our comparisons to the corresponding parts of any two leaves that are being investigated. For this reason it has become customary to select certain arbitrary positions in the leaf when making microscopical preparations for comparative purposes. For example in relatively narrow, linear leaves such as those that occur in grasses, comparisons are usually made in sections and epidermal preparations taken from a position midway between the apex and base of the lamina, excluding the sheathing leaf base.

Connor (1960) has, very usefully, drawn attention to the difficulties and dangers that can arise when attempting to distinguish between closely related species of the grass genus *Festuca* from the structure of the leaf. Whilst the difficulties to which he has drawn attention can be very real ones when dealing with minor taxa in a critical genus such as *Festuca*, in the writer's experience it would be wrong to jump to the conclusion that all histological characters are of doubtful value for taxonomic purposes. The writer has developed this subject, with reference to the grasses, more fully elsewhere (Metcalfe, 1960).

We are still comparatively ignorant concerning the taxonomic value of data relating to the structure of stems and roots in monocotyledons. It is, however, at once apparent that in a group as morphologically diverse as the monocotyledons the criteria of stem anatomy can only be of practical value when comparing plants that are similar in habit. One cannot, for example, very usefully compare the structure of the pseudo-stem of a banana with the aerial stem of a palm. On the other hand a palm trunk is comparable with the woody stem of a screw-pine or with the aerial part of the axis of a bamboo or even of an herbaceous grass. In making such comparisons, however, it is essential to make sure that we are referring to portions of stem that correspond to one another as nearly as possible, for there are marked variations in the distribution of sclerenchyma and in the structure and distribution of vascular bundles as one passes from the apex downwards in a single stem. Then again there is the practical difficulty that, although it is relatively simple to take sections of grass culms, when one comes to the trunks of palms the mixture of exceedingly hard sclerenchyma and of relatively soft ground tissue between the vascular bundles, together with the damage to a knife edge that can be caused by a mass of silica-bodies, all combine to ensure that the preparation of good sections is by no means easy. One remedy is to select young stems in which the tissues are less hard, but here the structure may well differ considerably from that of the mature trunk. A possible approach to this difficult problem would be to cut thin slices of material embedded in a thermoplastic resin with a diamond wheel and then to grind them down in the manner that is customary for rock sections. Although this has already been achieved by laborious hand grinding the possibility of mechanical cutting and grinding is now being explored at the Jodrell Laboratory.

Our knowledge of root structure in monocotyledons, like that of stems, is far from perfect. This is partly because roots are not always readily available. Owing to the labour involved, a field botanist passing through an interesting region of tropical vegetation seldom has the time or inclination to collect root material on more than a very small scale. It seems likely, however, that root structure will remain of limited taxonomic interest because it so often turns out that monocotyledons of diverse affinities are alike in having a polyarch stele surrounded by a well-developed endodermis, xylem marked by a circle of conspicuously large metaxylem vessels alternating with phloem strands, the ground tissue of the stele being more or less sclerozed in proportion to the "woodiness" of the species concerned. There are, however, some notable exceptions to this generalized pattern of root structure, e.g. in the palms where the structure of the roots is related to the mode of attachment of the root traces to the stem bundles (Tomlinson, 1961c, p. 60).

IV. RELATIONSHIP OF SYSTEMATIC ANATOMY TO OTHER BOTANICAL DISCIPLINES

A. DELIMITATION OF STRUCTURAL VARIATION BY HEREDITY

Earlier in this article (pp. 104–107) we have discussed some of the reasons why an anatomical survey is becoming increasingly essential to the more complete understanding of the classification of the flowering plants, with special reference to the monocotyledons. We have also seen how modern facilities for world travel and communication are broadening the scope and increasing the value of investigations of this kind. Our next task must be to ask ourselves which anatomical characters are the most important for systematic purposes. Before doing so, however, we must briefly indicate why a comparative anatomical survey is not something detached from the rest of botany and therefore of interest only to the student of plant classification. This at once becomes self-evident when we remember that, in the course of his work, the systematic anatomist cannot fail to ask himself whether the characters on which he relies for taxonomic purposes are distinct from those which are related to the environment in which the plants grow and from those which satisfy the physiological requirements of the plants in which they are to be found. Other anatomical characters are correlated with the sizes of the plants in which they are to be seen, or serve to provide mechanical compensation for morphological peculiarities.

In other words the study of systematic anatomy is intimately related to ecology and physiology as well as to plant size and morphology. The way in which a plant responds to its physiological needs and hazards of its environment is, however, limited by its hereditary make up. Thus although we shall find that similarity of structure can arise in plants which are in no way closely related to one another, nevertheless the plants concerned are always so stamped by their past evolutionary history that, in the writer's experience, their true affinities can usually be determined, provided that adequate material is available for this purpose. To take some familiar examples, although the leaf of a dead nettle (*Lamium*) may resemble that of a stinging nettle (*Urtica*) in superficial appearance, the two types of leaf could never be confused when differences in their histology have been noted. Similarities between members of the Cactaceae and some of the Euphorbiaceae (spurges) are also well known, and yet, quite apart from their floral differences, the true affinities of these plants can be determined from the structure of their vegetative organs alone. We might go further and recall that succulent members of families such as the Geraniaceae and Compositae sometimes bear a superficial resemblance to members of both the Cactaceae and Euphorbiaceae. With all of these examples, however, despite the similarities that we have noted, other characters that are less obvious are also to be found by which the family affinities of the plants concerned can be recognized. Another excellent example of similarity of structure in species belonging to two distinct families, this time from amongst the monocotyledons, is provided by *Gahnia radula* Benth., a member of the Cyperaceae from Australia, and the Marram grass (*Ammophila arenaria* (L.) Link.) of our English coasts. A student familiar with the appearance of a transverse section of the leaf of Marram grass, if shown a similar section of a leaf of *Gahnia radula* could readily be excused if he were to confuse the two plants because the leaf structure is so much alike (see Fig. 10, A–C). On examining the epidermis of the leaves of both species in surface view, however, the true situation would be at once revealed for it would now be seen that *Gahnia radula* has silica-bodies of a type that is common in the sedge family but unknown amongst the grasses (see also p. 137).

Considerations such as these allow us to see that structural characters revealed by the microscope result from demands made by the environment and physiological requirements on a developing plant body whose behaviour is limited and largely predetermined by heredity. In consequence it is by no means easy to assess the relative significance for taxonomic purposes of one character in relation to others. At present there is no way out of this difficulty except by making prolonged continuous and painstaking observations on a really wide range of material.

Conclusions based on long experience and much thought are our most useful guide, which means that a worker in this field of enquiry can scarcely expect to reach the apex of his research career until relatively late in life.

We must, however, now pass on to consider some examples of the ecological, physiological and morphological as well as the taxonomic aspects of this subject.

B. ECOLOGICAL CONSIDERATIONS

Certain ideas concerning the relationship of anatomical structure to environment are now so widely held that they are accepted as axiomatic, and most students are familiar with at least some of them. These include concepts such as that stomata in xerophytes are often sunken in the epidermis or in some other way protected from excessive water loss; that the epidermis of xerophytes is further protected by a thick layer of cutin on the outer surface. In plants from a marshy environment, on the other hand, intercellular air-spaces are usually well developed so that the ground tissue is spongy or aerenchymatous, the cuticle is usually thin, and stomata are not specially protected against loss of water either by special structures or by their position in the leaf. Then again we know that salt-loving plants, or halophytes, are often succulent. Many, or indeed probably most, of these generally accepted ideas are true or mainly correct, but, nevertheless the systematic anatomist engaged on a broad survey cannot fail to notice that this is not always so, or that at least some further explanation is needed. For example, if we consider for a moment the view that elaborate systems of intercellular air-spaces or canals are characteristic of hygrophytes, when we turn to the monocotyledons we may well ask ourselves if this represents the whole truth. There are indeed many monocotyledons from aquatic or marshy environments in which well-developed air-spaces and canals are to be found, as for example, in the Alismataceae. It must, however, also be remembered that extensive intercellular systems may likewise arise in plant organs that attain large dimensions by rapid growth. Tomlinson, for example, has pointed out that this may apply in organs such as the enormous leaves of bananas and their relatives the heliconias. It should also be noted that it is quite common to find radiately arranged, cortical, intercellular air cavities in grass roots, and this often applies to grasses from dry as well as from moist localities. Then again we are apt to think of superficial, non-glandular hairs on plant surfaces as being protective against water loss. Here the interesting fact may be noted that the capacity to produce hairs is much less well developed in monocotyledons than in dicotyledons. Furthermore although there are, for example, xerophytic grasses, e.g. *Spinifex hirsutus* Labill, which conform to tradition by having stomata that are well protected by a dense indumentum of hairs, the capacity to produce hairs is much more universally developed in the pan-tropical Commelinaceae than in most monocotyledons, although this family is by no means confined to arid localities.

Because plants from dry localities are sometimes characterized by assimilatory stems that bear very small or no leaves we may be led to suppose that reduction in leaf area and the possession of assimilatory stems is a xeromorphic character. Indeed this frequently applies both in dicotyledons and monocotyledons. We must, however, also remember that other plants such as many of the rushes (*Juncus* spp.) from marshy localities have assimilatory stems.

It is also well known that the leaves of many grasses from dry localities tend to become folded or rolled longitudinally in such a way that the furrows in their adaxial surfaces in which the stomata are situated are more adequately protected from desiccation. It has frequently been asserted that certain longitudinal bands of specially large translucent cells (bulliform cells) of the adaxial epidermis of grass leaves play an important part in these folding movements by expanding and contracting and so serving as hinge-cells. This explanation of the function of bulliform cells seems so immediately obvious that it has come to be commonly if not generally accepted. There are objections to it, however (Metcalfe, 1960, p. xxx), and the subject clearly needs further investigation. More pertinent to our present discussion is the fact that the leaves of sedges (*Carex* spp.) like those of the grasses are capable of becoming folded longitudinally along the midrib under certain conditions so that, just as in grasses, the two adaxial surfaces become pressed together. The real point of interest, however, lies in the fact that in most of the species of *Carex* examined by the writer, the adaxial surface is free, or almost free, from stomata whilst on the abaxial surface they are abundant. The result of the folding is, therefore, to give additional protection to the adaxial surface where there are no stomata and additional exposure to the abaxial surface where they are usually abundant. How then is this to be explained? Are we to say that *Carex* normally grows in marshy localities where maximum exposure of stomata is unlikely to do much harm? This seems plausible enough until we remember that some species of *Carex* grow in relatively dry localities where stomatal exposure would seem to be undesirable. The explanation of this anatomical difference between the leaves of sedges and grasses seems worthy of further investigation.

C. PHYSIOLOGICAL CONSIDERATIONS

The relationship of plant structure to physiological function raises many questions to which answers are not immediately evident. The following are some examples. In the dicotyledons calcium oxalate is deposited in members of many diverse families, chiefly as solitary crystals of various forms and as clusters or druses, but very much less frequently as raphides or needle-shaped crystals. Indeed the presence of raphides in dicotyledonous families is often a useful diagnostic microscopical character because of the restricted occurrence of this type of crystal. In the monocotyledons, there are families in which crystalline deposits are very rare or unknown. It is doubtful, for example, whether they occur in the Gramineae. In other families such as the Orchidaceae, and in many of the Scitamineae and Liliaceae, crystals are present, but, unlike those in the dicotyledons, the dominant type consists of raphides. In the Iridaceae crystals are mostly of the massive, solitary, elongated type known as styloids. Here the anatomist might well enquire of his physiologically-minded colleagues why crystals should exhibit these differences of form and taxonomic distribution. Are there differences of metabolism which correspond to them, and, if so, what is the nature of these differences. Somewhat similar problems arise in connection with the deposition of silica (see also p. 121). Firstly it may be noted that silica deposits are more common in, and characteristic of, monocotyledons than of dicotyledons, which seems again to denote some difference between the two groups at the metabolic level. There are also variations in the form in which silica is deposited as well as in the position in the plant in which deposition takes place. In the Cyperaceae (sedges) and Gramineae (grasses), for example, silica is deposited in epidermal cells as silica-bodies of various forms, but the forms of the bodies in the two families are totally different. On the other hand the forms of silica-body appear to be constant for any one species, and, in some instances, a particular type of silica-body may be characteristic of a genus or even of a whole tribe. In other families, such as those comprising the Scitamineae, as well as in the palms, silica is often deposited in longitudinal files of special cells that lie adjacent to strands of fibres or next to fibres at the periphery of vascular bundles. Silica-bodies in specialized cells accompanying the fibres or vascular bundles are commonly known as stegmata (see also p. 125). No matter whether we are dealing with epidermal silica-bodies or with stegmata, a given species often contains silica-bodies of more than one type, and here again the same combination of silica-body types appears to be constant for a species, and sometimes for taxa of higher rank. With all of these silica-bodies one cannot fail to ask oneself why they are so much more common in monocotyledons than in dicotyledons, and what determines the shape and location of these bodies within the plants in which they are to be found.

D. STRUCTURE IN RELATION TO SIZE OF THE PLANT BODY

Having noted that there are some histological characters that appear to serve primarily as adaptations to ecological conditions or to provide for the physiological requirements of the plants in which they occur, we should remember that plant structure is also bound up with the total dimensions and form of the plant body. Plants of large size need more physical support than small ones and, as the plant body increases in size, the provision of all its parts with adequate supplies of foodstuffs and water calls for certain minimum structural requirements. It is in fact necessary for the plant to be structurally and physiologically stable. If we turn our attention to monocotyledons we find that additional skeletal support in those of large size is very commonly provided by an increase in the amount of fibrous and other sclerenchymatous tissue in the plant body. A different method achieving the same end is found in plants which grow in length by intercalary growth at the bases of the stem internodes, e.g. in the grasses. Such a plant body would become structurally unstable without special provision because not only would each (mechanically weak) intercalary tissue have to carry the weight of the more mature tissue at the upper end of the internode of which it forms a part, but the intercalary tissue of the more basal internodes would have to carry the whole weight of the internodes above them. In the grasses and other plants of similar habit and mode of growth, collapse is prevented because the intercalary tissue is protected by the overlapping bases of the leaves. If the lower leaves of a grass are removed, the culm falls down because the intercalary tissue is too weak to hold it up.

Another interesting example of a type of plant that would be morphologically unstable without a special form of mechanical support is provided by the screw-pines (*Pandanus*) which are so top-heavy that they would immediately fall down unless supported by stilt roots. This same type of structure also occurs occasionally amongst the grasses, e.g. in *Ischaemum santapaui* Bor and *Triplopogon spathiflorus* (Hook. f.) Bor (Metcalfe, 1960, pp. 265 and 509), and in the Iriartoid group of palms (Tomlinson, 1961c, p. 11, Fig. 6).

Turning now to the question of supplying a large plant body with adequate food it is noteworthy that in monocotyledons, which do not, like the dicotyledons, increase the diameters of their trunks by means of secondary xylem derived from a cambium, additional conducting tissue is made available by the production of more vascular bundles.

V. CHARACTERS OF TAXONOMIC VALUE

A. GENERAL CONSIDERATIONS

From what has been said it is clearly not desirable to rely for taxonomic purposes primarily on characters that are wholly called forth to meet the demands of the environment, or on characters that have become necessary in consequence of the size or morphological peculiarities of the plants in which they occur. Such characters can, however, possess diagnostic value when they are of restricted occurrence and they can be of taxonomic significance provided they have arisen only once in the course of evolution. The difficulty of deciding when we are dealing with parallel evolutionary tendencies is one of the greatest obstacles to the use of the histological approach to the taxonomy of monocotyledons.

Because of the morphological diversity exhibited by monocotyledons it is not surprising to find that characters that are of taxonomic or diagnostic value in one family are not necessarily of equal value in another. Furthermore a family that consists of species of varying external form is often very uniform from the histological standpoint. Tomlinson has found this to be true for example in the Marantaceae (1961b). On the other hand there are families in which similarity of external form is combined with histological variations of considerable interest. This applies for example in the Cyperaceae, Gramineae and Restionaceae.

The characters that are known to be of taxonomic value in various monocotyledonous families have already been listed in the books and papers mentioned in the bibliography. For the sake of completeness a selection of these characters is repeated below.

B. TYPES OF VASCULAR BUNDLE

One of the most obvious ways in which the different types of vascular bundle differ from one another is in the number and arrangement of conspicuously large metaxylem vessels or tracheids. In using this character it must be remembered that a single vascular bundle may appear very different in transverse sections taken at various levels throughout its length (Cheadle and Uhl, 1948a; Cheadle and Whitford, 1941a). Vascular bundles also vary in the extent to which they are oval, rounded or angular in outline, and these differences can be of considerable diagnostic value (Metcalfe, 1960, Fig. VIII, p. 675). For

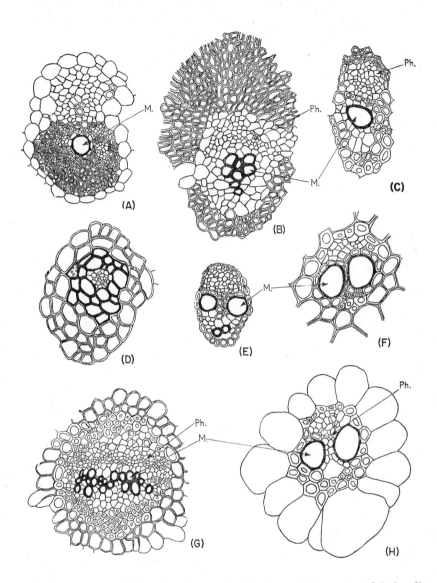

Fig. 1. Types of vascular bundle. (Here, and in Fig. 2, note different sizes of the bundles themselves and differences in the number and distribution of conspicuously large metaxylem elements and phloem strands.) All ×130. M. Metaxylem. Ph. Phloem. A, *Freycinetia banksii* Cunn. Leaf bundle. B, *Agave fourcroydes* Lem. Leaf bundle. C, *Orchidantha* sp. probably *O. longiflora* H. Winkl. Leaf bundle. D, *Achlyphila disticha* Maguire and Wurdack. Bundle from inflorescence axis. E, *Carex pendula* L. Leaf bundle. F, *Cannamois virgata* Steud. Stem bundle. G, *Hosta* sp. Leaf bundle. H, *Xyris indica* L. Large bundle from leaf.

example amongst the grasses angular vascular bundles are to be found chiefly amongst the panicoid group. In considering vascular bundles it must also be remembered that the conducting elements may be vessels or tracheids. The end walls of vessels range from transverse to very oblique and the pores in these end walls vary from simple to scalariform with numerous bars. Tracheids also have tapering ends with scalariform pitting in which the number of bars shows a considerable range of

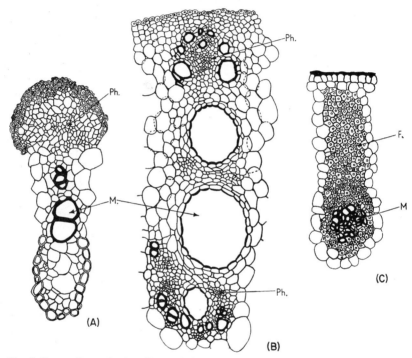

Fig. 2. Types of vascular bundle *contd.* All × 130. F. Fibres. M. Metaxylem. Ph. Phloem. A. *Heliconia metallica* Planch. and Lind. Leaf bundles. B, *Dioscorea bulbifera* L. Stem bundle. C, *Nietneria corymbosa* Kl. and Schomb. Leaf bundle.

variation. The morphology of conducting elements, as has been shown by Cheadle (1942a, 1943a, 1943b), provides characters that can be used as indicators of the level of phylogenetic specialization. The tracheid is pictured as being ancestral to the vessel element and the oblique scalariform perforation plate is thought to have given rise to the simple perforation in a transverse end wall. There is also the added point of interest that, as Cheadle has shown, the phylogenetic change from tracheid to vessel is thought to have occurred first in the root and then at successively higher levels in the plant body, ending with the leaves.

Amongst monocotyledons there are some species in which vessels are confined to the roots, others in which they occur in roots and aerial stems, whilst in the highly evolved monocotyledons such as the grasses the conducting channels throughout the plant consist of vessels. Although Cheadle was the first to make these important generalizations.

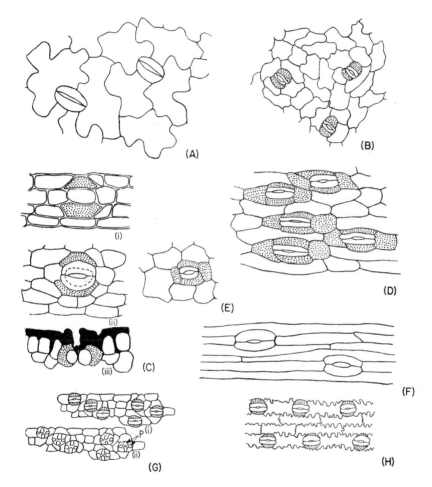

Fig. 3. Types of stomata in surface view. A–F × 130 ; G–H × 180.   Subsidiary cells marked with dots. A, *Ranunculus lingua* L. Anomocytic. B, *Alisma plantago-aquatica* L. Paracytic. Subsidiary cells dome-shaped. C, *Doryanthes excelsa* Correa. (i) high, (ii) low focus, surface view, (iii) transverse section. Paracytic, sunken. D, *Abolboda macrostachya* Spruce ex Malme var. *angustior* Maguire. Tetracytic. E, *Carludovica insignis* Duch. Tetracytic. F, *Campynema lineare* Lab. Anomocytic. G, *Lepironia articulata* Domin. Stomata from culm. Paracytic with papillae (P) overarching the stomata. (i) low, (ii) high focus. H, *Carex. polyphylla* Kar. and Kir. Paracytic.

the work at Kew has extended and confirmed his views without, so far, leading to the discovery of any species to which they do not apply.

C. TYPES OF STOMATA

It is well known that in the stomata of grasses the lumina of the guard cells are narrow throughout most of their length but wider at either end, the guard cells thus being somewhat dumb-bell shaped in surface view. This type must be distinguished from that which is more general amongst the monocotyledons where the guard cells are sausage-shaped. Whilst this well-known difference is taxonomically important, it is probably less well known that the shapes of the subsidiary cells and their positions in relation to the guard cells are of undoubted taxonomic value. We have, on the one hand, stomata of the paracytic type in which each is accompanied on either side by a lateral subsidiary cell. This type occurs in grasses (Metcalfe, 1960, Fig. IV), sedges, lilies and in plants belonging to many other families of monocotyledons. Further-more within the Gramineae themselves we find that in some species the subsidiary cells are triangular in outline in surface view and in others dome-shaped, the domes being low or tall in different species. In many grasses there are mixed stomata with triangular and dome-shaped subsidiary cells, and transitional types of subsidiary cells also occur (Metcalfe, 1960, Fig. IV). A similar range of types occurs also in the Cyperaceae. If now we turn to the Commelinaceae, to the palms (Tomlinson, 1961c), to the Scitamineae or to such genera as *Carludovica* (Cyclanthaceae) (Fig. 3, E) and *Abolboda* (Xyridaceae) (Fig. 3, D), we find stomata with two lateral and two polar subsidiary cells, making four subsidiary cells in all. The writer has called stomata of this kind tetracytic at the suggestion of his colleague Mr. H. K. Airy-Shaw (Metcalfe, 1961). The occurrence of paracytic and tetracytic stomata in the monocotyledons has also been noted quite independently by Stebbins and his associates at the University of California (Stebbins and Khush, 1961). Indeed it was not until the 9th International Botanical Congress was held at Montreal in 1959 that Stebbins and the writer discovered that they had both recognized stomata of these two types as being of taxonomic value.

D. HAIRS AND PAPILLAE

Hairs are, on the whole, much less common in monocotyledons than in dicotyledons, so their usefulness for taxonomic purposes is limited to families in which they occur. Hair structure is important, for example in grasses (Metcalfe, 1960, Figs. II, IIA and VII), but here we must distinguish between the large or macro-hairs which are usually unicellular and form an obvious indumentum on leaves and the much smaller, usually two-celled, micro-hairs. Macro-hairs are generally of diagnostic value only at the species level, whereas micro-hairs of one type or another are often characteristic of tribes and they occur mainly in panicoid grasses. Hairs are more than usually well developed in the Commelinaceae where current investigations by Tomlinson (unpublished) indicate that they are of taxonomic value; they are rare in the Palmae (Tomlinson, 1961c) but useful for diagnostic purposes wherever they occur. Prickle-hairs consisting of a short but inflated base bearing a short barb which is directed towards the leaf apex are so common in grasses (Metcalfe, 1960, Fig. VI) that they are seldom of much diagnostic value. The "roughness" of grass leaves and stems as the fingers are drawn over their surfaces from the apex of the leaf downwards is due to these hairs. Prickle-hairs also occur in Cyperaceae and other families.

Papillae of many types are common on the surface of the leaves of certain grasses (Metcalfe, 1960, Fig. III) and sedges, and they often overarch and therefore tend to protect the stomata (Metcalfe, 1960, Fig. XXVIII, C). They occur sporadically and are often of diagnostic value only at the species level.

E. LONG-CELLS AND SHORT-CELLS

In most of the monocotyledons the epidermis of both stem and leaf is made up of cells of more or less uniform size, although the cells overlying the veins frequently differ in appearance from those of the intercostal zones. It is noteworthy, therefore, that the epidermis of grasses usually differs from that of other monocotyledonous families in being made up of cells of two very distinct sizes (Metcalfe, 1960, Figs. XX, XLI, XLV). The genus *Joinvillea* of the family Flagellariaceae also resembles the grasses in this respect (Smithson, 1956).

F. ERGASTIC SUBSTANCES

1. Silica

Ergastic substances, that is to say visible products of metabolism which are deposited in cells or cavities, are often highly characteristic of the plants in which they are produced and therefore of taxonomic value. Reference has already been made to silica (p. 114), but the forms in which silica is deposited are of such great interest that the subject needs to be amplified. In grasses silica is deposited particularly in short-cells of the leaf epidermis, the silica-cells in which it occurs usually being accompanied by and frequently alternating with other epidermal

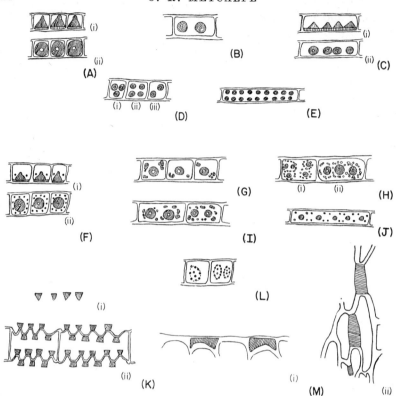

Fig. 4. Diagrams of principal types of silica-bodies in epidermal silica-cells of the Cyperaceae. The silica-bodies are marked with sloping lines. A, Bodies conical, 1 per cell. (i) in leaf sections, (ii) in surface view. B, Bodies conical, 2 per cell in horizontal rows: surface view. C, Bodies conical, 4 per cell in horizontal rows. (i) in leaf sections, (ii) in surface view. D, Bodies conical, 2–4 per cell. (i) in a triangular group (triad), (ii) in two horizontal pairs (tetrads), (iii) in an oblique pair. All in surface view. E, Bodies conical, more than 4 per cell, in two horizontal rows. In surface view. F, Bodies conical, 1 per cell, each surrounded by small silica particles. (i) in leaf sections, (ii) in surface view. G, Bodies conical, 1 per cell, each accompanied by larger silica particles than those in F. In surface view. H, Bodies conical, 3–4 or more per cell, each body, or sometimes a pair of bodies, surrounded by silica particles. (i) 5 per cell, (ii) 3 per cell. In surface view. I, Bodies conical, 2 per cell, accompanied by relatively large silica particles. In surface view. J, Bodies dome-shaped, more than 4 per cell, in a single horizontal row, accompanied by small silica particles. (A similar arrangment but with conical particles may also occur.) K, Bodies wedge-shaped, and appearing as triangles in transverse sections of the leaf. Bodies of this kind seen only in leaves with coarsely sinuous anticlinal walls, one body being attached to the apex of each sinuation as seen in surface view. (i) Triangular appearance of the bodies in sections, (ii) surface view showing bodies attached to sinuous anticlinal walls. (In surface view the observer is looking down on the surface of the wedges where they are attached to the external wall of the epidermal cell.) L, Bodies spherical and each made up of an aggregate of small particles; 1–2 per cell. (Bodies of this kind are often very translucent and the spherical outline of the whole body is often difficult to see. The particles appear to consist of a form of silica that is denser than that of which the remainder of each body is composed.) M, Bodies arched or vaulted and usually embedded in the outer wall of the epidermal cell. (i) two bodies embedded in the outer cell walls of 2 epidermal cells, (ii) two large and one small bodies in surface view. (Bodies of this type occur in epidermal cells with coarsely sinuous walls and extend across the cells so as to unite 2 (or more) sinuations. See also Fig. 6, D–G.)

cells which differ from their neighbours in having suberized walls. The silica-bodies of grasses include types that are commonly referred to as saddle-shaped, cross-shaped, dumb-bell shaped and so on (Metcalfe, 1960, Figs. I, IA and II). Silica-bodies of these particular types are highly diagnostic within the family Gramineae, but they appear, so far as we know at present, to be confined to this one family. Silica is also deposited in many other forms in other monocotyledonous families. In the Cyperaceae or sedge family, the silica-bodies occur chiefly over the veins, and they may be present in long or short rows depending on the species. They also vary in frequency from species to species, and their frequency on one surface of a single leaf is more often than not quite different from the frequency on the other. The dominant type of silica-body in the Cyperaceae is conical (Figs. 4–6). There may be from one to many cones per cell according to the species (Fig. 4, A–J). Where there is more than one cone per cell they are variously arranged (see Fig. 4, B–E). Close examination of these cones shows that their bases usually rest on the inner walls of the epidermal cells in which they occur, the apices of the cones being directed towards and often coming into contact with the outer epidermal cell wall (Fig. 4, A(i)). Sometimes a number of cones arise from a single siliceous base so that, in transverse sections through the epidermis, they resemble a row of miniature volcanic peaks comprising a single mountain range (see Fig. 4, C(i); 5, H). In many species the apices of the cones when observed in surface view can be seen to be encircled by small, independent particles of silica (see Fig. 4, F–J). Another variation is for the silica-bodies to consist of spherical aggregations of particles, e.g. in *Diplacrum longifolium* (Griseb.) Clarke (Fig. 5, P). Sometimes a single epidermal cell contains a central row of relatively large silica-bodies accompanied by peripheral bodies that are distinctly smaller, e.g. in the culm of *Eleocharis multicaulis* (Sm.) Sm. (Fig. 5, K). In *Fintelmannia* the bodies are of more uniform size and arranged in two ranks. In *Macrochaetium hexandrum* (Nees) Pfeiffer there are numerous silica-cones in each cell, but the cones are so small that they appear as sand-like particles (Fig. 6, C). In some of the silica-cells on the adaxial surface of the leaf of the same species a single row of small cones of the type just described is accompanied at one end of the cell by a single dome-shaped papilla often with a silica deposit in the inner part of the wall (Fig. 6, A–B). It is rare in the Cyperaceae to find a species from which epidermal silica-bodies are absent, but this appears to be so in certain species of *Chorizandra*, *Hypolytrum* and *Lepironia*. A very remarkable situation is to be seen in *Mapania* and *Scirpodendron*. Here the anticlinal walls of the epidermal cells are prominently and coarsely sinuous in surface view. In surface-view preparations of *Mapania wallichii* there sometimes

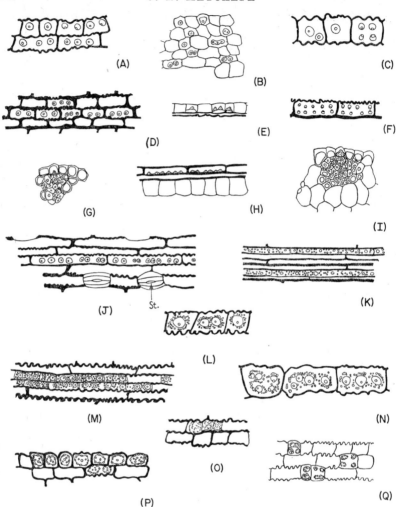

FIG. 5. Silica-bodies of Cyperaceae. Examples selected to illustrate the various types shown diagrammatically in Fig. 4. A, *Everardia neblinae* Maguire and Wurdack. Abaxial epidermis ×270. Bodies conical, 1–3 per cell. B, *Cephalocarpus* sp. (Maguire and Wurdack 42107). Adaxial epidermis ×180. Bodies conical, 1–2 per cell and sometimes compound (cf. E). C, *Ficinia indica* (Lam.) Pfeiffer. Abaxial epidermis ×270. Bodies conical, 1–4 per cell, including oblique pairs and tetrads: horizontal pairs and short rows also occur in this species. D, *Cladium mariscus* (L.) Pohl. Adaxial epidermis ×180. Bodies conical, 2–3 per cell, in horizontal rows. E, *Cephalocarpus* sp. (Maguire and Wurdack 42107). Epidermis in L.S. leaf ×180. Bodies conical, sometimes compound (cf. B). F, *Fintelmannia lhotzkyana* (Nees) Pfeiff. Abaxial epidermis ×270. Bodies conical, more than 4 per cell, in two horizontal rows (bodies viewed somewhat obliquely in the figure). G, *Cladium mariscus* (L.) Pohl. Conical body (dotted) in epidermal cell adjoining fibre strand ×180. H, *Asterochaete glomerata* (Thunb.) Nees. Epidermis in L.S. leaf with up to 6 conical bodies (dotted) per cell, some tendency to be compound ×180. (Sometimes there are more than 6 bodies per cell cf. J). I, *Asterochaete glomerata.* Conical body (dotted) in

appear to be small blocks of silica based on the apices of the sinuosities and extending into the lumina of the cells (Fig. 6, D, S.B. (i)). Where pairs of these bodies are attached to sinuosities facing one another on opposite sides of a single cell they may become united to form a single body (Fig. 6, D–E, S.B. (ii)). Sometimes two (Fig. 6, D, S.B. (iii)) or even three (Fig. 6, D, S.B. (iv)) such bodies are inconspicuously united where indicated by the broken lines in the drawing. Examination of the leaves in section shows that bodies of type S.B. (i) are more or less wedge-shaped, so that, in surface view, the observer is looking down on the base of the wedge. Bodies of type S.B. (ii) are arch-shaped when seen in section (Fig. 6, f) and those of type S.B. (iii) are similar, but the silica that forms the span of the arch is a very thin layer and therefore inconspicuous. Bodies such as that represented in S.B. (iv) resemble vaults. In *Scirpodendron costatum* (Fig. 6, G) the situation is somewhat similar to that of *Mapania*, but here the silica-bodies are mostly if not all, wedge-shaped rather than arch-like, and, in transverse section of the leaf, slices of the wedges appear as triangles (Fig. 6, G (i)). The fact that the silica-bodies of *Mapania* and *Scirpodendron* are rather different from those in most members of the Cyperaceae is of considerable taxonomic interest, and the subject will be further investigated.

If we turn now to the families examined by Tomlinson, i.e. to the Scitamineae, Commelinaceae and palms etc. we find that silica is most commonly deposited in stegmata (see p. 114). Tomlinson recognizes several categories of silica-body in stegmata and refers to them as hat-shaped (in some Palmae (Tomlinson, 1961c), Marantaceae (Tomlinson, 1961b), Lowiaceae); spherical or druse-like (in some Palmae, Strelitziaceae, Zingiberaceae-Costoideae, but not in Zingiberaceae-Zingiberoideae (Tomlinson, 1956)). He also refers to stellate or druse-like bodies in Cannaceae (Tomlinson, 1961a), rectangular bodies each with a deep

---

epidermal cell adjoining fibre strand ×180. J, *Asterochaete glomerata*. Adaxial epidermis ×180. cf. H. Bodies conical, 8–9 per cell, in a single horizontal row. More bodies per cell may also occur in this species. St. Stoma. K, *Eleocharis multicaulis* (Sm.) Sm. Epidermis of culm ×180. Bodies of two distinct sizes; a few large bodies and numerous silica particles per cell. L, *Cyperus diffusus* Vahl. Abaxial epidermis ×270. Bodies conical, 1–2 per cell, accompanied by numerous fine particles of silica. M, *Carex echinata* Murr. Abaxial epidermis ×180. Bodies conical, 2–6 per cell in a single horizontal row and each accompanied by numerous silica particles. N, *Cymophyllus fraseri* (Andr.) Mack. Abaxial epidermis from over a vascular bundle ×270. Bodies conical, 3–5 per cell, variously arranged, accompanied by numerous silica particles. O, *Cymophyllus fraseri* (Andr.) Mack. Adaxial epidermis ×180. Bodies warty. (These bodies are infrequent and can be easily overlooked.) P, *Diplacrum longifolium* (Griseb.) Clarke. Adaxial epidermis ×180. Bodies spherical, each made up of an aggregate of small particles. (cf. Fig. 4, L. and legend, and Fig. 5, O–P.) 1–2 bodies per cell. Q, *Carex pendula* L. Adaxial epidermis ×150. Bodies spherical, compound, somewhat similar to those in P but smaller; about 4 bodies per cell.

central depression in Heliconiaceae and to trough-like silica-bodies in Musaceae (Tomlinson, 1959). Of these types those described as hat-shaped appear to be not unlike the conical bodies of the Cyperaceae. Tomlinson also mentions silica-sand in the epidermis of the Zingi-beraceae-Zingiberoideae which serves to separate the Zingiberoideae

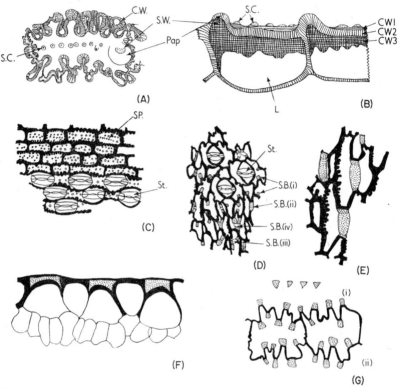

Fig. 6. Silica-bodies of Cyperaceae *contd.* A, *Macrochaetium hexandrum* (Nees) Pfeiffer. Single cell from adaxial epidermis in surface view × 270. With a single row of conical bodies (S.C.) and a single papilla (Pap) at one end of the cell with a silica deposit (S.W.) on the inner part of the wall. B, *Macrochaetium hexandrum.* A whole cell and part of a second cell from the adaxial epidermis in section × 270. CW1, CW2, CW3, outer, middle and inner layers of the cell wall. L. Lumen of cell. Other letters as per A. C, *Macrochaetium hexandrum.* Abaxial epidermis × 180. Showing numerous silica particles (SP.) in some of the cells, and stomata (St.). D, *Mapania wallichii* C. B. Clarke. Abaxial epidermis × 180. Showing stomata (St.) and arch-like silica-bodies marked with dots. Some of the bodies do not traverse the lumen of the cell completely. (The layer of silica forming the span of the arch is often very thin and difficult to detect in surface view). E, *Mapania wallichii.* Abaxial epidermis × 270. Showing 3 arch-like silica-bodies (dotted) in surface view. F, *Mapania wallichii* × 270. Abaxial epidermis and subjacent layer of cells in transverse section showing 3 arch-like silica-bodies (dotted) embedded in the outer cell wall of the epidermis. G, *Scirpodendron costatum* Kurz. × 270. (i) Wedge-shaped silica-bodies appearing triangular in T.S. leaf. (ii) Two epidermal cells showing wedge-shaped silica-bodies (dotted) in surface view.

from the Costoideae. In the Commelinaceae (unpublished) he finds spinulose silica-bodies in the thickened walls of the epidermal cells.

2. Crystals

The dominant type of crystal in the monocotyledons consists of bundles of needles known as raphides. These occur in cells that are usually parenchymatous and often contain mucilage as well as raphides. Raphides occur in a number of monocotyledonous families including Orchidaceae, Liliaceae, Palmae, Commelinaceae (in longitudinal files of cells resembling laticifers) as well as in some of the Scitamineae. On the other hand there are families of monocotyledons such as the Gramineae and Cyperaceae in which raphides have never been recorded.

Crystals of other types are far less common in the monocotyledons, but the presence in the Iridaceae of solitary, elongated crystals, of the type known as styloids, is noteworthy.

3. Tannin, oil-cells, laticifers and starch

The presence of amorphous substances, often referred to by anatomists as tannin or tanniniferous substances, is common in monocotyledons, especially in certain families such as the Cyperaceae and Palmae.

Oil-cells are sometimes present and may be of taxonomic interest. For example in the Zingiberaceae they occur in the Zingiberoideae but not in the Costoideae (Tomlinson, 1956).

Laticifers are also of interest because of their restricted occurrence. They have been noted for example in Musaceae (Tomlinson, 1959) and certain species of *Allium*.

The morphology of starch grains needs further examination before its value in the taxonomy of the monocotyledons can be fully assessed. The distinction between simple and compound grains has been extensively used in the Gramineae, and no doubt this character is very valuable when we can be quite sure whether the grains are simple or compound. The danger seems to be that loose aggregates of grains sometimes occur and it is not then always easy to say whether the grains are simple or compound. When this is so there is always a risk of misinterpretation. Starch grains of a morphologically distinct type are, however, sometimes of diagnostic value, e.g. the starch grains of ginger which are oval to oblong, flattened and transversely striated with the hilum in a terminal protuberance.

G. THE LEAF IN TRANSVERSE SECTION

In transverse sections of grass leaves information of diagnostic or taxonomic value is provided by (i) the heights and shapes of ribs on the leaf surface; (ii) the arrangement of the assimilatory tissue of the mesophyll in a radiate or non-radiate manner; (iii) the presence of translucent fusoid-cells in the mesophyll of bamboos; (iv) the number of bundle-sheaths and the nature of the cells of which they are composed; (v) the distribution pattern of sclerenchyma both as independent fibre strands and as fibres providing mechanical support for or serving as buttresses to the vascular bundles; (vi) the number, sizes and arrangement of the vascular bundles; (vii) the vascular structure of the midrib, especially when this part of the lamina is well developed; (viii) the distribution of bulliform and other translucent cells (Metcalfe, 1960, pp. xxxvii–l). These characters are frequently useful in other families of monocotyledons as well, and to them may be added the presence of hypoderm; the occurrence and distribution of intercellular spaces; the presence of transverse diaphragms and the nature of the cells of which they are composed. It may be noted in passing that a number of interesting variations in the distribution and arrangement of tissues in transverse sections of the leaves of Cyperaceae have been seen (see Figs. 7–11) many of which differ considerably from the corresponding tissue arrangements in the Gramineae (Metcalfe, 1960, pp. xxv, xlix and Fig. XVIII) (see also p. 132).

H. VALUE OF COMBINATIONS OF CHARACTERS

In ending this discussion of the histological characters of taxonomic value it must be emphasized that the systematic anatomist is most likely to arrive at valid taxonomic conclusions if he relies on combinations of characters rather than on single characters considered in isolation from one another. It is very difficult to lay down any list of precepts that mark the royal road to progress. The information given here is intended to give a rough indication of the characters that have proved to be valuable in those families of monocotyledons with which the writer is directly or indirectly familiar. We have still a long way to go before anything like a complete picture of the anatomy of the monocotyledons can be given. Further investigations may lead to changes of view concerning the taxonomic importance or relative significance of any one of the characters discussed above. There may be other important characters which at present remain overlooked. In short, the reader should realize that we are still hindered from further advance by the very great need to accumulate factual information concerning the wealth of species, belonging to nearly all of the families of monocotyledons, that still await anatomical examination.

VI. Taxonomic Conclusions Based on Anatomical Evidence

A. GENERAL CONSIDERATIONS

For the reasons given in the previous paragraph it would clearly do more harm than good to attempt, at this stage, to draw any all-embracing taxonomic conclusions for the whole of the monocotyledons. Nevertheless the rapidly accumulating anatomical evidence is becoming an increasingly important tool in elucidating the taxonomy of the group.

B. CONCLUSIONS CONCERNING INDIVIDUAL FAMILIES AND GENERA

For the purpose of the present discussion it will be most convenient to take certain families and genera concerning which the anatomy seems to point towards definite conclusions and to discuss them in turn.

1. Gramineae

The writer's work on the anatomy of grasses has already been fully described (Metcalfe, 1960). Here it will suffice to remind readers that histological characters of grass leaves generally provide good characters for the separation of panicoid and festucoid grasses and to recognize the bamboos as a distinct group within the Gramineae. There are, however, a few genera of grasses that have the same special type of leaf structure as the bamboos, but which taxonomists do not accept as undoubted members of the Bambuseae. Anatomical evidence is also of value in the separation of tribes, but the anatomical differences between genera that are generally accepted as being closely related to one another are seldom clear cut. Histological evidence can also sometimes be used to separate species of a known genus but the characters that serve for this purpose differ from those that are of value in the recognition of tribes. Very great caution must, however, be exercised when using anatomical characters to separate closely related species as the writer has already emphasized (Metcalfe, 1960). Connor (1960) has also drawn attention to this need for care by anatomists when he describes the variation in leaf anatomy that he noted in two related species of *Festuca*.

2. Cyperaceae

In spite of the external similarity of the leaves of the Gramineae to those of the Cyperaceae they can in general be easily distinguished on histological characters, and the morphology of the silica-bodies in the epidermal cells taken alone is generally sufficient for this purpose

6

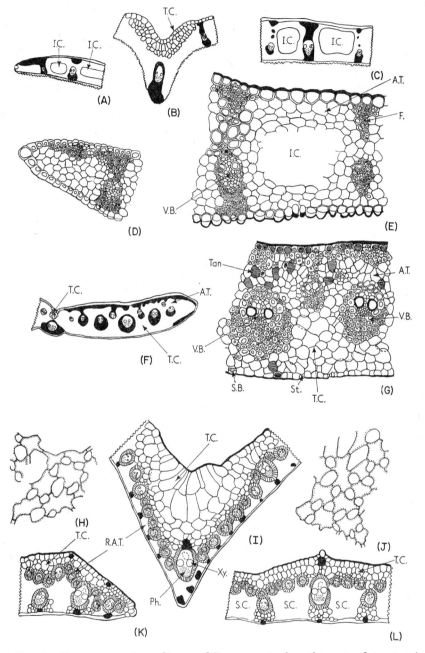

Fig. 7.—Transverse sections of leaves of Cyperaceae to show the range of structure in the family. Note especially the occurrence and distribution of air-cavities, differences in the structure and distribution of assimilatory tissue, translucent cells and in the fibre

(see p. 123). It seems possible on the evidence at present available that the Cyperaceae are more closely related to the Juncaceae than to the Gramineae.

An interesting range of leaf structure is to be seen in different members of the Cyperaceae. The writer's investigations have not yet gone far enough to show how closely these differences correspond to existing concepts of tribes, genera and species. It is becoming increasingly clear, however, that the leaf structure in the Cyperaceae provides a field of study that nobody who is concerned with the classification of the genera and species within the family can afford to ignore. Hitherto most of the anatomical work on Cyperaceae has been on genera such as *Carex, Scirpus, Cyperus* etc. of which there are many well-known species from north temperate regions. The points of interest that are now emerging, however, refer particularly to Cyperaceae from tropical regions or from the southern hemisphere. It is impossible to describe these in detail here, but some idea of what is being discovered is given in the following notes on certain species of a few selected genera and also by studying the illustrations.

We will start this survey by turning first to *Carex*, of which the leaf structure of a considerable number of species is already well known. The essential characters of the leaf of *Carex* (see Fig. 7, A–E) are (i) the presence of conical silica-bodies in the silica-cells of the leaf epidermis, the conical bodies sometimes being accompanied by silica particles (see Fig. 5, M, Q); (ii) the adaxial epidermis commonly composed of cells conspicuously larger than those of the abaxial epidermis as seen in transverse sections through the lamina; (iii) paracytic stomata, more frequent on the abaxial than on the adaxial surface and often confined to the abaxial surface; (iv) vascular bundles of unequal sizes but always arranged in a single row as seen in transverse sections; (v) conspicuous air-spaces, sometimes with the partly disorganized remains of translucent cells in them, forming a single row in transverse sections of the leaf and situated in the mesophyll between the vascular bundles; (vi) a well-marked, median, adaxial band of bulliform cells. Individual species vary in the distribution and frequency of the silica-cells; in the

---

strands and buttresses. A.T. Assimilatory Tissue. F. Fibres. I.C. Intercellular Cavity. Ph. Phloem. R.A.T. Radiate Assimilatory Tissue. S.B.Silica-body. S.C. Stellate Cells. St. Stoma Tan. Tannin Cell. T.C. Translucent Cells. V.B. Vascular Bundle. Xy. Xylem. A, *Carex pendula* L. Leaf margin × 36. B, *Carex pendula* L. Midrib × 36. C, *Carex pendula* L. Lamina × 36. D, *Carex pendula* L. Leaf margin × 130. E, *Carex pendula* L. Lamina × 130. F, *Cephalocarpus* sp. (Maguire and Wurdack 42107) Lamina × 36. G, *Cephalocarpus* sp. Lamina × 130. Stomata in abaxial surface away from the assimilatory tissue. cf. *Ficinia and Fimbystilis* in Fig. 6. H, *Cyperus longus* L. Stellate cells from air cavity × 130. I, *Cyperus longus* L. Midrib × 36. J, *Cyperus longus* L. Stellate cells from air cavity × 130. K, *Cyperus longus* L. Leaf margin × 36. L, *Cyperus longus* L. Lamina × 36.

amount and distribution of fibre strands and their relationship to the vascular bundles; in the presence or absence of epidermal papillae; in the size of the epidermal cells in surface view and in the extent to which their anticlinal walls are sinuous.

Some idea of the very striking range of structure that is to be seen in transverse sections of the leaves of different members of the Cyperaceae can be gained by now passing on to consider the following selected types and to compare them with the structure in *Carex* which has just been described.

    i. Cephalocarpus sp. *aff. C. linearifolius* Gilly. Material examined: Maguire and Wurdack, 42107. British Guiana (Fig. 7, F–G).

Intercellular cavities absent. Assimilatory tissue more or less homogeneous with no special orientation of cells. Adaxial surface and vascular bundles strongly supported by sclerenchyma, the median and nearly all other vascular bundles being completely sheathed by sclerenchyma. Vascular bundles in more than one horizontal row. Bulliform cells (Fig. 7, F, T.C.) in the midrib in the form of a single U-shaped, adaxial group.

    ii. Cyperus longus Linn. Material examined: cultivated at Kew (Fig. 7, H–L).

Adaxial part of the mesophyll consisting of translucent cells, the assimilatory tissue being restricted to radiately arranged cells surrounding the vascular bundles. Large intercellular cavities between the veins filled with a loose tissue of somewhat stellate cells. Bulliform translucent tissue well developed over the midrib. Median vascular bundle supported by an adaxial strand of sclerenchyma next to the xylem.

    iii. Didymiandrum stellatum (Boeck.) Gilly. Material examined: Maguire and Wurdack, 42110 (Fig. 8, A–D).

Mesophyll with assimilatory tissue clearly differentiated into palisade and spongy portions. Cells of the spongy portion interconnected by narrow, projecting lobes appearing circular to oval in transverse section. Groups of large translucent cells present in the spongy mesophyll between the vascular bundles. A well-defined hypodermis beneath the adaxial epidermis with fine strands of fibres embedded in it. Cells with amorphous, probably tanniniferous contents present in the palisade tissue and in the inner bundle sheaths. Median vb. not accompanied by adaxial sclerenchyma. Translucent tissue well developed in the midrib, but component cells not markedly bulliform. (Structure in some respects resembling that of *Everardia neblinae*.)

    iv. Evandra aristata R. Br. Material examined: from Sydney, N.S.W., supplied by J. W. Green (Fig. 8, E–F).

Assimilatory tissue more or less homogeneous. Intercellular spaces absent from the mesophyll between the veins. Large vascular bundles

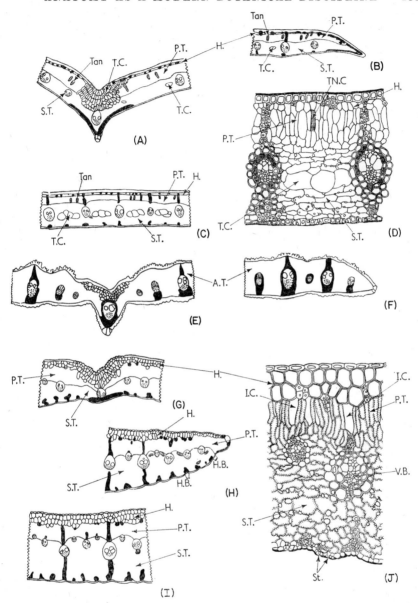

FIG. 8. Transverse sections showing leaf structure in the Cyperaceae (*contd.*). H. Hypodermis. P.T. Palisade tissue. S.T. Spongy tissue. Other letters as for Fig. 7. A, *Didymiandrum stellatum* (Boeck.) Gilly. Midrib × 36. B, *Didymiandrum stellatum* (Boeck.) Gilly. Leaf margin × 36. C, *Didymiandrum stellatum* (Boeck.) Gilly. Lamina × 36. D, *Didymiandrum stellatum* (Boeck.) Gilly. Lamina × 130. E, *Evandra aristata* R. Br. Midrib and lamina × 36. F, *Evandra aristata* R. Br. Leaf margin × 36. G, *Everardia neblinae* Maguire and Wurdack. Midrib × 36. H, *Everardia neblinae* Maguire and Wurdack. Leaf margin × 36. I, *Everardia neblinae* Maguire and Wurdack. Lamina × 36. J, *Everardia neblinae* Maguire and Wurdack. Lamina × 130.

ensheathed and girdered with sclerenchyma both adaxially and abaxially. Median vascular bundle surrounded and strongly supported adaxially and abaxially by sclerenchyma. Midrib with numerous adaxial, but not markedly bulliform, translucent cells.

v. *Everardia neblinae* Maguire and Wurdack. Material examined: Maguire and Wurdack, 42108 (Fig. 8, G–J).

Mesophyll clearly differentiated into palisade and spongy portions, the spongy tissue consisting of cells interconnected by narrow projecting lobes. Large intercellular cavities absent from the mesophyll between the veins. A two- or three-layered adaxial hypodermis of translucent cells present. Large vascular bundles with narrow adaxial and abaxial girders of sclerenchyma. Abaxial strands of sclerenchyma independent of the vascular bundles. Other strands of sclerenchyma at the boundary between the hypodermis and palisade sclerenchyma and abutting on the adaxial surface of some of the smaller vascular bundles. Median vascular bundle supported abaxially by parenchyma. Adaxial part of the midrib composed of translucent, but not markedly bulliform, cells. (Structure in some respects similar to that of *Didymiandrum stellatum*.)

vi. *Ficinia indica* (Lam.) Pfeiffer. Material examined: supplied by M. Levyns from Rondebosch Common, S. Africa (Fig. 9, A–D).

Leaf narrow and somewhat fleshy, the median adaxial part being composed of translucent cells with moderately thick walls. Mesophyll with assimilatory tissue restricted to the abaxial half of the mesophyll and the wings, the assimilatory cells being mostly palisade-like as seen in transverse section, but with occasional narrow lobes connecting adjacent cells. Large intercellular cavities present at the internal boundaries of the chlorenchyma. (The structure recalls that of *Gymnoschoenus sphaerocephalus* but lacks substomatal cavities lined with sclereids.)

vii. *Fimbristylis diphylla* (Retz.) Vahl. Material examined: collected by Metcalfe in Trinidad (Fig. 9, E, G, I).

Cells of the locally double-layered adaxial epidermis outstandingly large. Mesophyll with more or less homogeneous assimilatory tissue consisting of cells of which some are palisade-like, and many interconnected by narrow projecting lobes. Large intercellular cavities between the vascular bundles absent. Sclerenchyma present as strands at the outer edge of the assimilatory tissue towards both surfaces. Some secretory, probably tanniniferous cells scattered in the chlorenchyma. (Structure in some respects similar to that of *Kyllinga monocephala*.)

viii. *Fintelmannia lhotzkyana* (Nees) Pfeiffer. Material examined: A. C. Smith 3643, British Guiana (Fig. 9, F, H, J).

Structure similar to that of *Carex pendula* but with the sclerenchyma

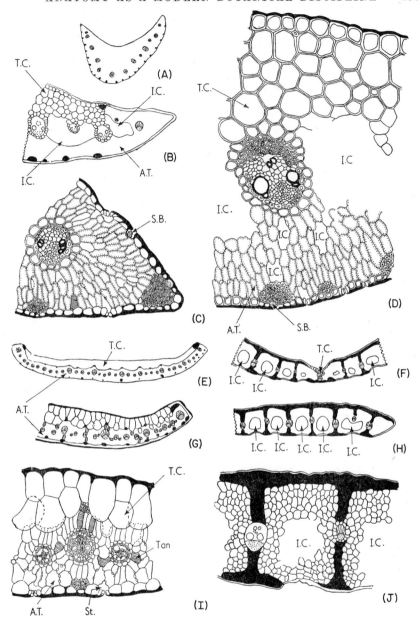

Fig. 9. Transverse sections showing leaf structure in the Cyperaceae (*contd.*). A, *Ficinia indica* (Lam.) Pfeiffer. Lamina ×10. B, *Ficinia indica* (Lam.) Pfeiffer. Leaf margin ×36. C, *Ficinia indica* (Lam.) Pfeiffer. Lamina ×130. D, *Ficinia indica* (Lam.) Pfeiffer. Lamina × 130. (Note stomata next to the assimilatory tissue. cf. *Cephalocarpus* in Fig. 7 and *Fimbristylis* in Fig. 9, I.) E, *Fimbristylis diphylla* (Retz) Vahl. Lamina ×20. F, *Fintelmannia lhotzkyana* (Nees) Pfeiffer. Lamina ×36. G, *Fimbristylis diphylla*. Lamina ×36. H, *Fintelmannia lhotzkyana*. Lamina and leaf margin ×36. I, *Fimbristylis diphylla*. Lamina ×130. (Note stomata next to the assimilatory tissue. cf. *Ficinia* in Fig. 9,D and *Cephalocarpus* in Fig. 7.) J, *Fintelmannia lhotzkyana*. Lamina ×130.

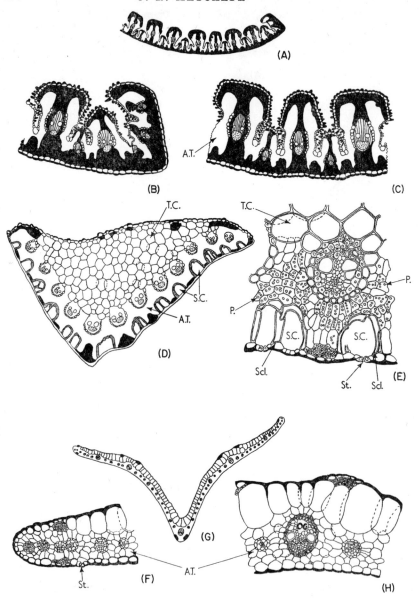

FIG. 10. Transverse sections showing leaf structure in the Cyperaceae (*contd.*). P. Pitted assimilatory cells. S.C. Substomatal cavity. Scl. Sclereid. Other letters as for Fig. 7. In F–H cells of adaxial epidermis specially large. A, *Gahnia radula* Benth. Lamina ×10. B, *Gahnia radula* Benth. Leaf margin ×36. C, *Gahnia radula* Benth. Lamina ×36. D, *Gymnoschoenus sphaerocephalus* (R. Br.) Hook. f. ×36. E, *Gymnoschoenus sphaerocephalus* (R. Br.) Hook. f. ×130. F, *Kyllinga monocephalus* Rottb. Leaf margin ×130. G, *Kyllinga monocephala* Rottb. Lamina ×20. H. *Kyllinga monocephala* Rottb. Lamina ×130.

very strongly developed, especially beneath the adaxial epidermis and in association with the vascular bundles. Mesophyll with homogeneous chlorenchyma and well-developed intercellular cavities between the vascular bundles. Differs from many species of *Carex* in having the cells of the adaxial not appreciably larger than those of the abaxial epidermis. Translucent cells in the adaxial part of the midrib not very numerous and not bulliform.

ix. Gahnia radula Benth. Material examined: supplied by W. M. Curtis from Blackman's Bay, Hobart, Tasmania (Fig. 10, A–C).

Leaf characterized by alternating tall and low ribs on the adaxial surface, the ribs being massively supported by T-shaped girders of sclerenchyma, the vascular bundles being embedded in the stems of the T's. Assimilatory tissue homogeneous and confined to the sides of the ribs. Adaxial ribs covered with cutinized papillae. (Structure immediately recalling that of Marram grass (*Ammophila arenaria*) but distinguished by the small, inconspicuous, mostly more or less circular silica-bodies in the abaxial epidermis, there being one body in each cell.)

x. Gymnoschoenus sphaerocephalus (R. Br.) Hook. f. Material examined: supplied by W. M. Curtis from Blackman's Bay, Hobart, Tasmania (Fig. 10, D–E).

Leaf narrow and fleshy, the structure somewhat recalling that of *Ficinia indica* but differing notably in having abaxial substomatal cavities lined with sclereids, and mesophyll cells interconnected by small, projecting, pitted lobes.

xi. Kyllinga monocephala Rottb. Material examined: collected by Metcalfe in Trinidad (Fig. 10, F–N).

Leaf structure recalling that of *Fimbristylis diphylla*, the specially large cells of the adaxial epidermis being alike in the two species. Adaxial sclerenchyma strands in *Kyllinga monocephala* situated immediately below the outer surface of the leaf in contrast to those of *Fimbristylis diphylla* at the boundary between the cells of the epidermis and the subjacent assimilatory tissue.

xii. Lagenocarpus sp. Material examined: Maguire and Wurdack, 41761, British Guiana (Fig. 11, A–D).

Leaf exhibiting a two-layered, partly sclerosed hypodermis; vascular bundles arranged in two rows, the individual bundles in the two rows being opposite one another; adaxial chlorenchyma palisade-like; abaxial part of the mesophyll composed of lobed cells; two-celled prickles present; stomata tetracytic.

(Tetracytic stomata appear to be unusual in the Cyperaceae judging from the species that have so far been examined.)

xiii. Lepidosperma laterale R. Br. Material examined: supplied by the National Herbarium, Sydney, New South Wales (Fig. 11, E).

6§

xiv. Lepidosperma tortuosum F. Muell. Material examined: supplied by W. M. Curtis from Electrona, Tasmania (Fig. 11, F).

Leaf of *L. laterale* isobilateral, equitant, with a set of vascular bundles beneath each surface, with the xylem towards the centre of the leaf, the xylem strands of the opposed pairs of bundles facing towards one another. Leaf of *L. tortuosum* similar but much narrower, with large intercellular cavities in the central ground tissue.

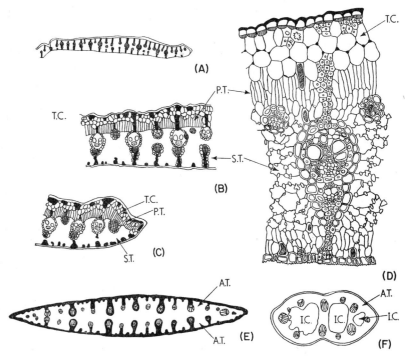

Fig. 11. Transverse sections showing leaf structure in the Cyperaceae (*contd.*). A, *Lagenocarpus* sp. (Maguire and Wurdack 41761). Lamina ×10. B, *Lagenocarpus* sp. (Maguire and Wurdack 41761). Lamina × 36. C, *Lagenocarpus* sp. (Maguire and Wurdack 41761). Leaf margin ×36. D, *Lagenocarpus* sp. (Maguire and Wurdack 41761). Lamina ×130. E, *Lepidosperma laterale* R. Br. Whole leaf ×20. F, *Lepidosperma tortuosum* F. Muell. Whole leaf × 36.

By some botanists the isobilateral, equitant type of structure is believed to indicate that the leaf is really a petiole. The leaf could equally be interpreted as a true lamina in which the two halves have phylogenetically become folded together on either side of the median vascular bundle in such a way that the adaxial surfaces of the two halves have not only fused together but the tissues have merged so as to form the central ground tissue of the lamina. The possession of iso-bilateral equitant leaves does not, by itself, indicate that there is necessarily any close affinity between plants that possess them because they are known to occur in a number of families which, on other grounds, do not appear to be closely related to one another (see p. 143).

3. Palmae

The following notes are based on observations by Tomlinson (1961c). The family is somewhat isolated from the other monocotyledons, the Cyclanthaceae being the one family that is in some respects similar. The subdivision of the palms into those with pinnate and palmate fronds respectively is not entirely satisfactory taxonomically. In Tomlinson's experience, classification based on the distinction between fronds in which the segments are folded in a V-shaped manner and those in which the folding is inversely V-shaped leads to a more satisfactory result. The vascular structure of ribs where the folds occur is also important in this basic subdivision of the family. Tomlinson also feels that the Palmae should be divided into a number of sub-families, although he has not actually taken this step, but in his book refers to them under group names such as Arecoid, Borassoid and Caryotoid palms. Some of these groups, such as the three just mentioned, consist of more than one genus, but other groups of equivalent taxonomic status consist of one genus only. This applies for example to the Phoenicoid palms, all of which belong to *Phoenix*. In certain of the groups with the status of sub-families and which consist of more than one genus it may happen that one particular genus stands out rather clearly from the remainder. This applies, for example, to *Bactris*, *Borassus* and *Chamaerops*. Histological interspecific differences of leaf structure are usually small, only quantitative and therefore not of much practical taxonomic value. Tomlinson concludes that it is possible from the anatomy of the fronds to identify palms at the level of the tribe or genus but not to the species level.

It is not possible here to give any further particulars of Tomlinson's study of the anatomy of palms or to describe in detail the histological characters on which his conclusions are based, but his ideas are fully set forth in his recently published book to which reference is made above.

4. Scitamineae

For information concerning the families in this order we are again indebted to Tomlinson (1956, 1959, 1960, 1961a, 1961b). Some interesting conclusions that have emerged from his observations include the following.

i. Zingiberaceae. In this family the Zingiberoideae and Costoideae are morphologically and histologically distinct, and these differences are repeated also in their cytology and pollen morphology.

ii. Heliconiaceae. The hundred or more species included in the genus *Heliconia* are sufficiently distinct in their morphology and anatomy to merit elevation to family rank. The family is allied to Musaceae, especially through *Musa coccinea* of which the leaf structure is very similar to that of *Heliconia.*

iii. Musaceae. No reliable differences of leaf structure could be found whereby most species of *Musa* can be separated from one another. A notable exception is *M. coccinea* (see last paragraph). These conclusions were reached after examining leaf material of *Musa* supplied by N. W. Simmonds from the College of Tropical Agriculture in Trinidad.

iv. Cannaceae. The combination of characters exhibited by members of the Cannaceae suggest that the family is somewhat isolated from the remainder of the Scitamineae.

v. Strelitziaceae (Ravenala, Phenakospermum, Strelitzia). These three genera consisting of small trees of palm-like habit form a natural group in which *Heliconia* should not be included.

5. *Commelinaceae*

This pan-tropical family, unlike most monocotyledons, is noted by Tomlinson (unpublished) as characterized by diverse types of hairs. Specially interesting are the uniseriate, glandular hairs, two to three cells long, each with a delicate club-shaped terminal cell. Hairs of this type, if reduced in length, would resemble the two-celled micro-hairs of the Gramineae. Two-celled spines which occur at the leaf margins of certain species recall the one-celled spines or prickle-hairs that are common in the Gramineae and Cyperaceae. The present writer found that *Lagenocarpus* is particularly interesting because it has two-celled prickles similar to those of the Commelinaceae, although it is a member of the Cyperaceae. On other anatomical grounds its inclusion in the Cyperaceae is also somewhat unexpected.

The Commelinaceae differ from the Gramineae in having tetracytic stomata. They also differ from most of the Cyperaceae in this respect, although tetracytic stomata have been noted in a few cyperaceous genera such as *Lagenocarpus* and *Lepironia.*

6. *Alismataceae* (Metcalfe, 1961)

It has sometimes been suggested that the monocotyledons have a ranalian ancestry, the similarity of the Alismataceae to the Ranunculaceae being cited in support of this suggestion. The anatomical evidence does not appear to support this suggestion, however, as indicated by the following important differences. In the Alismataceae the stomata are paracytic and tetracytic in contrast to the anomocytic type (Fig. 3) found throughout the Ranunculaceae. It is well known that the spongy tissue of the Alismataceae consists of a complex parenchymatous network and intercellular cavities, the cavities being traversed at intervals by plates of special cells. The writer has noted that this contrasts with the spongy tissue of aquatic members of the Ranunculaceae which consists of loosely arranged cells with irregular intercellular spaces between them. In the xylem of the Ranunculaceae the main conducting channels throughout the plant consist of vessels, whereas in the Alismataceae vessels are confined to the roots, the conducting elements of the xylem elsewhere in the plants consisting of scalariform tracheids with sloping end walls. Occasional raphides have been noted by the writer in a few species of Alismataceae but none have been recorded or noted in the Ranunculaceae. The fact that, in the Alismataceae, vessels are restricted to the roots is very significant, because tracheids are ancestral to vessel elements, which indicates that the vegetative organs of the Alismataceae have reached a less advanced evolutionary level than those of the Ranunculaceae. This being so it is difficult to see how the Alismataceae can have been derived from the Ranunculaceae as has often been stated or implied. The situation just described does not necessarily exclude the possibility of the Ranunculaceae and the Alismataceae having had a common ancestry, but, if this is so, it seems that the two families must have been evolving along independent lines for a very long time.

7. Flagellariaceae

Work by Smithson (1956), whose results have been confirmed by Tomlinson (unpublished), suggests that this family is not a natural group, in the sense that *Hanguana* exhibits many points of histological difference from *Flagellaria* and *Joinvillea*, the two other genera included in the family. *Flagellaria*, and even more so *Joinvillea*, resemble the Gramineae in leaf characters. Evidence provided by pollen morphology also indicates that *Flagellaria* and *Joinvillea*, but not *Hanguana*, have affinities with the Gramineae.

8. Coleochloa (Metcalfe, 1961)

Taxonomists have been uncertain whether this genus, and a few others to which it is closely allied, should be included in the Gramineae or Cyperaceae. Smithson (unpublished), who examined the structure of the leaf and ovary, was able to show quite conclusively that the genera in question are rightly to be regarded as members of the Cyperaceae.

The type of silica-bodies in the epidermis and the structure of the lamina in transverse sections are especially valuable in elucidating this problem.

9. *Haemodoraceae* (Metcalfe, 1961)

The writer has noted that this family is not anatomically homogeneous. For example secretory canals occur in *Dilatris* and *Haemodorum* (Haemodoreae), but none have yet been noted in the Conostyleae or in *Lanaria* and *Wachendorfia*. Here again pollen morphology (Erdtman, 1952) also indicates lack of uniformity within the family.

10. *Sparganiaceae and Typhaceae* (Metcalfe, 1961)

The writer finds that leaf anatomy provides convincing evidence that these two families are related to one another. Transverse sections of the leaf are especially interesting since they show the vascular bundles to be variously orientated for no apparent reason. Here again a close relationship between the two families is supported by similarities in pollen morphology (Erdtman, 1952).

11. *Xyridaceae* (Metcalfe, 1961)

The writer has found that *Abolboda* differs from *Xyris* particularly in the arrangement of the subsidiary cells around the stomata and also in the structure of the bundle sheaths. This distinction is also supported by palynological evidence (Erdtman, 1952). This family is now receiving more detailed attention from Carlquist (1960) who finds *Xyris* to be rather isolated within the family.

12. *Iridaceae* (Metcalfe, 1961)

The widespread occurrence of elongated, bar-like crystals (styloids) within the family is noteworthy.

C. GENERAL CONCLUSIONS

Although many more observations will have to be made before we are in a position to give anything like a complete picture of the taxonomy of the monocotyledons as it appears to an anatomist, it is clearly evident that considerable progress has already been made.

We are now in a position to say that anatomy, and particularly leaf anatomy, enables non-flowering specimens to be identified, but the taxonomic level to which this identification can be carried varies from family to family. We can also begin to see when families are closely related to one another and when they are not, and to detect possible weaknesses in classification when a supposedly natural taxonomic group turns out to be anatomically heterogeneous. When we turn from taxonomy to phylogeny we at once find ourselves on more difficult ground, but the accumulating evidence is still confirming Cheadle's discovery that vessels were evolved from tracheids first in the root and then at progressively higher levels in the plant body. The task of studying the morphology of the conducting channels of the xylem throughout a single plant is from its very nature slow and tedious, and this in turn means that comprehensive phylogenetic conclusions based on the structure of the xylem are still a long way off. Nevertheless we can already see that throughout the monocotyledons various levels of tracheal specialization have been reached. We can for example recognize the Gramineae as an advanced family and the Cyperaceae as one that is slightly less so. There is also evidence of different levels of specialization within many individual families. We are, however, not yet in a position to decide with authority whether the monocotyledons are a monophyletic group or not. Furthermore the writer is not quite certain whether we are absolutely clear in our own minds precisely what we mean by this term. Whether or not one envisages a taxon as being monophyletic depends on how broadly based one supposes the starting point to have been. However this may be, in considering the phylogeny of monocotyledons there is the great difficulty of deciding when we are dealing with evolutionary convergence as opposed to similarities that are due to true taxonomic affinity between plants that have evolved from a common starting point. There is ample evidence that, in the monocotyledons, many points of similarity must have been evolved along independent lines. As examples it may be mentioned that secretory cells and raphides occur in so many different families of which some are clearly not closely allied to one another that it is hard to believe that they have been evolved only once. Then again there are leaves of the isobilateral equitant type, i.e. the type of leaf that occurs in many of the irises, where there are two opposed rows of vascular bundles with the xylem groups of the opposed bundles facing one another (Fig. 11, E–F). The morphological interpretation of this type of structure has already been discussed (p. 138), but it is not important for our present discussion to accept one rather than the other of the two alternative explanations that have been suggested. The fact to be emphasized is that, no matter how the isobilateral equitant leaf has been evolved, leaves of this type, besides being present in the Iridaceae, have been noted in plants that are wholly unrelated such as *Juncus xiphioides* E. Mey. var. *triandra* Engelm. of Juncaceae, in *Narthecium* and *Nietneria* of Liliaceae, in *Abolboda* and *Xyris* of the Xyridaceae, and in a modified form in *Philydrum lanuginosum* Banks of the Philydraceae. Isobilateral equitant leaves have been noted in species of *Lepidosperma*

(Fig. 11, E–F) of the Cyperaceae and also occasionally in the Orchidaceae. The fact that this distinctive type of leaf has appeared in so many families at least raises the possibility that it has been evolved more than once and should put us on our guard against accepting all similarities as evidence of taxonomic affinity without very careful consideration.

VII. General Discussion

This article has been written at a moment in the history of botany when the study of plant science has become fragmented into a number of specialized branches of knowledge. This fragmentation is not wholly harmful for it has led and is still leading to spectacular advances in some directions, particularly in the realms of physiology, biochemistry and cyto-genetics. In addition there is the all important study of ultra-structure which has been so greatly stimulated by the use of the electron microscope for the study of vegetable cells. Nevertheless one may perhaps be forgiven for asking all who are concerned with the scientific study of plants and plant life, and, after all this is the true purpose of botany, to stand back for a few moments and consider this edifice of knowledge as it exists today.

One of the dangers of the present position is that specialized workers develop their own language which is not widely understood. This is perhaps inevitable, and it need do no harm provided the specialist realizes that his colleagues outside his own immediate circle cannot understand him without translation. Very often a specialist in one field does not like to admit his ignorance of the meaning of technical terms when discussing a subject with a colleague outside his own immediate province and this inevitably leads to lack of mutual comprehension on the broader issues that concern them both. Because of the mental effort needed to overcome this language difficulty, and because of the urge to become eminent in some field no matter how small it may be, there is a danger that many will seek to attain superiority over their fellows by narrowing their enquiries to a pin-point. Thus we find, for example, that the discussions of those who are concerned with the minutiae of botanical classification or nomenclature arouse no interest whatsoever in the minds of those concerned with the biochemical approach to plant classification, the ultra-structure of a cell wall or the behaviour of a meristem. This tendency towards splinter movements increases the danger that in the search for new knowledge we may forget the essentials of what we already know, rather as if we were studying the behaviour of a disembodied growing point whilst forgetting about the morphology of the parent plant from which it has arisen.

The classical approach to morphology and anatomy is one that is at present passing through a rather stagnant period. The writer hopes, however, that the content of this article will help to show that there is still a lot to learn in this field, and that it will be a serious loss to the cohesion of botanical science as a whole if this branch of knowledge is allowed to lapse any longer. Our need is to arrest the fragmentation of botanical knowledge and to concentrate attention on the frontiers between the different compartments into which it has become divided. Just as nature abhors distinct lines of demarcation, so should these boundaries not only become more nebulous, but provision should be made for the two-way diffusion of ideas across these membranes and cell walls that have become or are becoming too impervious. It is precisely because we have now reached the stage at which further advances in botanical knowledge are likely to stem from "fringe subjects" that the classical approach to morphology and anatomy is so important. Nowhere better than here can we find an approach that possesses such good unifying properties, for the subject interconnects all of the specialized branches into which botany has become divided. Here we find a common meeting ground for students of evolution, taxonomy, physiology, genetics, ultra-structure and other specialized approaches to the scientific study of plants. Here the comparative anatomist can also join forces with the palaeobotanist and the archaeologist. Students should no longer be brought up in ignorance of the fact that plant anatomy has its part to play in helping to solve everyday problems concerning the identification of economic plant products such as the fibres, foodstuffs and timbers that are so important in the world's economy. Nor should we ever be allowed to forget that the physiology of growing crops cannot be fully understood whilst we are ignorant of the structure of the crop plants themselves.

If we may conclude this discussion by taking a final look at the anatomist's contribution to taxonomy, the writer cannot emphasize too strongly that the study of systematic anatomy should never be conducted in an ivory tower by a group of narrow-minded specialists. It must be remembered that histological data are in no way superior to other criteria that are used for taxonomic purposes. What the anatomist does claim is that he is providing additional characters of which full use should be made both in general taxonomy and in the identification of fragmentary vegetable material and economic plant products. The extent of our ability to help with these tasks is at present limited by our imperfect knowledge of the distribution of histological characters throughout the flowering plants. This situation provides a challenge which we should be able to meet, for we live in an age in which modern techniques enable us not only to obtain but also to handle and analyse large masses of data. We must, however, work hand in hand with those whose interest lies in the traditional herbarium approach to taxonomy, and also with those concerned in plant classification from the respective standpoints of the cytologist, palynologist, biochemist and fossil botanist. To synthesize a broad picture of the evolution and classification of the flowering plants based on these several considerations will be no mean task, and it will call for very capable and broad-minded botanists to achieve it. Nevertheless, in the writer's view, this synthetic approach could enrich the whole of botany by giving us a deeper and more perfect understanding of the evolution of the flowering plants and of their classification on a natural basis.

REFERENCES

Carlquist, S. (1960). *Mem. N.Y. bot. Gdn* **10**, 65–117. Anatomy of Guyana Xyridaceae: Albolboda, Orectanthe and Achlyphila.

Carlquist, S. (1961). *Aliso* **5**, 39–66. Pollen morphology of Rapateaceae.

Cheadle, V. I. (1937). *Bot. Gaz.* **98**, 535–555. Secondary growth by means of a thickening ring in certain monocotyledons.

Cheadle, V. I. (1942a). *Amer. J. Bot.* **29**, 441–450. The occurrence and types of vessels in the various organs of the plant in the Monocotyledoneae.

Cheadle, V. I. (1942b). *Chron. Bot.* **7**, 253–254. The role of anatomy in phylogenetic studies of the Monocotyledoneae.

Cheadle, V. I. (1943a). *Amer. J. Bot.* **30**, 11–17. The origin and certain trends of specialization of the vessel in the Monocotyledoneae.

Cheadle, V. I. (1943b). *Amer. J. Bot.* **30**, 484–490. Vessel specialization in the late metaxylem of the various organs in the Monocotyledoneae.

Cheadle, V. I. (1944). *Amer. J. Bot.* **31**, 81–92. Specialization of vessels within the xylem of each organ in the Monocotyledoneae.

Cheadle, V. I. (1948). *Amer. J. Bot.* **35**, 129–131. Observations on the phloem in the Monocotyledoneae. II. Additional data on the occurrence and phylogenetic specialization in structure of the sieve tubes in the metaphloem.

Cheadle, V. I. (1953). *Phytomorphology* **3**, 23–44. Independent origin of vessels in the monocotyledons and dicotyledons.

Cheadle, V. I. (1956). *Amer. J. Bot.* **43**, 719–731. Research on xylem and phloem—progress in fifty years.

Cheadle, V. I. and Uhl, N. W. (1948a). *Amer. J. Bot.* **35**, 486–496. Types of vascular bundles in the Monocotyledoneae and their relation to the late metaxylem conducting elements.

Cheadle, V. I. and Uhl, N. W. (1948b). *Amer. J. Bot.* **35**, 578–583. The relation of metaphloem to the types of vascular bundle in the Monocotyledoneae.

Cheadle, V. I. and Whitford, N. B. (1941a). *Amer. J. Bot.* **28**, suppl. to No. 10. A discussion of some factors which influence the form of the vascular bundle in the Monocotyledoneae.

Cheadle, V. I. and Whitford, N. B. (1941b). *Amer. J. Bot.* **28**, 623–627. Observations on the phloem in the Monocotyledoneae. I. The occurrence and phylogenetic specialization in structure of the sieve tubes in the metaphloem.

Connor, H. E (1960). *New Zealand J. Sci.* **3**, 468-509. Variation in leaf anatomy in *Festuca novae-zelandiae* (Hack.) Cockayne and *F. Matthewsii* (Hack.) Cheeseman.

Erdtman, G. (1952). "Pollen Morphology and Plant Taxonomy", xii + 539 pp. Uppsala.

Fahn, A. (1954). *J. Linn. Soc. (Bot.)* **55**, 158–184. The anatomical structure of the Xanthorrhoeaceae Dumort.

Mann, L. K. (1952). *Hilgardia* **21**, 195–231. Anatomy of the garlic bulb and factors affecting bulb development.

Mann, L. K. (1959). *Amer. J. Bot.* **46**, 730–739. The *Allium* inflorescence: Some species of the section *Molium*.

Mann, L. K. (1960). *Amer. J. Bot.* **47**, 765–771. Bulb organization in *Allium*: some species of the section *Molium*.

Metcalfe, C. R. (1956). *Bot. Mag., Tokyo* **69**, 391–400. Some thoughts on the structure of bamboo leaves.

Metcalfe, C. R. (1959). A vista in plant anatomy. *In* "Vistas in Botany" (W. B. Turrill, ed.), pp. 76–99. Pergamon Press, London.

Metcalfe, C. R. (1960). "Anatomy of the Monocotyledons. I. Gramineae", 731 pp. Clarendon Press, Oxford.

Metcalfe, C. R. (1961). "Recent Advances in Botany", 146–150. University of Toronto Press. The anatomical approach to Systematics. General introduction with special reference to recent work on monocotyledons.

Metcalfe, C. R. and Chalk, L. (1950). "Anatomy of the Dicotyledons", 1500 pp. (2 vols.). Clarendon Press, Oxford.

Smithson, E. (1956). *Kew Bull.* 491–501. The comparative anatomy of the Flagellariaceae.

Stebbins, G. L. and Khush, G. S. (1961). *Amer. J. Bot.* **48**, 51–59. Variation in the organization of the stomatal complex in the leaf epidermis of monocotyledons and its bearing on their phylogeny.

Tomlinson, P. B. (1956). *J. Linn. Soc. (Bot.)* **55**, 547–592. Studies in the systematic anatomy of the Zingiberaceae.

Tomlinson, P. B. (1959). *J. Linn. Soc. (Bot.)* **55**, 779–809. An anatomical approach to the classification of the Musaceae.

Tomlinson, P. B. (1960). *J. Arnold Arbor.* **41**, 287–297. The anatomy of *Phenakospermum* (Musaceae).

Tomlinson, P. B. (1961a). *J. Linn. Soc. (Bot.)* **56**, 467–473. The anatomy of *Canna*.

Tomlinson, P. B. (1961b). *J. Linn. Soc. (Bot.)* **58**, 55–78. Morphological and anatomical characteristics of the Marantaceae.

Tomlinson, P. B. (1961c). "Anatomy of the Monocotyledons. II. Palmae", 453 pp. Clarendon Press, Oxford.

Palynology

G. ERDTMAN

Palynological Laboratory, Stockholm-Solna, Sweden

I. INTRODUCTION

A. SCOPE OF PALYNOLOGY

Palynology—the science of pollen grains and spores—is a relatively new weft in the complicated web of *Scientia amabilis*. It is particularly connected with palaeobotany (pollen analysis of Quaternary and Pre-Quaternary deposits) and taxonomy.

The following does not pretend to be a detailed presentation of the whole of palynology. Such an approach would be difficult since palynology is at present penetrating into many other domains of botany and allied sciences, from which it is not separated by distinct lines of demarcation. Attention will, instead, be focused upon the nucleus of palynology, namely the pollen grains and spores themselves and the morphology, fine structure etc. of their wall (sporoderm), particularly of its outermost layer, the exine. This layer is usually extraordinarily resistant and possesses at the same time so many characteristic details that it can be used for the identification of plants, recent as well as fossil, much as criminals can be identified from finger-prints. The resistance of grains and spores is so great that they can be preserved as fossils in peat and lake sediments etc. and thus provide a record of climatic and vegetational history stretching from our own days right back to Cambrian or, maybe, even Pre-Cambrian times. A gram of peat may contain 100 000 fossil pollen grains, making 100 000 million or more per cubic metre. Moreover, pollen grains and spores are found almost everywhere. It has been said that they have probably a wider distribution, both with regard to time and space, than any other organisms or parts of organisms.

The word "palynology" was coined by Hyde and Williams (1945) as a substitute for "the science of pollen grains and spores". It comes from the Greek verb *palynein*, meaning "to spread, to distribute" in recognition of the fact that many pollen grains and spores are easily carried by the wind. Palynology is an adequate term although somewhat difficult to pronounce. "Sporology" has been mentioned as a possible alternative, and the cytological difference between "microspores" and "pollen grains" is not so great that these two groups cannot be dealt with under a common heading.

B. SUBDIVISIONS OF PALYNOLOGY

Tentatively, palynology can be subdivided as follows:

1. Basic palynology

i. Pollen and spore morphology.

ii. Theoretical aspects of applied palynology. Output and dissemination of pollen grains and spores, etc.; resistance to decay, etc.

2. Applied palynology

i. Palynotaxonomy. Pollen and or spore morphology and plant taxonomy.

ii. Geo- or palaeopalynology. The study of fossil pollen grains and spores (pollen-analytical investigations of Quaternary deposits; pollen and spore floras of Pre-Quarternary deposits).

iii. Melittopalynology. The study of pollen grains etc. in honey.

iv. Pharmacopalynology. The study of pollen grains etc. in drugs, tablets, etc.

v. Iatropalynology. The study of pollen grains etc. in connection with allergies etc. (Some pollen grains causing hay fever are shown in Pl. I opposite).

vi. Copropalynology. The study of pollen grains etc. in excrements.

vii. Forensic palynology. Palynology as an aid in criminology.

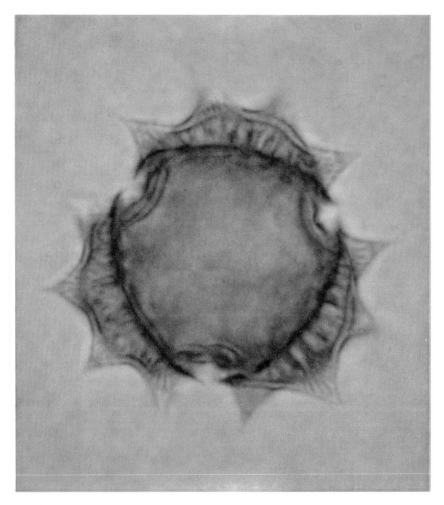

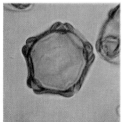 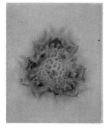 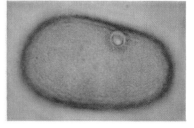

Upper figure: *Chrysanthemum leucanthemum* (polar view, optical cross section; × 3000).
Lower row (× 750): *Alnus incana* (polar view, optical cross section); *Chrysanthemum leucanthemum* (polar view, surface); *Secale cereale* (surface).

C. HISTORY

A detailed historical account has been given by Wodehouse (1935, 1960), and mention will be made here of only six outstanding names.

Hugo von Mohl (1805–72), well known for his classification (1834) of pollen grains according to their apertures, aperture membranes, etc.

Carl Julius Fritzsche (1808–71), whose paper "Ueber den Pollen" was published in 1837—the great scientific value of which was not properly realized until more than 50 years later—and who still stands almost unrivalled with regard both to the beauty of his illustrations and the accuracy of his descriptions.

Hugo Fischer (1865–1939), who made painstaking investigations, among other things, of the exine details in a great number of plants. The great value of his thesis "Beiträge zur vergleichenden Morphologie der Pollenkörner" (Breslau, 1890) was not properly recognized and Fischer did not attain a position worthy of his ability.

Lennart von Post (1884–1951) who, in 1918, read a paper embodying the principles of modern pollen analysis. As a student he learned from Gustaf Lagerheim, "the spiritual father of pollen analysis", how to identify the pollen grains of common Swedish trees and shrubs. Rutger Sernander instilled in him a keen interest in the history and development of the Swedish peat deposits. Building upon the knowledge acquired from these botanists, Lennart von Post, who was a geologist and in fact never studied botany at the university, laid down the principles of modern pollen analysis. He, and with him a great number of pupils, applied these principles in the investigation of many Swedish peat bogs. Since 1922 pollen analytical investigations have been carried out in increasing numbers outside Scandinavia as well.

Robert Potonié (b. 1889) who, in the early thirties, wrote a number of important papers on the morphology of fossil pollen grains and who, ever since, has been an active authority on palaeopalynological matters.

Roger P. Wodehouse (b. 1889), whose monograph "Pollen grains" (1935, reprinted 1960) is a classic from which a wealth of information and inspiration can be obtained.

Other palynological text-books and atlases have been published by Erdtman (1943, 1952 etc.), Firbas (1949–52), Faegri and Iversen (1950), Pokrovskaja (1950), Ikuse (1956), Dyakowska (1959), Wang (1960), Pla Dalman (1961), and others.

II. Palynological Research Centres and Their Organization

A. PALYNOLOGICAL RESEARCH CENTRES

Palynological research centres either exist or are being established in various parts of the world from Calgary, Stockholm and Alma Ata in the north to Concepción (Chile), Bloemfontein, and Melbourne in the south. A short list of centres was published in *Grana Palynologica* some years ago (Vol. 1, No. 2, pp. 136–139, Stockholm, 1956). Many centres have been established by oil and coal companies. Other centres, probably less numerous, have been set up at universities as special institutes or as sub-departments of botany or geology. Only very few centres are wholly concerned with basic palynology and its taxonomical and (at least as far as the theoretical aspects go) pollen statistical implications. One of these is the Palynological Laboratory, Nybodagatan 5, Stockholm-Solna, which is briefly described here.

B. THE PALYNOLOGICAL LABORATORY, STOCKHOLM-SOLNA

The Laboratory was established by the Swedish Natural Science Research Council in 1949. The present staff includes, beside the director, three senior assistants, a secretary, two technical assistants, and a clerk. In addition there is room for up to four visiting scientists. Here is housed a very representative range of material which may be classified as follows.

1. The Palynological Training Collection

This is a collection of elementary pollen and spore slides, simple models of pollen grains and spores, explanatory drawings and diagrams, photomicrographs, text-books, etc.

2. The Sporothecae. (Main collections of pollen and spore slides.)

i. The Scandinavian Sporotheca. This consists of one cupboard containing about a hundred boxes of pollen slides of plants chiefly from Sweden, Norway, Finland, Denmark, and Iceland. A hundred slides can be placed in each box and there are usually several slides of each species. The slides are arranged in the same way as those in the International Sporotheca (see below). Photomicrographs of pollen grains and spores in this collection have been published by Erdtman *et al.* (1961). This part of the Sporotheca is essentially intended as a reference collection for the benefit of pollen analysts dealing with the pollen statistics of North European Quaternary deposits.

ii. The International Sporotheca. This is housed in six cupboards and totals about 400 slide boxes in all. The boxes are now from 50–75% full and arranged in the following way.

(a) Angiosperms (*Acanthaceae-Zygophyllaceae*). The families are arranged according to Engler and Diels (1936) though quoted in alphabetical order to make it possible to find any slide as quickly as possible. Within the families the genera, and within the latter the species are arranged in alphabetical order. If slides have been made from different collections of a certain species, these slides are likewise arranged in alphabetical order of collectors' names. (b) Gymnosperms (*Abies-Zamia*). Genera in alphabetical order. (c) Pteridophyta (*Abacopteris-Xiphopteris*). Genera in alphabetical order. (d) Bryophyta (*Acrocryphaea-Zygodon*). Genera in alphabetical order.

There are also a number of boxes with spores of thallophytes and boxes with pollen grains and spores of *plantae incertae sedis* as well as with hystrichospherids and similar cysts etc.

In building up the International Sporotheca emphasis has been placed largely upon the following points. Every angiosperm family, sub-family, tribe etc., should, if possible, be represented by at least the more important genera. With regard to gymnosperms and pterido-phytes every genus should, if possible, be represented, together with any species, or group of species, of particular interest. With regard to the bryophytes efforts have been made with the kind assistance of many bryologists, to have all families and the more important genera and species represented. Special emphasis has been laid upon plants from the Tropics and Sub-Tropics, since the flora of these countries is usually less known than that in the rest of the world. Brazil, the Canary Islands, the Cap Verde Islands, Madagascar, Mauritius, Ceylon, New Caledonia and part of the tropical areas of the mainland of Africa, Asia, and Australia are thus fairly well represented.

This Sporotheca has been established for the study of the importance of palynological data in plant taxonomy. It includes slides of type specimens of new species and of a number of rare plants believed to be extinct. The total number of slides in the Scandinavian and International Sporothecae is estimated (January, 1962) at about 25 000.

3. The Graphotheca

Parallel with the Sporotheca and its subdivisions is a "graphotheca", an archive of descriptions of the slides. The descriptions are written or typed on special forms. They include dimensions, notes on sporoderm stratification (cf. p. 165), sketches illustrating aperture details (see Fig. 32), rough palynograms (Fig. 26), and finally—so far in certain species only—a pollen or spore diagnosis. There is a special folder of notes for each angiosperm family. At the end of each folder is a collection of the literature on the pollen morphology of the family in question with excerpts from various papers. There is also a list of the relevant literature references. In a special folder, marked "SY–PO", is a collocation of the ideas of various botanists, from A. de Jussieu (1789) onwards, concerning the position of the family in the system.

4. The Tomotheca, Glyptotheca etc.

The collections also comprise a tomotheca (collection of the thin section slides), a glyptotheca (collection of models, in plaster of Paris, clay, etc., of pollen grains or exine details), an iconotheca (collection of photomicrographs, ultra-violet micrographs and electron micrographs), and finally a bibliotheca (library).

5. The Library (Bibliotheca)

The library contains old papers by Ehrenberg, Fritzsche, Guillemin etc., and an extensive collection—particularly as far as the Scandinavian countries go—of literature on modern palynology. Papers in Russian, Ukrainian etc. (not provided with summaries in English, French, or German) are placed in a special collection. From the palynological laboratory and its forerunner, catalogues of literature on pollen statistics, palynology, and finally only basic palynology were issued in the years 1928–60. They comprise in all about 580 pages with approximately 10 000 titles, about 90% of which are to be found in the library.

III. POLLEN AND SPORE MORPHOLOGY

A. METHODS

Critical observation of pollen grains can be made in an ordinary binocular microscope with periplanatic eyepieces ($\times 10$) and two apocromatic objectives, one $\times 10$ (num. ap. $0 \cdot 25$) and the other $\times 100$ (num. ap. $1 \cdot 32$). Recourse to the more powerful phase contrast microscopy is very useful, however, and a polarizing microscope is often helpful (Sitte, 1960). Fluorescence microscopy has, so far, only been sporadically applied in palynology.

It is often necessary to determine the refractive index of the sporoderm and to measure, for instance, the thickness of sections. This can

PLATE II. Photomicrographs taken with interference microscope (phase differences appear as different colours). a. Section through part of osmium-fixed pollen grain of *Althaea rosea*. The background has been made equal in phase over the whole field. The unevenness in colour of the cytoplasm is due to thickness variations of the section. Geometrical thickness 4800 Å. b. Part of isolated pollen wall of same object as in Fig. a. The optical path difference (O.P.D.) can be determined by matching the colour of the object against the colour of the background seen in the hole of the rotatable compensation wedge. The difference in O.P.D. between the exine (1260 Å± 50 Å) and the intine (435 ± +60 Å) is clearly shown. c. Section through part of megaspore of *Selaginella selaginoides*. The resistant part of the spore wall has been isolated by acetolysis. d. Section through acetolyzed pollen wall of *Lavatera pallida*. The variation of colour of the exine is due to structural differences of the wall. Geometrical thickness around 1500 Å. Microscope adjustment same as in Figs. a and c. e. Section of acetolyzed pollen wall of *Malape trifida*. Background field crossed with interference bands. Geometrical thickness around 3000 Å. f. The unevenness of a common microscope coverslip (16 × 16 mm) chosen at random is shown. Variations of the gradient are seen as divergences from the colour of the background (situated in the "sensitive colour" of the first order). The red band of the object is of the second order and the O.P.D. thus around one wavelength. This picture was taken in a similar macroscopic set-up.—From L. Johansson and B. Afzelius *in* Statens naturv. forskningsråds årsbok, 9, Stockholm 1955 (p. 136).

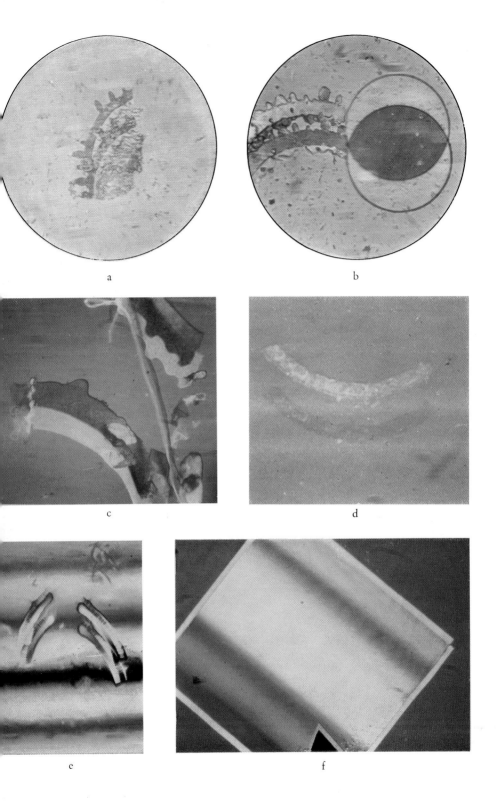

a

b

c

d

e

f

conveniently be done under an interference microscope, and examplet of the usefulness of this type of microscopy are illustrated in Pl. II. In special circumstances resort must also be made to ultra-violes microscopy (see Pl. IV). The finest details of structure can, of course, be seen only in an electron microscope, using the methods of ultra-thin sectioning and of surface replication. Plates V–VIII show some of the important observations which can be made in this way.

When making pollen and spore slides the material is usually subjected first to acetolysis, treatment with a mixture of acetic anhydride and concentrated sulphuric acid (9 : 1) (Erdtman, 1960). The pollen grains and spores are then embedded in glycdrine jelly and sealed under a cover-slip with paraffin wax. Photomicrographs are, as a rule, taken of aceto-lyzed and/or chlorinated pollen grains and spores suspended either in distilled water or in silicone oil (the latter is often better on account of the difference between the indices of refraction of the pollen grains and spores and that of the liquid). Plate III, 11–17 is given as an example.

By means of acetolysis most of the substance of the spores is dissolved away, leaving the resistant parts of their walls. Acetolyzed spores look very much like fossil spores in peat and sediments. Other, not so drastic methods are used in the investigation of less resistant parts of the spores, such as the intine and the aperture membranes, the study of which has, hitherto, been much neglected. Staining is not necessary. Large, acetolyzed spores often turn dark and have to be bleached by chlorination. It is advisable to make slides with a mixture of acetolyzed grains (on which measurements should be made) and grains that have been chlorinated after acetolysis (which then can be used for the study of fine details). In dark spores the use of a red light helps to distin-guish details which would otherwise be difficult to observe. Thin sections are usually stained and studied with phase contrast.

Fine details, particularly of the cytoplasm and the intine, are best studied in the ultra-violet microscope and this is used for preference in the study of thin sections ($0 \cdot 1$–$0 \cdot 5 \, \mu$; see Pl. IV) and of minute spores, e.g. *Lycoperdon*. Ultra-violet radiation can be converted into visible green light by means of a vidicone or ultrascope and observation can then be carried out in the same way as in ordinary light microscopy.

The fine relief of the pollen and spore surfaces can be studied only by electron microscopy, as briefly mentioned above. (See Pls. V–VIII; also Rowley, 1959, 1960; Yamazaki and Takeoka, 1962, and the literature quoted in these papers.) The development of this method means that a very important step has been taken towards the better knowledge of fine details. Several details, distinctly seen in electron micrographs, can also be seen, although less distinctly, in ordinary photomicrographs if suitable techniques are used.

An example of the powers of this method is presented in Pl. IX, 4, where the white dots are small spinules at high focus; the occurrence of such spinules—now known to be common in many types of pollen grains—was largely unknown to pollen analysts only a few years ago.

For the elucidation of subsurface details (fine details of the sporoderm stratification, etc.) EMGs of ultrathin sections are necessary (see Pl. VIII; also Afzelius, 1955, 1956; Afzelius *et al.*, 1954; Mühlethaler, 1953, 1955; Rowley, 1959, 1960; Sitte, 1957).

B. MORPHOLOGY

1. Polarity and symmetry

Cormophyte spores are usually produced in fours (tetrads) by spore mother cells. They have two poles at opposite ends of what may be called the polar axis. The proximal pole is situated in the centre of the proximal face, which is directed towards the centre of the tetrad. The distal pole, which faces in the opposite direction, is situated in the centre of the distal face. Most spores are, as can be seen from Figs. 1–20, either hetero- or isopolar. Those in Figs. 21–25 are apolar.

These definitions are mentioned in order to stress the necessity—to facilitate comparison—of reproducing pollen and spore illustrations in a uniform way. The polar axis should always be perpendicular, with the distal pole ("the North Pole") at the top and the proximal pole at the base. Non-observance of this simple rule will cause confusion. Winged coniferous pollen grains of pine and spruce, for example, may be— and in fact often are—shown upside down with the wings in the lower instead of the upper part of the figure. Uniformity is necessary whether one is dealing with a dial, a floral diagram, or a "palynogram" (Fig. 26) i.e., a diagrammatic illustration of a pollen grain or spore.

2. Apertures

Some mosses, ferns, gymnosperms, and monocotyledons have *atreme* spores (Gr. *trema* = aperture), i.e. they seem to have no special aperture. Most spores are not atreme, however, but usually *monotreme*, i.e. provided with one aperture. This aperture is either long or short and rounded. Long apertures are referred to as *colpi* (sing. *colpus*), short as *pores*. Colpi are simple or three-slit. In moss (Fig. 3) and fern (Fig. 1) spores the aperture is proximal. In the pteridosperms there is an important transition from spores with a proximal to spores with a distal aperture. In recent gymnosperms, monocotyledons, and in the tulip tree (*Liriodendron*) and other monocotyledonoid dicotyledons, the aperture is distal (Figs. 7–13). In some monocotyledons there is, in contradistinction to the recent gymnosperms, a tendency towards more than one aperture. In *Colchicum* there are thus two lateral apertures resulting from the coalescence of the central part of a long colpus. In the pollen grains of the palm *Sclerosperma mannii* the central part of a three-slit colpus coalesces, with the appearance of three marginal apertures (the ends of the former branches of the colpus) as a result. The dicotyledons generally have pollen grains with more than one (3—∞) apertures (three is the most common number). With regard to the number of apertures the general evolutionary tendency is thus from one to several.

With regard to the *position* of the apertures there is also a special evolutionary trend, viz. from a proximal (Fig. 1) to a distal position (Fig. 11) and from there to an equatorial (Figs. 15–17, 19, 20) or, in certain cases, a ubiquitous ("panto-") position (Figs. 21–24). In mosses, ferns and some of the pteridosperms the aperture is proximal (Figs. 1–3). In the other plants with monotreme pollen grains it is distal (Figs. 6–13).

In the dicotyledons, with the appearance of more than one aperture, the apertures take a zonal position, usually along the equator of the spores (Figs. 15–17, 19, 20). This very important step, the transition from mono- to pleiotrematy, can be traced in some primitive dicotyledons, e.g. the Magnoliales, the Nymphaeales *s. lat.* (including the Nelumbonaceae), and the Chloranthaceae.

According to the shape of the amb (short for ambit; from L. *ambitus*), i.e. the outline of a spore viewed with one of the poles exactly uppermost, the following extreme types can be distinguished in spores with zonal arrangement of the apertures, all interconnected by transitional types:

Peritreme spores have the centres of the apertures more or less uniformly distributed along a roughly circular amb (Figs. 15, 17, 20). Spores with equatorial apertures and $\pm$ angular (or lobate) amb are either *gonio-*, *pleuro-*, or *ptychotreme*. In the first case the centres of the apertures are situated at the angles of the amb, in the second about half-way between the angles. In the third, and rare, case the amb is lobate with the apertures in the concavities between the lobes.

Peritreme grains in a wider sense are, as a rule, not rigid. When swollen they are exactly peritreme, when dry they often shrink and assume a habit similar to that of ptychotreme grains. Gonio- and pleurotreme grains, on the other hand, are more rigid and cannot undergo such changes of shape.

If the equatorial belt becomes, as it were, too narrow or short to accommodate the apertures, the zonal arrangement is superseded by a ubiquitous arrangement: the apertures are uniformly spread over the entire surface. Such spores are said to be *pantotreme* (Figs. 21–24). The Chenopodiaceae have typical pantotreme pollen grains (in some

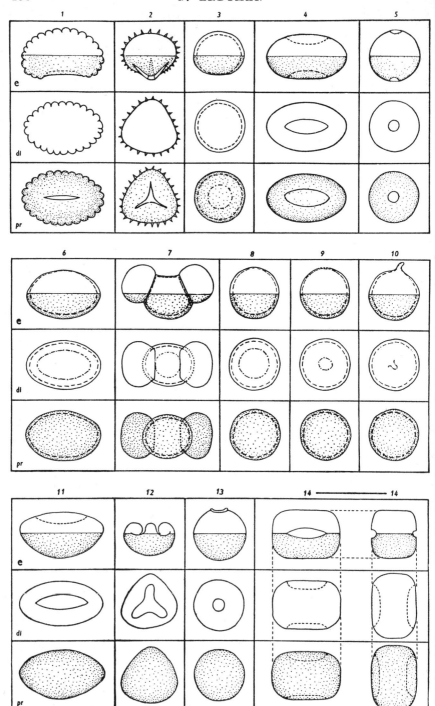

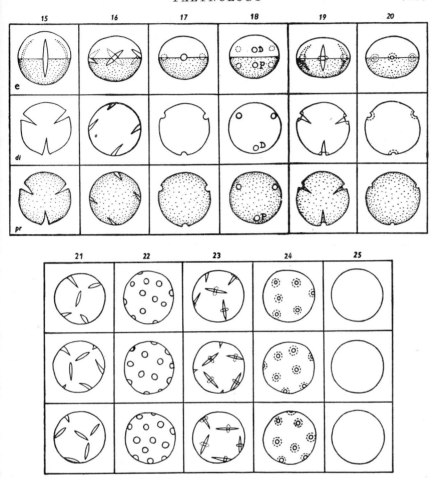

Figs. 1–25. *NPC*-classification (examples). *N*-classification.—Monotreme: 1–3, 6–13.
—Ditreme: 4, 5, 14.—Tritreme: 15, 17, 19, 20.—Tetrateme: 16.—Hexatreme: 18 (dizonotreme, 3+3).—Polytreme: 21–24.—Atreme: 25.

P-classification.—Catatreme: 1–3.—Anacatatreme: 4, 5.—Anatreme: 6–13.—Zonotreme: 14–17, 19, 20.—Dizonotreme: 18.—Pantotreme: 21–24.

C-classification.—Monolept: 3, 6–10.—Trichotomocolpate: 2 (referred to as trilete),
12.—Colpate: 1 (referred to as monolete), 4, 11, 14 (clinocolpate), 15 (orthocolpate), 16
(loxocolpate).—Porate: 5, 13, 17, 18, 22.—Colporate: 19, 23.—Pororate: 20, 24.

– . – . – . – . – · approximate outline of leptoma.

– – – – – – – – – – inner contour of exine in spores provided with leptoma.

e, equatorial view; *di*, distal face; *pr*, proximal face; D and P in Fig. 18, pore in dista
and proximal face, respectively.

NPC. 1: 113, catacolpate (*monolete*).—Example: *Polypodium vulgare*. 2: 112, cata-
trichotomocolpate (*trilete*).—*Lycopodium clavatum*. 3: 111, catalept.—*Polytrichum gracile.*
4: 223, anacatacolpate. 5: 224, anacataporate. 6: 131, analept, bilateral, asaccate. 7: 131,
analept, bilateral, saccate.—*Pinus silvestris*. 8: 131, *analept*, radiosymmetric, asaccate.—
Pseudotsuga taxifolia. 9: 131, *analept*, microlept. 10: 131, *analept*, micro- and prolept.—
Sequoia. 11: 133, anacolpate.—*Lilium bulbiferum*. 12: 132, ana-trichotomocolpate (*tri-*

species there may be as many as one hundred apertures). There is reason to believe that the poly- and pantotreme pollen types have been preceded by oligotreme-pantotreme (porate) pollen types, which, in their turn, have evolved from hexacolpate pollen grains with the colpi in panto-position. Hexapanto-colpate pollen grains are sometimes produced together with tricolpate pollen grains with the colpi in zonal arrangement. An increase in the number of apertures often occurs in connection with an increase in the chromosome number (see Maurizio, 1956).

A very rare spore type is that referred to as *dizonotreme*: here an increase in the number of apertures has been accomplished by the appearance of two aperture-bearing zones (on either side of the equator) instead of one (Fig. 18). It occurs in the Olacaceae (*Anacolosa*, *Cathedra*) among other families.

Apertures are either simple or composite. The latter (Fig. 32) ought to be considered as more "advanced" than the former. They are confined to the dicotyledons. They consist of an outer part, either colpoid or poroid, and an inner part, usually referred to as an "os" (L. *os*, mouth). Like a mouth it can vary in outline from the circular to the transversally or longitudinally elongated, etc. Colporate (colp-orate) spores (Figs. 19, 32) have composite apertures consisting of an os and an outer, colpoid part, whereas in pororate (por-orate) spores the outer part is poroid (Fig. 20).

Composite apertures are often provided with thickened, incrassate, margins (crassimarginate). The margins of the colpoid part, or of the os, or of both can be thickened. The margins of simple apertures are often tenuous (tenuimarginate) and their contours, as a result, less distinct than those in many composite apertures (examples may be found in the Altingiaceae, Bromeliaceae, and Ranunculaceae). In some cases—and this may denote a more "advanced" stage—the margins are thick and reinforced as in the Gramineae (Fig. 13, e).

---

chotomocolpate s. str.).—*Acanthorhiza mocinni*. 13: 134, *anaporate.*—*Zea mays*. 14: 243, zonocolpate (*2-colpate clinocolpate*).—*Tigridia pavonia* (morphotype with orthocolpate pollen grains: *Hypecoum procumbens*). 15: 343, zonocolpate (*3-colpate*).—*Biscutella auriculata*. 16: 443, zonocolpate (*4-colpate loxocolpate*).—*Myriophyllum spicatum* (morphotype with orthocolpate pollen grains: *Desmostachys preussii*). 17: 344, zonoporate (*3-porate*).—*Roëlla ciliata*. 18: 654, dizonoporate (3+3).—*Anacolosa lutea*. 19: 345, zonocolporate (*3-colporate*)..—*Lathyrus vernus*. 20: 346, zonopororate (*3-pororate*).—*Betula tortuosa*. 21: 763, pantocolpate.—*Echinocactus tabularis* (polypantocolpate; the pollen grains in *Spergula arvensis*, 663, are 6-pantocolpate). 22: 764, *pantoporate.*—*Salsola tragus* (polypantoporate; the pollen grains in *Fumaria officinalis*, 664, are 6-pantoporate; those with the formula 464 are 4-pantoporate). 23: 765, *pantocolporate* (polypantocolporate). 24: 766, *pantopororate* (polypantopororate).—*Juglans rupestris*. 25: 000, atreme. —*Cinnamomum camphora*.

PLATE III

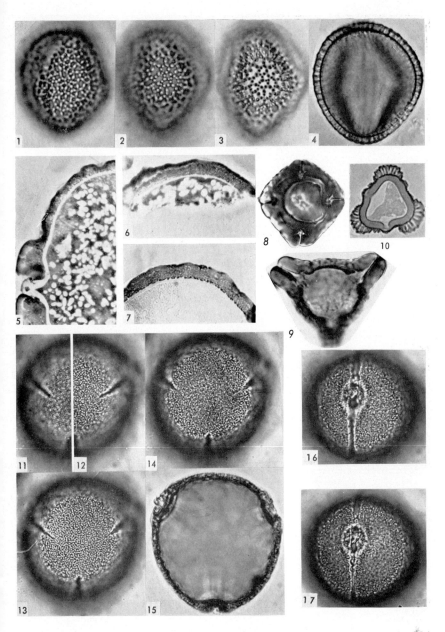

PLATE III. Various pollen grains (surface and sections).—Figs. 1–4, *Ligustrina amurensis*
(×850), pollen surface details at different foci from high to low.—Figs. 5–7, *Bouchea
fluminensis* (×650), showing compact, outer exine, and not compact inner ditto.—
Figs. 8–9, Upper Cretaceous pollen grains from Oebisfeld, Germany (slide made by
Dr. W. Krutzsch; ×850).—Fig. 10, *Drejerella (Beloperone) guttata* (median cross-section,
×600; from Raj, 1961).—Figs. 11–17: *Fagus orientalis*, pollen grain at different foci.—
Embedding medium glycerine jelly except in Figs. 11–17 where distilled water has been
used.

PLATE IV

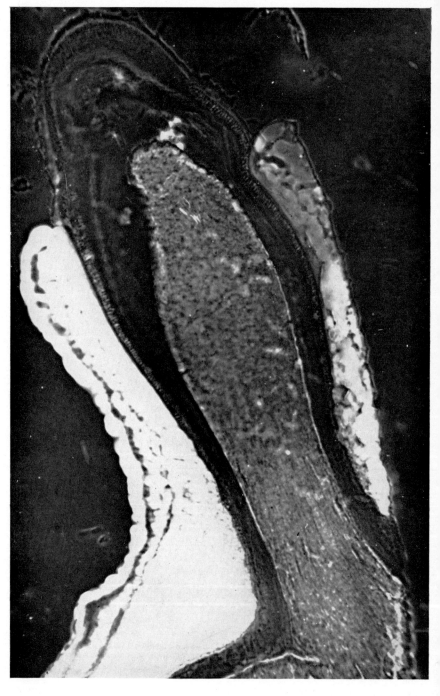

PLATE IV. *Morina longifolia*. Section through one of the three funnel-shaped apertures. UV-micrograph 2536 Å, × 2700 (negative print, original magnification × 1050). Diameter of small rodlike details immediately inside the exine in the outer part of the funnel about 0·1 μ.

PLATE V

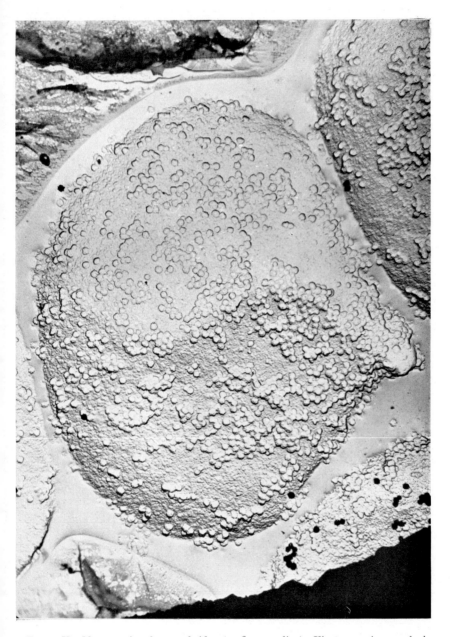

PLATE V. *Metasequoia glyptostroboides* (surface replica). Electron micrograph by M. Takeoka. ($\times$ 4000.)

PLATE VI

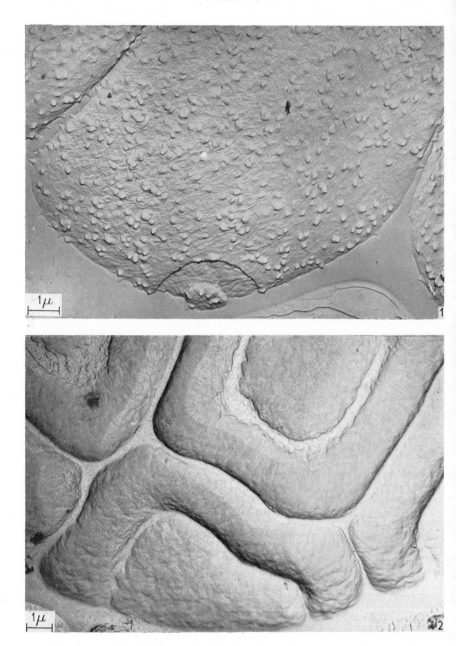

PLATE VI. Surface replicas. Electron micrographs by M. Takeoka. Upper figure: *Morus japonica*. (× 9000.) Lower figure: *Acacia longifolia* var. *floribunda*. (× 7000.)

PLATE VII

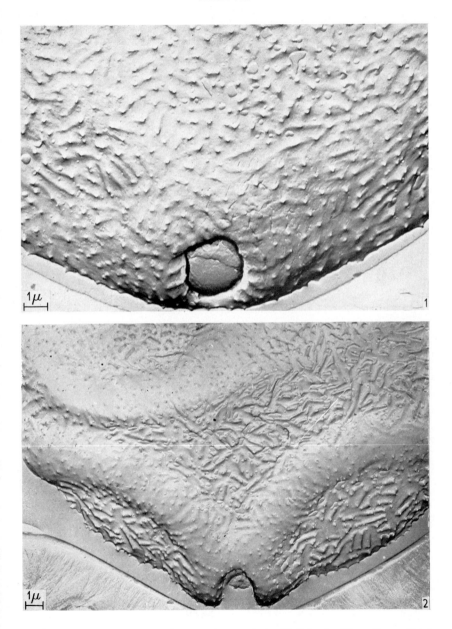

PLATE VII. Surface replicas. Electron micrographs by M. Takeoka. Upper figure: *Betula platyphylla* var. *japonica*. (× 6800.) Lower figure: *Alnus japonica*. (× 4800.)

Plate VIII

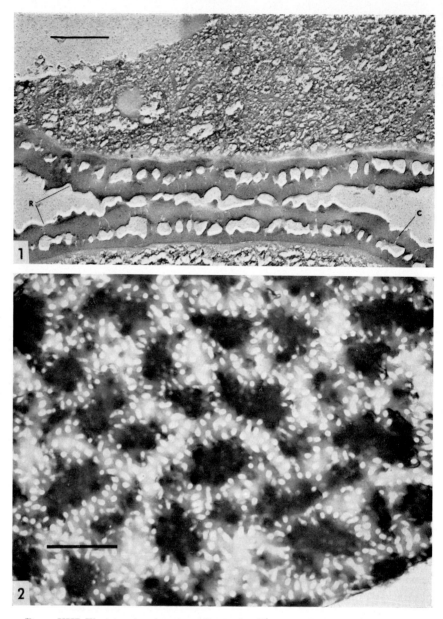

Upper figure: Section of the exine of two grains of *Poa annua* L. (stage prior to microspore mitosus). Fixation by 40% OsO_4, methacrylate embedding, removal of methacrylate with amyl acetate, and shadowing with platinum evaporated from two sources. Periodic depressions (R) in the ektexine are seen between some spinules (similar depressions form the bright reticulum in the lower figure). Channels (C) through the exine exposed in parallel and oblique setion. ($\times$ **17 000**.)

Lower figure: Electron micrograph (unshadowed) taken normal to the pollen grain surface of *Cynodon dactylon* (L.) Pers. using partially dissolved exine directly as a specimen. The white spots are channels traversing the exine perpendicular to its surface, the bright reticulum is an incized reticulum (compare R in the upper figure). The dark lacunae of the reticulum consist of spinules (black spots) and non-incized interspinule regions. The channels are larger here than in sections (compare upper figure at C), probably as the result of etching with concentrated nitric acid. ($\times$ **20 000**.)

PLATE IX

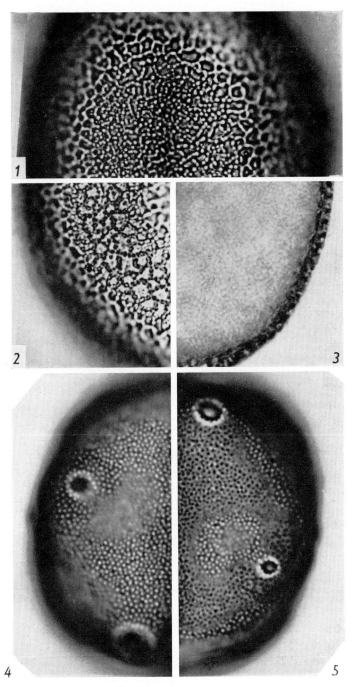

PLATE IX. Acetolyzed pollen grains in distilled water.—Figs. 1–3, *Iris pseudacorus.*—
Figs. 4, 5, *Juglans regia.* (×1500.)

PLATE X

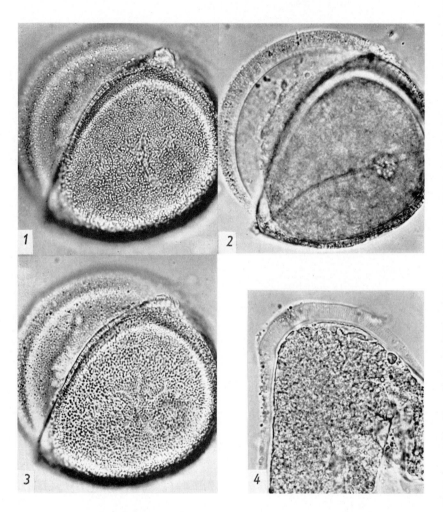

PLATE X. *Dracaena hookeriana*. Fresh, untreated pollen grains in distilled water (showing protruding intine). (×1000.)

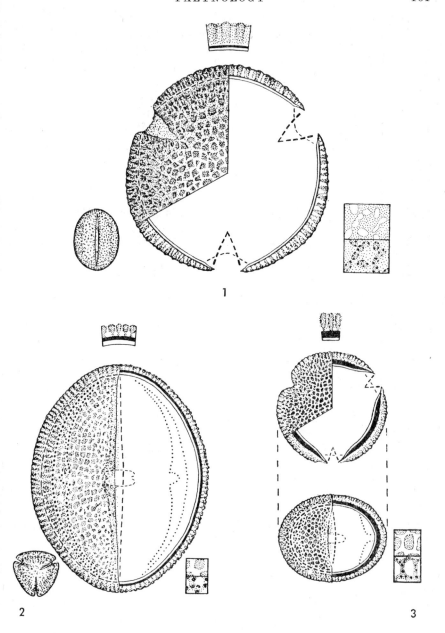

FIG. 26. Palynograms. *1. Jasminum humile, 2. Hypseocharis pimpinellifolia, 3. Panda oleosa.*

As shown in Fig. 27, upper row, spores can be subdivided into nine classes (N_0–N_8) with regard to the number (N) of their apertures. Anomotreme spores, i.e. spores with irregular apertures, are grouped as N_8. Those in N_1–N_7 are normal, nomotreme. With regard to the position (P) of the apertures there are seven classes (P_0–P_6; P_0 accommodates spores in which the position—distal, proximal, etc.—of the aperture or apertures is unknown). With regard to the character (C) there are likewise seven classes (C_0–C_6); see also Erdtman and Straka, 1961.

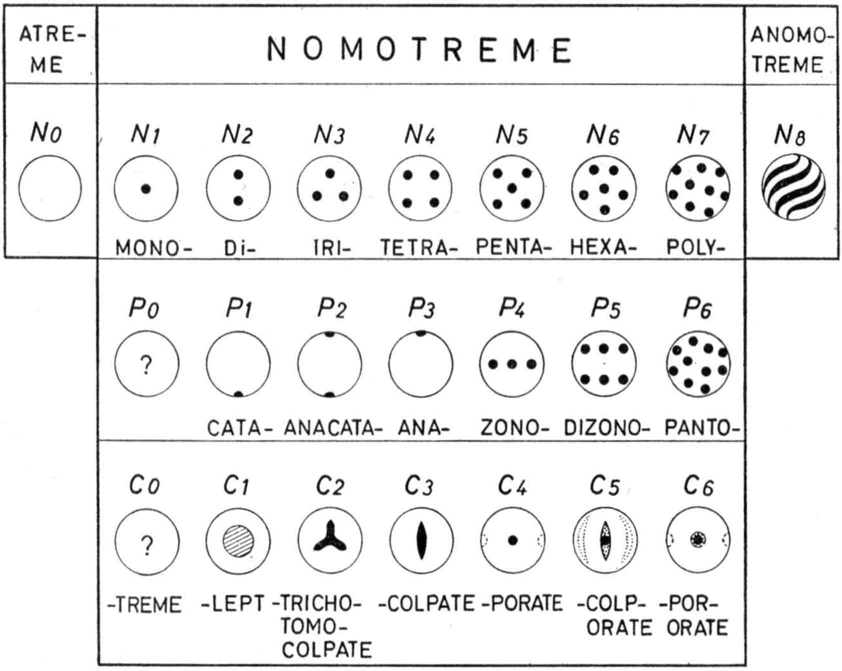

FIG. 27. The classification is based on the apertures, their number (N), position (P) and character (C). There are nine N classes (N_{0-8}), seven P (P_{0-6}) and seven C classes (C_{0-6}). N_0: atreme spores (spores without aperture); P_0: spores where the position of the aperture(s) is not known; C_0: spores where the character (shape etc.) of the aperture(s) is not known.

The classification N is one-dimensional. By combining it with the classification P we get a two-dimensional system; by combining all three we get in a sense, a three-dimensional classification, the NPC or number-position-character, classification. This can be used to classify any (individual) spore if the key characters (N, P, C), or one (N) or two (N and P, or N and C) of them are known.

Figure 28 shows a number of NPC-combinations. The large circles roughly illustrate the fact that spores with one (N_1) or three (N_3)

apertures are more common than for example spores with two (N_2) or five (N_5). Spores with apertures in zonal arrangement (P_4) are frequent, likewise catatreme (P_1) and anatreme spores (P_3), i.e. spores with one aperture in cata- or ana-position respectively (*ana*, up, denotes the distal, *cata*, down, the proximal pole or face). On the other hand dizonotreme spores (P_5) are rare. Colpate spores (C_3) are probably more frequent than porate spores (C_4), etc.

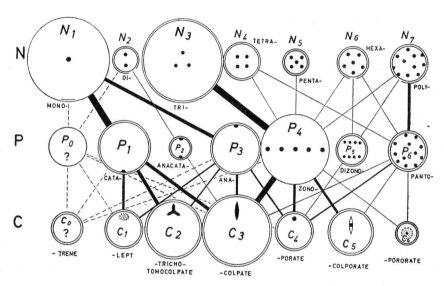

Fig. 28. The circles N_1 and N_3 are larger than the rest and thereby emphasize that mono- and titreme spores are more common than di-, tetra-, pentatreme etc. The size of the *P*- and *C*- circles is likewise approximately proportionate to the frequency of the features in question and so is also the thickness of the lines interconnecting the circles. Broken lines connote that the position or the character (or both) of the aperture(s) is not known. Single circles signify that the name of the feature in question is usually omitted. Example: a "tri-zono-porate" grain (classification number N_3-P_4-C_3, or simply 343) is referred to as "triporate", not as "trizonoporate" (P_4 is surrounded by a single circle, N_3 and C_3 by a double circle). 100 is a monotreme, 111 a catalept, 764 a polypantoporate and 654 a dizonoporate (3+3) spore.

N_3–P_4–C_3, or "343", is the number, or pigeon-hole, to which "tri-zono-colpate" (usually written tri-colpate or 3-colpate) spores should be referred. "133" are "mono-ana-colpate" grains (usually written anacolpate). Those in "663" are hexapantocolpate, with six colpi in pantoposition, i.e. placed in the same position as the six edges of a pyramid with a triangular base.

Figure 29 is a simplified chart of the NPC-system for use in routine work. An application of the NPC-system is given on pp. 171, 172.

3. Shape and size

With regard to radiosymmetrical dicotyledonous pollen grains it is sufficient to stress that their shape can vary from very flat (peroblate) with a short polar axis, to very elongate (perprolate), with the polar axis more than twice as long as the equatorial diameter. The shapes and their variations—from peroblate to oblate, suboblate, sphaeroidal, subprolate, prolate, and perprolate—are often quite characteristic but, from a taxonomic-phylogenetic point of view, less important than the apertures. To refer certain Cretaceous and Tertiary spore groups to "Longaxones" or to "Breviaxones", as has sometimes been done, may,

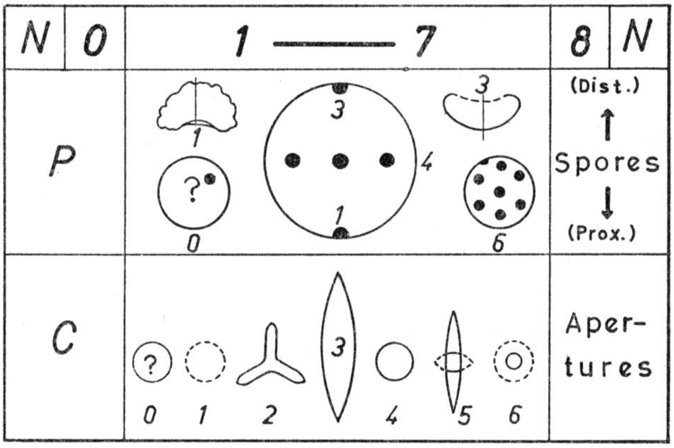

Fig. 29. The NPC-system (simplified chart).

therefore, not be of very great value. Thus the Tiliaceae include both flattened pollen grains of the *Tilia* type and elongate pollen grains of the *Grewia-Triumfetta* type. In the Elaeagnaceae, *Shepherdia argentea* has oblate, *S. canadensis* prolate pollen grains and so forth.

The size of the pollen grains in recent flowering plants varies from about $5 \times 2 \ \mu$ in some species of *Myosotis* to 200 μ or slightly more in *Orectanthe*, a new genus in the Abolbodaceae (near the Xyridaceae) and some cucurbitaceous and nyctaginaceous species. Much larger however, are the spores in certain ferns: the megaspores in *Selaginella exaltata* have a diameter of approximately 1·5 mm (1500 μ). In Carboniferous deposits megaspores measuring up to about 6–7 mm have been encountered (e.g. *Triletes giganteus*).

4. Absolute number of pollen grains

In order to calculate the number of pollen grains per flower or stamen material collected immediately before the opening of the anthers is acetolyzed and counted with the aid of a counting chamber (e.g. a chamber taking 0·1 cm³ of the liquid—glycerine and water, in equal parts—in which the exines are suspended; the chamber used in the Palynological laboratory, Solna, is 80 μ deep).

Pollen output per stamen has been recorded (Pohl, 1937) as follows: *Rumex acetosa* 30 000, *Secale cereale* 19 000, *Fraxinus excelsior* 12 500, *Acer platanoides* 1000, *Calluna vulgaris* 2000 (=500 tetrads). In January 1962 twenty-five dark, withered flowers of *Trifolium medium* which had been thawed out of snow after having been frozen for a few weeks, were subjected in our laboratory to acetolysis. It was calculated that there were 15 230 pollen grains in all, thus about sixty well-preserved pollen grains per stamen. In another specimen there were still about 110 grains per stamen. This shows that pollen grains retained in flowers of entomophilous plants may still be present in a number and condition sufficient for making pollen slides even if the plant and the flowers themselves are dead and have repeatedly been subjected to very unfavourable weather conditions, etc.

C. SPORODERM STRATIFICATION

1. Perine and exine

Sporoderm, "spore skin" or spore wall, is a term which was probably introduced by Bischoff (1842). After having been forgotten for a long time it has recently been revived and is now in common use. The main parts of the sporoderm are the exine (exosporium) and the intine (endosporium). Another layer is the perine or perisporium, which occurs in many fern spores.

It would be impossible to enumerate, evaluate and reconcile here, in brief, the still somewhat disparate ideas concerning sporoderm stratification. More definite information can only be derived from further painstaking and penetrating morphogenetic investigations.

Figure 30 gives some idea of the intricate sporodermal conditions that are sometimes met with in higher plants. Typical perine (P in Fig. 30) is said to occur in ferns only, e.g. in *Dryopteris*, *Asplenium*, and *Athyrium*. It does not occur in *Polypodium*, *Osmunda* and *Botrychium*. Its occurrence or non-occurrence is of taxonomical significance. If it is difficult to decide whether a layer is perinous or exinous—and such cases do occur—the layer in question may be referred to under the · neutral heading "sclerine" which indicates perine and/or exine.

If some small, faintly spinulose globulets had been inserted in Fig. 30 between layers 1 and 2 they would have corresponded, with regard to size and place, to the "orbicules", or "con-peito-grains", in certain gymnosperms. These have been known for a long time and can be seen in many old illustrations of pollen grains in *Juniperus* and other cupressaceous plants. They have recently been studied in greater detail by Japanese palynologists (Yamazaki and Takeoko, 1962) who consider that they are perinous and of great taxonomic value. The perinous nature of the orbicules may be correct but this has not yet been definitely proved. The taxonomic value of the orbicules, on the other hand, is quite

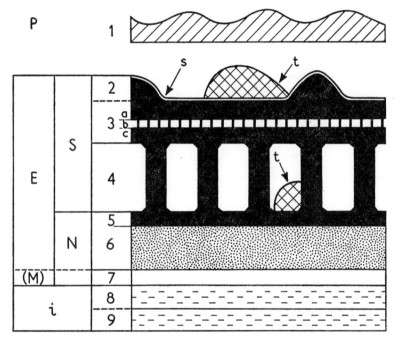

FIG. 30. Sporoderm stratification (diagram; explanation of figures etc. on p. 166-168).

apparent (orbicules are, for example, present in the Cupressaceae but absent from the Pinaceae and Podocarpaceae, including the isolated genus *Saxegothaea*).

The letter *E* on the left in Fig. 30 indicates the exine. In higher plants there is usually a marked difference between its outer part, the sexine (*S*) (also referred to as ektexine), and the inner, more solid part, the nexine (*N*) (or endexine). The sexine is usually more dissected than the nexine, which has taken its "n" from a Latin verb, *nexere*, to connect, join, bind together ("s" in sexine is from *secare*, to cut).

A differentiation of the exine into sexine and nexine is, so it seems, a prerequisite for the occurrence of composite apertures (at the same time the differentiation by no means precludes the occurrence of simple apertures). The differentiation is thus an "advanced" character, which occurs only faintly or sporadically, if at all, in mosses and ferns. That sexine and nexine both stem from a uniform exine is, at least to some extent, indicated by the fine, "granular" structure or both, and by the absence of a baculate, separating layer in certain Cretaceous pollen grains (Pl. III, 8, 9).

The t in Fig. 30 indicates "tryphine", a provisional comprehensive name for oil etc. on the surface of certain pollen grains and in the interstices of the sexine. The s refers to a poorly known layer, provisionally, and perhaps unnecessarily, referred to as "stegine" (Gr. *stegein* = to cover). It is clearly seen in an electron micrograph of the sporoderm stratification in a moss (Wettstein, Pl. V in Erdtman, 1957).

The outermost layer of the sexine (if the "stegine" is not considered) is the ectosexine (Fig. 30, 2–3), in Fig. 30 developed as a continuous layer, the tegillum (3). The tegillum is provided with suprategillar processes (2) and—only very occasionally—intrategillar rods (b). It is supported by the endosexine (4), which in Fig. 30 consists of coarse, infrategillar bacula (L. *baculum* = a rod) (also referred to as columellae); they usually stand at right angles to the tegillum and/or the general surface of the nexine. They are sometimes branched-digitate in their upper part, but not or only seemingly so (from the inclusion of small hollows) in their lower part.

The nexine (Fig. 30, N and 5 + 6) is sometimes differentiated into several layers, still imperfectly known and as yet much debated. Layer 5 in Fig. 30 can be referred to as "nexine 1", or "basosexine". It may chemically and physically be of the same kind as the sexine, yet topographically it undoubtedly forms part of the nexine proper. Electron micrographs of acetolyzed sporoderms of *Epilobium* have failed to show any difference in fine structure between layers 5 and 6, whereas in ultra-violet micrographs of non-acetolyzed grains of *Epilobium* these layers seem to be well differentiated. According to Raj (1961) the presence of layer 5 in the Acanthaceae (Pl. III, 10) and other plants is to be regarded as an "advanced" feature. Undoubtedly there are several facts in favour of this idea, which, however, at the present state of our knowledge must be considered as equally hypothetical as many, or most of the guesses concerning "advanced" and "non-advanced" features.

In some plants, e.g. *Alluaudia humberti*, *A. montagnacii* and *Alluaudiopsis fiherenensis* (Didiereaceae), the nexine seems to consist of very densely packed rods. In the pollen grains of other plants the nexine is partially or entirely laminated. This can be seen in acetolyzed grains of *Schismatoglottis* (Araceae, Fig. 31), in thin sections through the exine of *Lapageria* and *Myriophyllum*, *Cedrus*, *Pinus*, as well as—and more

7§§

distinctly—in electron micrographs of *Ephedra*, *Welwitschia*, *Picea*, *Poa annua* and other plants (according to statements—only partially published—by Afzelius, Praglowski, Rowley, Ueno, and others).

In the spores of *Lycopodium clavatum* there is a very fine lamellation also in the "sexinous" part of the sporoderm (Afzelius *et al.*, 1954).

Layer 7 (Fig. 30) is intensively debated. Many authors do not recognize it at all. Some authors consider that it is an intermediate

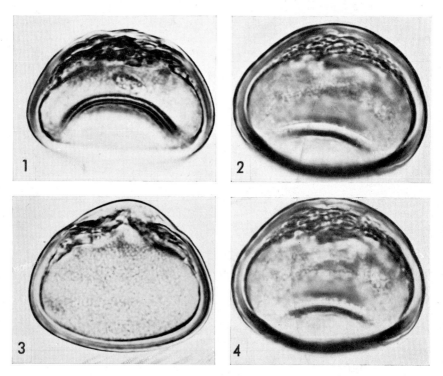

FIG. 31. Acetolyzed pollen grain of *Schismatoglottis* sp. (Araceae). (× 1700.)

layer between the exine and the intine, less resistant (and more refracting?) than the former but more resistant than the latter layer. It has been referred to as "mesine" by Rowley. This is however a term which cannot be used without certain reservations. Thus Fitting (1900) once distinguished a layer which he called mesosporium. If shortened in the same way as, e.g., exine (from exosporium), mesosporium would be mesine, a term which has also been used, although in another sense, by Gagnepain (1903). Recently the term "medine" has been suggested. The study of the intine (Fig. 30, $i = 8 + 9$) has hitherto, as already mentioned, been much neglected. In *Dracaena hookeriana* (Pl. X) the intine instantaneously swells and becomes very bulged on addition of water. The fine baculation, faintly seen in the upper part of Pl. X, 4, may possibly correspond to the fine baculation seen in the ultraviolet micrographs of a section through an aperture in *Morina longifolia* (Pl. IV).

The fine structure and physiological properties of the intine will be better interpreted when better knowledge of its chemical characteristics is acquired. According to Dr. Hans Bouveng, Swedish Forest Products Research Laboratory, Stockholm (private communication), pollen grains contain large amounts of low-molecular weight carbohydrates, though only a little is known of the nature of the high-molecular weight carbohydrates (polysaccharides) in pollen except that starch is recognized as a normal constituent.

The swelling properties of pollen grains are usually attributed to the presence of "pectic substances" in the intine. The term "pectic substances" refers to a complex mixture, difficult to separate, of apparently three different polysaccharides. These are present in various plant tissues, and on acid hydrolysis yield L-arabinose, D-galactose, and D-galacturonic acid. The latter essentially originates from a polygalacturonic acid (pectic acid), which is esterified with methyl alcohol in the native state. No polysaccharide of this type has, however, been found in the intine of *Pinus mugo*. The pollen grains were first extracted with hot monoethanolamine following Bailey (1960), who claimed that exine is selectively soluble in that solvent. The exine was, however, virtually unaffected by this treatment. Only 6–8% of the non-exinous part of the pollen grains, excluding starch and low-molecular weight material, went into solution. About two-thirds of the extract consisted of high-molecular weight carbohydrates, the rest of unidentified substances. Subsequent extraction with cold and hot water, and with cold, approx. 8% sodium hydroxide dissolved away a further quantity of high-molecular weight material (corresponding to about 4% of the original pollen material), mainly polysaccharides. A microscopic examination of the residue revealed that the intine had altogether disappeared leaving the cytoplasm as a more or less intact entity within, and free from, the exine.

An examination of the polysaccharide part of the extract showed the presence of a protein-like compound combined with D-xylose (xylose could not be separated from the accompanying protein by means of several well-established methods) together with a polysaccharide built up of L-arabinose, D-galactose residues and varying amounts of L-rhamnose and uronic acid. Pectic acid was virtually absent.

The xylose-protein complex had a positive optical rotation ($[a]_D$ of one preparation $+160°$). Xylans known hitherto from plant sources contain exclusively β-D-xylose residues and consequently have negative rotations (usually -60 to $-90°$). The "xylo-protein" which, in contrast to normal plant xylans, is readily soluble in water, providing solutions of low viscosity, seems to contain D-xylose residues of the α-linked type. The other polysaccharide seems to belong to the rather ill-defined group of polysaccharides usually referred to as "plant gums". It forms viscous solutions and may therefore be at least partly responsible for the swelling of the intine.

The investigation is so far (January 1962) in a preliminary stage. Dr. Bouveng hopes to publish a full account in *Acta Chemica Scandinavica*.

IV. POLLEN AND SPORE MORPHOLOGY AND PLANT TAXONOMY

A. INTRODUCTION

This theme has been dealt with by Wodehouse (1935, 1960), Erdtman (1952, 1957), Wagenitz (1955), Stix (1960), Raj (1961), and many others. Pollen diagnoses are increasingly added to the diagnoses of new species and it is gratifying to note that the results or suggestions obtained from pollen and spore morphology usually conform with those obtained from anatomy (Metcalfe, 1961) or other branches of botany. In the following only a few examples of the application of pollen morphology in plant taxonomy will be given. They comprise a note on the application of the "*NPC*-system" in the gross classification of the dicotyledons, followed by an exposé of pollen morphological possibilities in the classification of taxa below family rank. Finally a fairly extensive list of suggestions for further investigations is put forth.

B. APERTURE CHARACTERISTICS AND GROSS CLASSIFICATION OF THE DICOTYLEDONS

The average aperture conditions in one of the orders of the dicotyledons—the Parietales are chosen as an example—are as follows (Tab. 1). The families according to Engler and Diels (1936) are enumerated in alphabetical order; the figures before the name of each individual family constitute the average "*NPC* formula" (see Figs. 27, 28) of the family.

If the families with a query and the two families with two sets of figures are not taken into consideration, the table emphasizes the great apertural uniformity within the Parietales: The pollen grains are 3-treme ($N3$), zonotreme ($P4$), and colporate ($C5$), the "*NPC*-formula" thus generally being 345. One striking exception occurs, however, viz. the Canellaceae, where the pollen grains have one (distal) colpus only. The *NPC*-formula here is 133. Then there are two less striking exceptions, viz. the Frankeniaceae and Tamaricaceae, both with *NPC* 343.

TABLE I

NPC in Parietales

345	Achariaceae	345	Flacourtiaceae
345	Actinidiaceae	343	Frankeniaceae
343	?Ancistrocladaceae	345	Guttiferae
345	Begoniaceae	345	Loasaceae
345	Bixaceae	345	Malesherbiaceae
133	Canellaceae	345	Marcgraviaceae
345	Caricaceae	344	?Medusagynaceae
345	Caryocaraceae	345	Ochnaceae
345	Cistaceae	345	Passifloraceae
345	Cochlospermataceae	345	Quiinaceae
345	Datiscaceae	345	Strasburgeriaceae
343, 345	Dilleniaceae	343	Tamaricaceae
343, 345	Dipterocarpaceae	345	Theaceae
345	Elatinaceae	345	Turneraceae
245	?Eucryphiaceae	345	Violaceae

From a purely palynological point of view the following taxonomical conclusions may be suggested. 1. The Canellaceae (having 'monocotyledonoid" pollen grains) cannot be referred to the Parietales. 2. If the Frankeniaceae and Tamaricaceae also in some other respects deviate from the rest of the order they should perhaps more appropriately be referred to another group of plants. 3. The same may possibly also apply to the Dilleniaceae, Dipterocarpaceae, and Eucryphiaceae.

The Canellaceae are now, in fact, for general taxonomical reasons, including the indication of the "palynological compass needle", often separated from the Parietales and referred to the Magnoliales (e.g. by Hutchinson, 1959). The Frankeniaceae and Tamaricaceae are referred by Hutchinson to the Tamaricales. This order, however, also comprises the Fouquieriaceae (*NPC* 345), which are referred by Engler-Diels to the Tubiflorae-Convolvulineae. With regard to *NPC* the latter family thus constitutes an aberrant element within the Tamaricales *sensu* Hutchison and is probably better placed in another order.

It may be added that the Elatinaceae are referred by Hutchinson to the Caryophyllales. This is not supported by palynological evidence: the pollen grains have composite apertures (*NPC* 345), whereas those in the other families of the Caryophyllales *sensu* Hutchinson (i.e., the Molluginaceae, Caryophyllaceae, Ficoidaceae, and Portulacaceae) have simple apertures (*NPC* 343, 464, 663, 763, 764, etc.).

7§

It would be beyond the scope of the present paper to analyse all orders to which the families of the Parietales *sensu* Engler have been referred by Hutchinson. It may be pointed out, however, that the Bixaceae and seven other families are referred to a special order, the Bixales. The *NPC* formulae in this order are as follows.

TABLE II

NPC in Bixales

664 Achatocarpaceae	345 Flacourtiaceae
345 Bixaceae	345 Hoplestigmataceae
345 Cistaceae	345 Lacistemaceae
345 Cochlospermataceae	345 Scyphostegiaceae (?)

The Achatocarpaceae evidently do not fit in. The Hoplestigmataceae fit with regard to *NPC*, but not with regard to other palynological characters. They should, as suggested by Hallier (1910), probably be referred to near the Ehretiaceae. The pollen grains in the Scyphostegiaceae are not tricolpate, as was formerly believed, but tricolporate, and this constitutes an additional reason for not referring this poorly known Bornean family to the Urticales, an order characterized among other things by pollen grains with exclusively simple apertures. Further investigations are necessary in order to elucidate whether the general morphology of the pollen grains distinctly favours the reference of the Scyphostegiaceae to the Bixales *sensu* Hutchinson or not. The Achatocarpaceae are referred by most authors to the Centrospermae *sensu* Engler and Diels. Here they fit well, also with respect to *NPC*, as seen from the following collocation.

TABLE III

NPC in Centrospermae sensu *Engler-Diels,* 1936
(with the addition of the Halophytaceae)

764 Achatocarpaceae	764 Dysphaniaceae
343 Aizoaceae	343 Gyrostemonaceae
764 Amaranthaceae	664 Halophytaceae
663 Basellaceae (also 664, 764)	764 Nyctaginaceae (also 343, 443, 663,
764 Caryophyllaceae (also 343, 344,	763)
464, 663, 664)	343 Phytolaccaceae (also 443, 663,
764 Chenopodiaceae	763, 764)
	763 Portulacaceae (also 343, 764)

If we speculate, we may assume that the pollen grains with the formula 343, i.e. tricolpate grains, characterize more or less primitive genera or families. From there they may have evolved in one or two directions: in the Gyrostemonaceae and part of the Aizoaceae towards composite apertures (345), in other genera or families towards a greater number of simple apertures in panto-position, either colpi (663, 763) or pori (664, 764).

We may also assume that the pollen grains with the formula 343, i.e. tricolpate grains, characterize more or less primitive genera or families. From there they may have evolved in one of two directions: in the Gyrostemonaceae and part of the Aizoaceae towards composite apertures (345), in other genera or families towards a greater number of simple apertures in panto-position, either colpi (663, 763) or pori (664, 764).

From the point of view of *NPC*—and for some other reasons—it would perhaps be possible to include the Didiereaceae and the Cactaceae in the Centrospermae, or in a still wider group with the Centrospermae as the dominant element. With regard to the *NPC* the Plumbaginaceae also conform with the Centrospermae (343, 763, 764 etc.). They seem, however, as described on p. 190, to be more closely related to the Linaceae. Another family with connections with the Centrospermae is the Polygonaceae. This family, however, also includes more advanced genera with composite apertures (see p. 191). In summary, we may say that the "Centrospermoid" dicotyledons seem when subjected to an "*NPC*-analysis", to have several aperture conditions in common. They do perhaps have roots in the Linaceae as well as (and chiefly) in primitive members of the centrospermae proper. There are several lines of evolution: one goes from the Linaceae via the Plumbaginaceae to the Armeriaceae, another, not dealt with here, from the Linaceae to other families in the Geraniales etc. (maybe also to the Euphorbiaceae, Buxaceae, and Thymelaeaceae), a third to the Polygonaceae and a fourth (with several branches) to the more advanced members of the Centrospermae proper.

C. POLLEN MORPHOLOGY AND CLASSIFICATION OF TAXA BELOW FAMILY RANK

The *NPC*-analysis can also be applied in classificatory investigations or checks of sub-families, genera, sections of genera, species etc. Particularly in the classification of species the results will probably be better if, in the future, *NPC*-analysis is combined with the checking of the occurrence of "*XYZ*" or some other so far imaginary series of reliable characters derived from sporoderm details, fine structure, chemical and/or physical properties, etc.

1. Sub-families

Here we may take, as one example, the Acanthaceae-Thunbergioideae. In the Thunbergioideae all genera except one have irregular (anomotreme) pollen grains with a spiral aperture (Raj, 1961). Pollen grains with the same kind of apertures do not occur in the other sub-families. In conjunction with certain details including those of the sporoderm this fact speaks in favour of the homogeneity of the Thun-bergioideae (the deviating genus *Meyenia* has seven meridional colpi—*NPC*-formula: 743—and should, according to Raj (1961), be referred to the Pedaliaceae). So great, in fact, is its homogeneity (also in non-palynological respects), that it has been suggested that the Thunbergi-oideae should be regarded as a special family.

Again, in the Ulmoideae-Celtidioideae, the demarcation line between these two sub-families of the Ulmaceae is probably not drawn correctly: the pollen grains in *Zelkova* (and *Hemiptelea*) are of the same character-istic type as those in the Ulmoideae and different from those in the Celtidioideae, to which these genera hitherto have been referred.

2. Genera, Sections

Let us take, as example, the genus *Morina*. Without knowing any-thing of the pollen morphology of this genus De Candolle subdivided it into two sections, Acanthocalyx with four stamens, and Diotocalyx with two fertile stamens only. The pollen grains in these sections are so different that it would be worth while to see if the sections—also on other grounds—should more rightly be referred to as genera. The pollen grains in Diotocalyx are particularly large and strange. Plate IV shows part of one of the three lateral funnel-shaped aperturiferous out-growths in *M. longifolia*. The pollen grains in a third and unnecessarily established section is of the same type as those in Diotocalyx.

The pollen grains in Acanthocalyx are less strange, "less advanced". Their sexine is uniform, without tegillum and bacula etc. This may be of importance when looking for plants possibly related to *Morina* (so far *Morina* is usually regarded as a genus anomalum in the Dipsacaceae). The genus *Zabelia*, East-Asian shrubs earlier referred to *Abelia* (or *Linnaea*), has pollen grains with some characters in common with those in *Morina*.

A palynologist must not suggest a relationship on so loose a ground. He can refer, however, to statements in the literature pointing in the same direction. Thus Philipson (1948) has mentioned that "if the *Morina* type of inflorescence were to be shortened a capitate inflorescence similar to that of *Lonicera* would be produced". There is certainly a long step, particularly as far as habit is concerned, from *Morina* to *Linnaea* or *Lonicera*. But even untrodden paths or narrow, discouraging roads sometimes lead to the goal.

With regard to *Abelia* it should be added that even the eminent taxonomist Alfred Rehder found it difficult to delimit the section

Zabelia from the rest of the genus. Now, however, not many years later, *Zabelia* is considered a good, distinct genus, to which *Abelia* is probably less closely related than it is to *Linnaea*.

3. Species

In *Geranium silvaticum* I have seen specimens with relatively small flowers and small pollen grains (about 79 × 85 μ; specimen from Järvsö, Sweden, 15 June 1961) whereas in other specimens, as a rule, the grains are considerably larger (equatorial diameter up to 120 μ; in a specimen from Röst, Norway, 105 μ).

V. POLLEN MORPHOLOGY AND PLANT TAXONOMY: PARTICULARLY INTERESTING TAXA. SUGGESTIONS FOR FURTHER INVESTIGATIONS

A. ANGIOSPERMS
(Families in alphabetical order)

ABOLBODACEAE. The South American genera *Abolboda* and *Orectanthe* have comparatively recently been segregated from the Xyridaceae and referred to a special family, the Abolbodaceae. The pollen grains are spheroidal and have no apparent apertures. They are, furthermore, usually large (diameter up to 200 μ, or occasionally more), and provided with a thin, spinuliferous exine underlain by thick intine. The pollen grains in the Xyridaceae proper are smaller, aperturiferous, and the exine is relatively thicker and destitute of spinules.

ACANTHACEAE. This family is known for its particularly beautiful pollen grains (Pl. XI). According to Raj (1961) the Nelsonioideae can for pollen-morphological and other reasons, be referred to the Scrophulariaceae, whereas the Mendoncioideae and Thunbergioideae (in conformity with the ideas expressed by Bremekamp (1953)) deserve the rank of separate families. *Gilletiella* should be removed from the Mendoncioideae and included in the Acantheae. *Meyenia*, it seems, must be excluded from the Thunbergioideae and referred to the Pedaliaceae in conformity with *Thomandersia*, previously placed among the Asystasieae. Fossil pollen grains from the Tertiary deposits have been found to resemble those of *Ruellia*, *Hulemacanthus*, *Sanchezia*, etc.

ALANGIACEAE. The large and conspicuous pollen grains in this interesting family, the taxonomic position of which is still unsettled, are worth further study. Fossil, alangioid pollen grains have been found in Pre-Quaternary deposits. The fact that the pollen grains in *Alangium chinense* (Lour.) Harms from Kilimanjaro (Verdcourt *anno* 1954) are very different from those in what is said to be the same species (China, Nanking, 1931; Luh-Teng 9678 etc.), poses an interesting problem.

AMARANTHACEAE. Pollen grains of two main types, one (the *Amaranthus*-type) of the same type as the grains in the Chenopodiaceae (the grains are small, polytreme with well-defined tremata), the other (the *Gomphrena*-type) is more closely similar to certain pollen grains in the Caryophyllaceae (they are larger than those of the *Amaranthus*-type, oligotreme, with usually not well-defined tremata). Further investigations into the distribution of these types within the family are desirable.

AMARYLLIDACEAE. A eurypalynous, not too well-defined family (eurypalynous: with several, well defined pollen types; families with a more or less uniform pollen morphology are "stenopalynous"). The occurrence of dicolpate pollen grains in *Crinum* and other genera is of taxonomical importance and should be followed up in greater detail. The pollen grains in the monocotyledons are usually monotreme. In the Amaryllidaceae this monopoly has been broken: some genera have pollen grains with more than two (up to eight?) apertures. Further study of these conditions, of the sporoderm stratification (including the intine!) and so forth is recommended.

ANACARDIACEAE. As has been known for a long time the pollen grains in *Pistacia* are markedly different from those encountered in the rest of the family. Kuprianova (1961) recently took this as a reason for referring it to a family of its own, the "Pistaceae" (Pistaciaceae?). Other deviating genera are *Dobinea* and the newly described *Campylopetalum* (Forman, 1953). *Dobinea* grows in Yunnan, *Campylostemon* in Thailand. They would constitute the family Podoonaceae Baillon ex Franchet (related to the Sapindaceae and Anacardiaceae) if that family, suggested in 1889, be revived (*Dobinea* and *Podoon* are considered to be congeneric). Pollen morphology and, as it seems also cytology, may be in favour of establishing a new family. On the other hand there are no significant anatomical discrepancies between the Anacardiaceae and "Podoonaceae".

ANNONACEAE. From the point of view of pollen morphology this is one of the most interesting (and most primitive?) angiosperm families (a monographical study of annonaceous pollen grains is under way in the U.S.A.). The prominent processes in the pollen grain of *Dasymaschalon clusiflorum* somewhat resemble the irregular processes of the Jurassic sporomorpha *Ricciisporites tuberculatus* Lundblad (Svensk bot. Tidskr., Vol. 48, 1954, Pl. IV : 9, at p. 409, and Fig. 10, p. 480).

ARACEAE. This family is perhaps as interesting as the Annonaceae and much in need of further study (apertures, sporoderm stratification including fine structure, etc.). In *Schismatoglottis* sp. (New Guinea,

1961; leg. K. Stopp) some pollen grains are "colpoidate" (Fig. 31), provided with a more or less colpoid or leptomatoid concavity. Other grains have no such colpoid area. In all grains the main, nexinous part of exine opposite to the colpoid streak (or the corresponding area) is more or less irregularly split into a series of transverse lamellae. This probably imparts a certain weakness to the exine, and the area in question may, therefore, be perforated and function as an exit at pollen tube formation (cf. the irregularly and faintly lamellate nexine surrounding the inner parts of the apertures in *Tilia* etc.). In Fig. 31 the thin, outermost exine layer, the sexine, has been locally separated from the underlaying irregularly lamellate nexine. If encountered in a fossil state, pollen grains of this type could perhaps be referred to as spores of non-angiospermous plants by palynologists unaware of the presence of this peculiar pollen type in the Araceae.

ASCLEPIADACEAE. The Periplocoideae are now usually referred to as a special family, the Periplocaceae. This is substantially supported by pollen morphology: the grains are united in tetrads (not in pollinia as in the rest of the Asclepiadaceae) of much the same type as in several members of the Apocynaceae (*Apocynum, Trachomitum,* etc.).

BALANOPHORACEAE. The pollen grains are relatively small (longest axis about 13–44 μ) but, as far as is known, very diversified with regard to apertures etc. Some grains—e.g. those in *Mystropetalon* (South Africa)—are of a unique type (*Mystropetalon* should possibly be referred to a family of its own). Practically nothing is known with regard to sporoderm stratification and details of aperture membranes (there seems to be at least a superficial resemblance between the aperture membranes in *Balanophora elongata* and the membranes in some genera of the Proteaceae).

BIGNONIACEAE. A monograph of this family along the same line as the monograph of the Acanthaceae mentioned above would be welcome.

BOMBACACEAE. With regard to pollen morphology the Bombacaceae are a somewhat difficult but very attractive family. The pollen grains are usually large and their morphology often of apparent taxonomical importance (e.g. in the South American genera *Matisia* and *Quararibea*). By means of fossil pollen grains the history of the bombacaceous plants can, it seems, be traced back to Late Cretaceous times. Attention should be drawn to the vague lines of demarcation, in need of revision, between the Bombacaceae and the other families in the Malvales (*Fremontia* should be referred to the Sterculiaceae; *Adansonia* may be malvaceous rather than bombacaceous, etc.).

BORAGINACEAE. A eurypalynous, heterogeneous family with many problems of great palynological interest. In some genera (*Myosotis* etc.) a detailed study of the pollen morphology is prevented—if only ordinary light microscopes are available—by the very small size of the grains (longest axis sometimes 5 μ or even less).

BROMELIACEAE. See the comment under the Bignoniaceae.

CALLITRICHACEAE. Pollen grains comparatively small, but diversified with regard to apertures, exine details etc. Fresh, carefully collected and determined material may, i.a., shed more light upon the question whether the Callitrichaceae are related to the Euphorbiaceae or maybe to the Scrophulariaceae or other families, etc.

CALYCANTHACEAE. According to pollen morphology this is a "primitive", slightly "monocotyledonoid" family which cannot be referred to Rosales or to Myrtales.

CAMPANULACEAE (s. lat.). This family presents an array of interesting problems. In the Campanuloideae-Campanulaceae there is a gradual transition from "primitive" genera, such as *Cyananthus* and *Leptocodon* (East Asia), with pollen grains with many colpi (up to nine or ten), via genera with colporate pollen grains (example: *Platycodon grandiflorum* with six oriferous colpi) to others with porate pollen grains with a reduced number of apertures (usually three) and more intricate exine patterns. Examples: *Campanula, Iasione, Phyteuma, Roëlla, Wahlenbergia*. Attention should be drawn to the fact that gradual transitions can also occur within a single genus (e.g. *Canarina*). A supposed saxifragaceous genus, *Berenice*, Réunion, possesses pollen grains of campanulaceous type. This genus is also for other reasons truly campanulaceous and should, it seems, be transferred to one of the existing campanulaceous genera in continental Africa.

In pollen morphology the Lobelioideae seem to have no very clear connections with the Campanulaceae-Campanuloideae. The position of the rare genus *Pentaphragma* (Indo-China, Celebes, etc.) remains obscure. Its peculiar flattened pollen grains are very different from those in the Begoniaceae, to which family the genus has sometimes been believed to be related.

CANELLACEAE. This small family, consisting of monotypic, widely scattered genera with restricted distribution, is decidedly primitive: the pollen grains are of the same type as those met with, e.g., in many members of the Magnoliales. It has been much discussed by taxonomists and has formerly been referred to, or near to, the following families: Bixaceae, Cistaceae, Cochlospermaceae, Flacourtiaceae, Guttiferae, Meliaceae, Moringaceae, Pittosporaceae, Theaceae, and Violaceae.

CAPPARIDACEAE. In its pollen morphology this is a very uniform family. There is however, at least one striking exception: the West Australian monotypic genus *Emblingia* has pollen grains of a very deviating type pointing towards the Polygalaceae, Leguminosae, or—perhaps—a new family. With more material at hand this question can probably be easily settled. *Emblingia* was originally described as a "very typical capparidaceous plant".

CAPRIFOLIACEAE (*s. lat.*). Pollen morphology and other features are in favour of referring *Adoxa* to near *Sambucus* and of regarding *Zabelia* —formerly a section of *Abelia*—as a well-defined genus. The pollen grains of *Alseuosmia* are not typically caprifoliaceous.

CARYPHYLLACEAE. The Caryophyllaceae are in need of revision. Detailed studies of genera, e.g. *Silene, Melandrium, Stellaria* etc. (species by species), would be rewarding. The pollen grains in *Stellaria holostea* referred to a section of its own, are thus very different from those in *S. nemorum*, for example (Erdtman *et al.*, 1961, Pl. 12), and the grains in *Melandrium noctiflorum* (earlier referred to as *Silene noctiflora*) are of much the same type as those in *Melandrium rubrum*. Attention should be drawn to the fact that Greguss in several papers published more than 30 years ago and now, as it seems, quite forgotten (e.g. Greguss, 1929), drew attention to the size variation in pollen grains in certain dioicious plants such as *Melandrium rubrum*. This item seems to deserve further study by means of refined methods.

CASUARINACEAE. There is a pronounced similarity between the pollen grains in *Casuarina* and those in *Betula* and, maybe, *Myrica* (see, for example, Praglowski, 1962). The pollen morphology of the Hamamelidaceae is different. Further studies of *Casuarina* pollen grains are recommended also with regard to the possibility of contributing to the knowledge of its geological history by means of pollen analysis.

CERCIDIPHYLLACEAE. The pollen grains are of a "primitive", singular type. They have been described as 3-colpate or "3-colpoidate". The "colpi", which are often not all of the same length, are, however, not similar to ordinary colpi. They can perhaps better be compared with the thin, furrow-like leptomata, which serve as apertures in many gymnosperms (*Cycas, Ginkgo* etc.). If exposed to lateral pressure the pollen grains often exhibit the leptomata in the following way: in the centre a wide, prominent leptoma; on the other side of the grain two, apparently, rather small streaks (representing two compressed leptomata of more or less the same size as the central, "large", leptoma). Fossil pollen grains, similar to grains of this type and known as "*Eucommiidites*", have been found in Jurassic and Cretaceous deposits. Their real nature—angiospermous or gymnospermous—is under debate.

CHLAENACEAE. See Rhodolaenaceae.

CHLORANTHACEAE. The pollen grains have been described as atreme, anacolpate, or more or less irregular ("polycolpoidate"). *Ascarina* has anacolpate grains with the slightly sunken colpus protected by a thick operculum. Fossil pollen grains of a similar type have been found in Early Tertiary and Cretaceous deposits in Europe. Now, the genus is restricted to Asia and Australia, from the Philippines to New Zealand. In *Chloranthus*, possibly also in *Hedyosmum*, there seems to be a transition—recommended for further study—from an anatreme, primitive apertural condition (with one aperture at the distal pole) to a pleiotreme status (with more than one aperture either at the equator or more or less uniformly distributed over the surface).

CNEORACEAE. The factors behind the striking difference between the 3-colporate, striate pollen grains in *Cneorum tricoccum* and the 4–6-colporate, verrucose grains in *C. pulverulentum* Vent. (also referred to as *Neochamaelea pulverulenta* (Vent.) Erdtm.) ought to be investigated.

COMPOSITAE. Attention should be paid to a recent paper by Erika Stix (1960) dealing with, for instance, sporoderm stratification (Fig. 32), index of refraction of the exine layers etc., and the importance of these characteristics in taxonomy. The pollen grains, particularly in Cynareae and Mutisieae, present many unsolved problems. The relationships of the family remain somewhat obscure, the Calyceraceae (and perhaps also the Goodeniaceae) excepted.

CONVOLVULACEAE. The large pollen grains of the Ipomaea type would be particularly suitable for studies in morphogenetics, sporoderm stratification, and fine structure.

CRUCIFERAE. A typical stenopalynous family. Unexpected pollen morphological findings have recently been made in *Rorippa silvestris* (Erdtman, 1958). Their cytotaxonomical background is being studied.

CUCURBITACEAE. A taxonomical revision of the family has been made at Kew by Jeffrey (1962). In this connection a palynological co-operation was to some extent established between Kew and the Palynological laboratory in Solna. From Kew polliniferous material of carefully determined species was delivered while pollen slides—for Kew and for the Palynological laboratory in Solna—were made in Stockholm-Solna. Owing to lack of time—and of palynologists interested in pollen morphology—a detailed investigation of these slides has not yet been made. Jeffrey, however, has made use of the main pollen morphological characters in his new treatment of the family.

CYCLANTHACEAE. In his monograph of this family Harling (1958) has made extensive use of the possibilities of pollen morphology in plant taxonomy earlier hinted at in connection with the family.

CYPERACEAE. The pollen morphology is probably more varied than was formerly believed (see, for example, Ikuse, 1956) and worth further investigation, particularly in connection with morphogenetic and cytological studies.

DEGENERIACEAE. Further studies, of fine structure etc., are desirable.

DIDIEREACEAE. A particularly interesting family, endemic in Madagascar. In some species the large pollen grains exhibit clearly a very

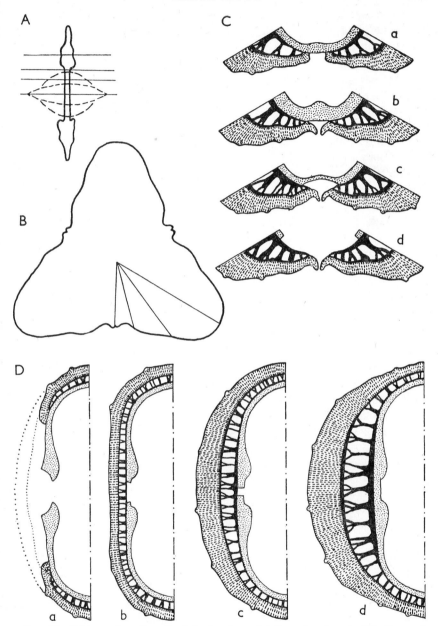

FIG. 32. *Echinops banaticus* Rochel. Aperture details and exine stratification.—A, aperture from outside. Sections along the four horizontal lines are shown in C, a–d.—B, median cross-section through a pollen grain showing the four lines along which the longitudinal sections in D, a–d, have been made.

Broken, thin lines: ectosexine.—Full black: endosexine (branched or unbranched bacula) and nexine 1.—Stippled areas: nexine 2. (From Stix, 1960).

rare feature: the nexine consists of densely packed vertical rods. This is specially evident in the nexinous part of the thick colpus membranes. On pollen morphological grounds an affinity between the Didiereaceae and Centrospermae (cf. particularly the Nyctaginaceae, e.g. *Phaeoptilum*) has been suggested (Erdtman, 1948). There is also reason for taking the Cactaceae into consideration, for Rauh of Heidelberg, has recently been able to graft didiereaceous shoots on cacti.

DIPSACACEAE. The pollen grains are usually very large and well suited for studies in morphogenetics, fine structure, etc. It goes without saying that collaboration with cytotaxonomists should be sought.

The taxonomic position of the genus *Morina* is unsettled. The grains, particularly in the members of the section Diotocalyx, are very large and apparently constitute a valuable material for studies in fine structure, and of chemical and physical properties such as index of refraction.

DIPTEROCARPACEAE. In this family pollen grains with very thin (almost absent) nexine have been found (e.g., in *Dipterocarpus crinitus*). A systematic comparison—with regard to morphogenetics and other features—between "tenui-nexinous" pollen grains (as in some Dipterocarpaceae as well as, occasionally, in other families) and "crassi-nexinous" grains (as in several malpighiaceous genera, e.g. *Aspicarpa*) has not yet been made.

DROSERACEAE. The pollen morphology in *Drosera* and *Dionaea* can be advanced in favour of an affinity—perhaps seemingly contradicted by other features—between the Droseraceae and Nepenthaceae. The sporoderm stratification and the fine structure of the pollen walls should—just as in the Nepenthaceae—be studied in greater detail.

ELAEAGNACEAE. The apertures, sporoderm stratification etc. deserve further study, both with regard to modern genera and also on account of the importance of this family in palaeobotany etc.

ELAEOCARPACEAE. The pollen grains in this family are as a rule 3-colporate and small (longest axis about 11–18 μ). Fossil pollen grains of much the same not very characteristic type as in the Elaeocarpaceae are often encountered in Tertiary and Cretaceous deposits. Studies leading to better possibilities for the determination of such "elaeocarpoid", "lacistemoid", "castaneoid", "elatinoid", etc. pollen grains are to be desired.

EMPETRACEAE. Many characters, including pollen morphology, make it necessary to refer the Empetraceae to Ericales. Linnaeus, among others, described *Empetrum nigrum* and *E. album* (now usually referred to as *Corema alba*). With regard to *Empetrum nigrum* he could hardly have anticipated that the *Empetrum* of Lapland etc. should come to be considered a special form or variety of *E. nigrum* or even as a special species (*E. hermaphroditum* Hagerup), or—*mirabile dictu* —a sub-species of the North American *E. eamesii* Fernald and Wiegand (Löve, 1960). This is not palynology, but the matter is mentioned here in order to stress the desirability of testing—in the Empetraceae as well as in other families—the bearing of pollen morphological characters in recognizing and delimiting taxa of various kinds, such as genera, sections, species, sub-species etc.

EPACRIDACEAE. The occurrence of heterodynamosporous tetrads, i.e. tetrads where all pollen grains are not of the same size (cf. also the Rhodolaenaceae), has, so far, been studied in greater detail from a cytological than from a pollen morphological angle (Smith-White, 1948).

ERICACEAE. Further investigations are desirable in order to elucidate the aperture details, the "viscin strands", the narrow grooves in the lower surface of the nexine as well as the conditions underlying the fact that some species, e.g., *Erica stricta*, have single pollen grains (monadse, whereas in other species, e.g., *Erica tetralix*, the grains are firmly unitod in tetrads.

ERIOCAULACEAE. The pollen grains in this family are of a morphalogically advanced "gyrotreme" type, i.e., the surface of the sphericl) pollen grains is provided with one (or several?) spiral, weak, narrow apertural area(s).

EUCOMMIACEAE. Further investigation, particularly of fresh pollen grains, of the interesting monotypic Chinese genus *Eucommia* are to be desired. The grains have been described as tricolporoidate (colpi as a rule of unequal length: either one long and two short, or one short and two long).

EUPHORBIACEAE. A eurypalynous, more or less heterogeneous family. Special interest is attached to the small, spherical, pantotreme pollen grains in *Phyllanthus*, to the zonotreme grains in *Breynia fruticosa* (grains colpate; each colpus provided with two ora), and, above all, to the occurrence of the so-called "Croton-pattern", a sexinous pattern with conspicuous processes in surface view usually triangular, verrucoid or gemmoid. Several palynologists are engaged in the study of the pollen morphology in this family and its bearing on the taxonomy.

EUPOMATIACEAE. Pollen grains probably of a primitive type. Detailed studies have hitherto been prevented on account of shortage of material.

FAGACEAE. There is a very remarkable difference between the pollen grains in *Fagus* and those in *Nothofagus*. The grains in the former genus are 3-colporate or, rarely, 4-colporate, subprolate to suboblate and destitute of typical spinules, whereas the grains in *Nothofagus* (including the New Caledonian genus *Trisyngyne*, earlier referred to the

Euphorbiaceae) are more flattened, spinuliferous, and, as a rule, provided with more than four apertures.

Electron micrographs of replicas have been of value in elucidating the fine details of the exine surface in Japanese species of *Quercus* (Yamazaki and Takeoka, 1959b).

There is a great difference between the pollen grains in the Fagaceae and those in the Betulaceae. Further investigations will possibly show that some features are common to the grains in some *Nothofagus* spp. and those in some *Alnus* spp.

Fossil pollen grains of the *Nothofagus* type have been found in Cretaceous strata. Stray grains have been encountered in the London Eocene Clay (Sein, 1961), though there has been some discussion about the reliability of the determination of these. The risks of confounding *Nothofagus* pollen with the nothofagoid grains of, for instance, *Stylidium* spp., does not, however, seem to be very great.

FLACOURTIACEAE. A monographic study of the pollen morphology in this eurypalynous, heterogeneous family is very much to be desired. The pollen grains in *Gynocardia odorata* R. Br. are united in tetrads in a specimen collected in India (Darjeeling) by Clarke (34904A), but free in a specimen from Assam collected by King (s.n.) in 1893 (Herb. Brit. Mus. Nat. Hist.). They are provided with large, loosely attached verrucae. The mode of formation of the latter is unknown.

FLAGELLARIACEAE. *Joinvillea* and *Flagellaria* have graminoid, monoporate grains whereas those in *Hanguana* (Susum) are non-graminoid, without aperture. *Hanguana* certainly does not belong to this family.

GOODENIACEAE. In pollen morphology, although apparently not in general taxonomy, *Leschenaultia* is very isolated. The pollen grains are united in large tetrads or polyads (the only example of polyads in the "Sympetaleae" so far known). Cytological investigations may shed light on this phenomenon.

GRAMINEAE. A typically stenopalynous family. There is, however, as Rowley (1960) and others have shown, a great diversity in fine details which may make pollen morphology a useful tool also in the taxonomy of the grass family. For the applicability of pollen grain characters in pollen analysis (history of cultivated grasses etc.) the work of Hinz-Rohde (quoted in Beug, 1961) should be consulted.

Attention should be drawn to the fact that pollen grains of much the same type as in the Gramineae also occur in a number of restionaceous genera (cf. also *Flagellaria* and *Joinvillea*).

HALORAGACEAE. The pollen grains in this family present several interesting, probably more or less primitive features. Some species have a lamellate nexine (thin, horizontal layers which can be seen even with an ordinary light microscope). The Gunneraceae deviate from this condition.

LABIATAE. Pollen grains of two main types: (a) 2-nucleate, usually 3-colpate (sometimes 4-colpate; examples: *Scutellaria*, *Marrubium*, *Lamium*); (b) 3-nucleate, usually 6-colpate (examples: *Rosmarinus*, *Lavandula*, *Nepeta*, *Salvia*, *Satureja*, *Ocimum*). Some genera, e.g., *Ajuga* (pollen grains 3-colporoidate), seem to belong to a transitional group between the Labiatae and Verbenaceae.

LEGUMINOSAE. The Mimosoideae comprise genera with monads (e.g. *Desmanthus*, *Entada* and *Leucaena*), tetrads (e.g., *Mimosa* and *Schrankia*), and polyads (8-, 12-, or 16-celled in the Acacieae and Ingeae). For fine details of exine surface, see Pl. VI (lower fig.). Some tetrads are calymmatous, i.e., the sexine is present as a common cover ("calymma") but absent from the dissepiments. In a-calymmatous tetrads it is present along the latter as well (van Campo and Guinet, 1961). The cytological etc. background to the occurrence of tetrads and polyads deserves further study.

Dizonotreme pollen grains, i.e., grains with apertures in two zones on both sides of the equator, occur in *Dumasia truncata* (Ikuse, 1956). Otherwise such grains are only seldom met with (in the Olacaceae they occur, e.g., in Anacolosa).

LILIACEAE. A eurypalynous, heterogeneous family in need of further study. Most of the species have anacolpate grains of the common monocotyledonous type. *Colchicum* has 2-porate grains (the pores seem to correspond to the ends of a furrow). Trichotomocolpate grains occur in *Dianella*, *Johnsonia*, *Arnocrinum*, *Phormium* etc.

The colpus in *Dracaena* and *Sansevieria* (probably also in other genera) is underlain by chemically not very resistant intine, part of which instantaneously swells if moisture is added. The colpus thereby changes from a slight depression to a strongly bulging surface (Pl. X). The exine in *Lapageria rosea* is, it seems, lamellate.

LIMNANTHACEAE. The pollen grains in *Limnanthes* and *Floerkea* markedly differ from those, for instance, in the Geraniaceae. Morphogenetic studies and investigations of the fine structure may provide clues to the puzzling relationships of this family.

LINACEAE. The important taxonomic position of this family, earlier emphasized i.a. by Hallier (1921), is substantiated by pollen morphology. In the main part of the family the pollen grains have simple or no apertures: the grains are atreme, colpate, pantocolpate, or pantoporate. They are, as a rule, beset with characteristic, often dimorphic processes. As shown by Saad (1961) the bacula or baculoid structures within the "sexine" (or within a stratum which looks as if it were the main part of the "nexine") are endogeneous. Saad's paper provides glimpses of the possibilities for further research, in order to check among other things the importance, if any, of the series: uniformly granulate exine → exine with granules locally agglomerated to form baculoid streaks → baculate, tegillate exine (with or without suprategillar processes).

There is an apparent pollen morphological similarity between certain linaceous plants (e.g. *Linum hirsutum*, a species with three colpi) and the Plumbaginaceae-Plumbagineae (including *Aegialitis* earlier usually referred to the Staticeae; cf. also under Plumbaginaceae).

The Hugonieae are closely related to the Anisadenieae and Lineae but slightly more "advanced" with regard to the apertures (the grains are colporoidate).

Colporate pollen grains occur in other groups including the Ixonanthoideae. Whether these plants are truly linaceous is another question.

LOGANIACEAE. Eurypalynous, heterogeneous. In need of revision.

LORANTHACEAE. In some species the pollen grains look like small tridents, the arms of which are sometimes provided with lateral, more or less deep concavities. If by the action of certain chemicals, for instance, the contents of the arms swell, the concavities disappear and are replaced by convexities. Pollen grains seen in polar view then appear triangular instead of trident-shaped.

MAGNOLIACEAE *s. lat.* Magnoliaceae *s. str.* (*Liriodendron, Magnolia* etc.) and, it seems, the Austrobaileyaceae, have "monocotyledonoid" pollen grains with a distal furrow. Grains of fairly similar type are also encountered in the Canellaceae, Bennettitales etc.

The grains in the Winteraceae are, as far as is known, associated in tetrahedral tetrads and have a distal, rounded aperture. In *Drimys winteri* the sexine forms a conspicuous reticulum in the distal face except in a narrow zone surrounding the aperture. The muri of the reticulum are distinctly carinate. In the proximal face, on the other hand, the sexine gradually diminishes (and ultimately disappears?) towards the proximal pole.

In *Schisandra* there are usually six colpoid streaks. Three of these units meet at the (distal?) pole and seem to serve as lines of dehiscence. These streaks differ from colpi of the usual type by having membranes with a median linear thickening.

The grains in the Illiciaceae are provided with three colpoid streaks. In some species of *Illicium* they meet at both poles. In *I. anisatum* the streaks are comparatively long but do not, however, reach the poles.

The grains in the Schisandracae and Illiciaceae are reticulate with, at least in the latter, carinate muri. Pending further investigations nothing can be said with certainty about whether there is some line of transition from the primitive monotreme monocotyledonoid pollen grains in *Drimys* to the more "dicotyledonoid" ones in the families just mentioned.

MALPIGHIACEAE. The Malpighiaceae often have characteristic, easily recognizable pollen grains with thin sexine and thick nexine, a comparatively rare feature recalling conditions met with, e.g., in certain Linaceae. The apertures are often, as in *Aspicarpa*, slightly irregular.

MALVACEAE. The pollen grains in this family are usually large and provided with very conspicuous, sometimes dimorphic spines. In *Malva pusilla* specimens with low, rounded verrucae instead of spines have been found.

The Bombacaceae, Sterculiaceae, and Tiliaceae have stray genera, or groups of genera, with pollen grains of a more or less malvaceous habit. Pending a thorough taxonomical study of the Malvales it seems better to refer these genera, or some of them, to the Malvaceae than, for instance, to transfer the Hibisceae from Malvaceae to Bombacaceae.

MARTYNIACEAE. The sporoderm stratification (fine structure included) as well as the germination of the pollen grains (there are no apparent apertures) need to be investigated.

MELIACEAE. The position of the genus *Nymania* S. O. Lindberg (*Aitonia* Thunb.) has been under debate. It was referred by Bentham and Hooker to the Sapindaceae as a genus anomalum. The pollen grains are, however, of the same, rather queer type as those met with in *Turraea robusta* (Meliaceae-Melioideae-Turraeae) and other genera.

MONIMIACEAE. This family is in need of further study. The pollen grains are usually thin-walled and some types at least are "monocotyledonoid". The pollen morphology corroborates the division of the family into three families (the Amborellaceae, Monimiaceae *s. str.*, and Trimeniaceae) as well as the subdivision of Monimiaceae *s. str.* into four sub-families: the Atherospermoideae, Hortonioideae, Monimioideae, and Siparunoideae.

MUSACEAE. In the pollen grains of this family, the exine is very thin and the intine thick. These structures should, like those in the pollen grains in the Zingiberaceae etc., be studied by means of electron microscopy.

MYRICACEAE. Here, the exine has very minute, densely and regularly distributed spinules. In this respect the pollen grains recall those in Juglandaceae more than those in the Betulaceae and Casuarinaceae. The spinules can be easily seen in grains mounted in silicone or in distilled water and, still better, in electron micrographs of surface replicas. The pollen grains of the Myricaceae and the Myrtaceae (or supposed forerunners to these families) are frequently mentioned in palaeobotanical-palynological papers. However, pollen grains of a myrtaceous habit also occur in the Santalaceae (*Fusanus* etc.) and Sapindaceae. The determination of fossil grains with a myrtaceous habit must therefore be accepted with caution.

NELUMBONACEAE. The pollen grains are tricolpate (or tricolporoidate), different from those in the rest of the Nymphaeaceae *s. lat.* and, at the same time, deviating from those of ordinary tricolpate grains (the nexinous floor of the colpi being thicker than the nexine in the mesocolpia).

NEPENTHACEAE. The pollen grains are spinuliferous, united in small tetrads of probably much the same shape as the tetrads in the Droseraceae.

NYMPHAEACEAE *s. lat.* The pollen morphology testifies to the heterogeneity of this family (cf. the Nelumbonaceae). The pollen grains are of a primitive, monocotyledonoid type. There may be a faint resemblance to the Pontederiaceae.

OENOTHERACEAE. This family has large pollen grains and is well suited for further palynological studies, preferably combined with cytological and genetic investigations. The apertures as well as sporoderm stratification present several problems of interest.

OLACACEAE. The pollen grains in this interesting, little known, probably very heterogeneous tropical family, are as a rule quite small (diameter ranging from about 10–45 μ) but often remarkable with regard to apertural features etc. The grains in the Old World genus *Anacolosa* and the American genus *Cathedra* have six pores, with two pores immediately in line along each of the three meridians through the rounded corners of the flattened grains. Fossil grains of this type have been found in European Pre-Quarternary deposits.

ORCHIDACEAE. Detailed studies of the morphogenetics and fine structure of the pollen walls are still awaited.

PALMAE. The pollen grains in *Nipa*, now often referred to a family of its own, deviate from those in other palms. Fossil pollen grains of the *Calamus* type have been found i.a. in European Tertiary deposits. An extensive monograph is needed, for taxonomical as well as for palaeobotanical reasons.

PANDANACAE. The taxonomy of the large *Pandanus* genus is difficult to disentangle but attention should be drawn to the fact that there are at least two distinct pollen types. In *Gnetum* there is a parallelism between pollen morphology and taxonomic subdivision and it may be suspected that the same be the case in *Pandanus* also.

PIPERACEAE (incl. Peperomiaceae). The pollen grains are small (longest axis 9·5–17 μ), of a primitive type. More detailed studies, with the application of advanced techniques such as electron microscopy, should be undertaken, to find out if, for instance, there are any palynological features pointing towards some sort of relationship with *Gnetum*.

PLANTAGINACEAE. This family may perhaps belong to, or be near, the Centrospermae, as forecast already by Jussieu. Electron microscopical investigations may perhaps help to show whether this suggestion is correct or not.

PLUMBAGINACEAE. Pollen morphology seems to cast some doubt about the assumption of a close relationship between the Plumbaginaceae and the Centrospermae (e.g. according to Friedrich (1956), the Phytolaccaceae, or, according to Agardh (1858), the Nyctaginaceae-Pisonieae).

The Plumbaginaceae *s. lat.* may be subdivided into the Plumbaginaceae *s. str.* (comprising the Plumbaginoideae and the rare, primitive genus *Aegialitis*) and the Armeriaceae (comprising the Armerioideae with the exception of *Aegialitis*).

The Plumbaginoideae have pollen grains of much the same type as the grains in several linaceous plants, for instance, the genus *Linum s. lat.* They are thus generally tricolpate, an exception being found only in *Ceratostigma willmottianum*, where the grains are pantocolpate and have about thirty colpi. In *Linum* pantocolpate grains occur in *L. monogynum* from New Zealand, likewise in *L. jamaicense*, although here the apertures are shorter, more irregular, almost pore-like. In *Linum bulgaricum*, *L. flavum* and *L. grandiflorum* the colpi consist of longitudinal slits or grooves in the sexine, surrounded by sexinous processes bent sideways towards the colpus parallel to the surface of the grain. Similar conditions are met with, e.g., in *Aegialitis*.

The Plumbaginoideae have been said to possess monomorphic pollen grains but on closer examination several characters indicating a pollen dimorphism may be found (possibly not in *Aegialitis*). The flowers in *L. breweri*, *L. californicum* and *L. catharticum* are homostylous and have pollen grains with monomorphic processes. *Linum usitatissimum*, however, has also homostylous flowers but the processes are dimorphic: a few relatively large processes are interspersed among densely spaced smaller processes. *Linum bulgaricum* and many other species have heterostylous flowers and pollen grains with dimorphic processes.

In the Armerioideae the pollen grains are, as a rule, dimorphic, a feature first described in *Statice limonium* by MacLeod (1887). The grains are, furthermore, "reticulate", in contradistinction to those in the Plumbaginoideae and Linaceae which are "tectate", without apparent signs of any reticulation.

From a pollen morphological point of view a part, at least, of the family Linaceae thus seems to form a platform from which evolution has proceeded in different directions: one branch towards the Plumbaginaceae *s. lat.*, viz. to *Aegialitis*, and from there to the rest of the Plumbaginaceae *s. str.*, and finally to the Armeriaceae. The other, well-known lines (to the Geraniaceae etc.) will not be discussed here.

The pollen morphological reasons for assuming a relationship between the Plumbaginaceae and Linaceae are supported by a number of other arguments. Thus in both families the leaves are usually simple, alternate, estipulate. The flowers are hermaphordite, actinomorphic, bracteate, with petals free, fugacious in the Linaceae as well as in *Aegialitis*. The anthers bithecious and introrse, opening lengthwise and the ovary is superior, with two integuments. The styles ((3–)5), are filiform and may be free or united and the embryo is straight. There are also certain similarities in the geographical distribution.

There are also differences, however: the calyx in the Plumbaginaceae is gamosepalous whereas in the Linaceae it is chorisepalous (or with the sepals partially united). In the Plumbaginaceae the corolla is gamopetalous (except in *Aegialitis*) with the five lobes or segments extending almost to the base. In the Linaceae the petals are free, fugacious. The ovary is 3–5-locular in the Linaceae, unilocular (5-capellary, usually 5-lobed or 5-ribbed) in the Plumbaginaceae. There are two ovules in each loculus in the former family, only a solitary anatropous ovule in the latter, pendulous from a basal funicle. The fruit in the Linaceae is a capsule, in the Plumbaginaceae a utricle or tardily circumscissaly dehiscent. What is perhaps the most obvious difference concerns the stamens. In the Plumbaginaceae these are five, hypogynous (Plumbaginae except *Aegialitis*) or perigynous (Staticeae), opposite the corolla lobes, whereas in the Linaceae they are of the same number as the petals and alternate with them, sometimes alternating with small staminodes; the filaments are connate at the base.

Do these differences, particularly the last one, really exclude the possibility, strongly indicated by certain pollen morphological details, of a relationship between the two families? According to Friedrich the petals and the superimposed stamina in *Aegialitis* are morphological unities, products of a cleavage of the outer set of stamina (the inner set is lacking). How are the conditions, in this respect, in the Linaceae, Geraniaceae, Malvaceae, Euphorbiaceae, Thymelaeaceae etc.?

POLEMONIACEAE (*s. lat.*). A monographical study of this family is much wanted. Stray observations in *Cobaea penduliflora* pollen on the tooth-like processes "fixed in the nexine like teeth in a jaw", the upper part of which later amalgamates and forms a large, conspicuous reticulum, may perhaps turn some ideas of the character of the "nexine" upside down (see Frontispiece Fig. 2, and Fig. 193, p. 330 in Erdtman, 1952).

POLYGALACEAE. The idea of Martius (1856) that *Diclidanthera*, re-

ferred by Gilg to a family of its own, should be referred to the Polygalaceae, has been confirmed by several investigators on different
grounds, including palynological characteristics.

POLYGONACEAE. The sub-division of the very heterogeneous genus
Polygonum (*s. lat.*) is one of the best and most striking examples of
parallelism between palynological characters on the one hand and
non-palynological characters on the other (Hedberg, 1957). Further research into the sporoderm stratification, e.g. in *Fagopyrum*, is to be desired.

In contradistinction to *Polygonum s. lat.* the genus *Eriogonum* is not
eury-, but stenopalynous with regard to apertures, shape etc. With
regard to pollen sizes there is, however, a great diversity: the grains
in *E. xanthum* Small, *E. flavum* Nutt., and *E. jamesii* Benth. measure
about 65 × 40 μ, whereas those in *E. mohavense* S. Wats., *E. dendroi-
deum* (Nutt.) Stokes, *E. baileyi* S. Wats. and *E. salsuginosum* (Nutt.)
Hook. only measure about 30 × 20 μ or slightly less. This, in conjunction with cytological studies, may shed new light on the history
and dispersal of the often more or less "apocratic" (see p. 196) members
of the genus *Eriogonum*.

PROTEACEAE. In the tetrad stage the apertures (in pollen grain with
three apertures) meet three and three in four places. This arrangement
is in accordance with "Garside's rule", whereas, in most non-pro-
teaceous plants, the apertures meet two and two in six places
("Fischer's rule"). In sporoderm stratification there is still much to be
investigated—cf. particularly the aperture membranes, which in *Hakea*,
Grevillea, and other genera are very conspicuous indeed.

RANUNCULACEAE. In *Anemone* and other genera pollen morphology
will prove to be of great value in taxonomic revisions. Pollen morphology and other features emphasize the isolated position of *Paeonia*.
The pollen grains in this genus are often slightly colporoidate whereas
in the other genera the apertures are simple, i.e. more "primitive".

RAPATEACEAE. In this interesting, and until recently but poorly
known family, the palynological possibilities in taxonomical context,
hinted at by Erdtman (1952) have been more fully followed up by Cronquist (1960).

RESTIONACEAE. Fossil pollen grains which no doubt belong to this
family, which in our days is mainly restricted to the southern hemisphere, have been found in Old Tertiary strata in Europe. There is a
gradual transition, within this family, from pollen grains of almost
centrolepidaceous habit to pollen grains of much the same type as those
in the Gramineae (there are also anatomical resemblances between the
Restionaceae and Gramineae; cf. Metcalfe, 1961).

RHODOLAENACEAE. Usually referred to earlier as Chlaenaceae, is an
interesting family, endemic in Madagascar, previously often placed in

the Malvales. Its members have pollen tetrads, often of a peculiar shape and once, by mistake, interpreted as polyads. With regard to its relationships it is believed that "the palynological compass needle" points towards the Ericales. Note, e.g., that some genera have heterodynamosporous tetrads (this is also the case in the Epacridaceae).

RHOIPTELECEAE. *Rhoiptelea chilianthea* is a monotypic Chinese tree genus, which, as far as pollen grains are concerned, seems to be more nearly related to, for instance, the Betulaceae than to the Ulmaceae.

ROSACEAE. The Neurada-group—with its very queer pollen grains (zonotreme with regard to the ora)—is very isolated. *Rubus chamaemorus* has spinulose pollen grains, a feature which is singular both with regard to the genus and to the family.

The genus *Sanguisorba* is of particular interest cytologically as well as from the standpoint of pollen morphology (Pl. XII). From the morphology of fossil (e.g. Late Glacial) pollen grains conclusions as to the number of chromosomes may be drawn (Erdtman and Nordborg, 1961).

RUBIACEAE. This is a heterogeneous family in need of further study. There is often a gradual transition from a colporate to a porate status, the "ora" in the former status finally standing out as true pores in the latter status (after the disappearance of the colpal "frames"). As a rule pores of this kind, e.g., in *Tocoyena*, are crassimarginate, i.e. have thickened borders, whereas "simple pores", e.g., in other families, in *Thalictrum*, *Altingia* and *Costus*, are "tenuimarginate", provided with very thin margins.

SANTALACEAE. This is, together with the Loranthaceae and Olacaceae one of the most interesting dicotyledonous families from a pollen morphological point of view. The pollen dimorphism in *Arjona* (and *Quinchamalium* ?) should be studied in greater detail, as well as the fossil record of santalaceous or santalaceoid pollen types in Cretaceous layers which may prove illuminating.

SAPINDACEAE. The pollen morphology of this family has by no means been so fully studied as would be required with regard to the size, taxonomy, and history of the family. Cronquist (1957) has suggested a relationship between the Sapindaceae and the Umbelliferae. It has been pointed out (Erdtman, 1960) that the pollen grains in the Australian species *Diplopeltis huegelii* are so remarkably similar to the pollen grains in the Umbelliferae that, if found in a fossil state, they would undoubtedly be referred to the Umbelliferae by palynologists unaware of the *Diplopeltis* pitfall. The pollen grains in the second species, *D. stuartii*, are less umbelliferoid. It is interesting to note that each species has been placed in a section of its own.

SAURURACEAE. The small pollen grains in *Anemopsis*, *Houttuynia* and *Saururus* are "monocotyledonoid", similar to those in *Ascarina* (Chloranthaceae) and *Piper* (Piperaceae).

SAXIFRAGACEAE. This is a very heterogeneous family. Pollen morphology has been instrumental in subdividing the family into smaller units and elucidating the relationships of these.

SCROPHULARIACEAE. Pollen grains with composite apertures (e.g. in *Hebenstreitia* and *Verbascum*) are less common than grains with simple apertures. The pollen grains in *Pedicularis* are zonotreme (2–3-colpate) or occasionally pantotreme (6-pantocolpate).

SIMAROUBACEAE. The pollen grains in such genera as *Irvingia*, *Suriana* and *Kirkia* deserve further study.

SOLANACEAE. *Mandragora* has pollen grains with very thin exine, the stratification and apertural conditions of which are not well known. In the pollen grains in the South American species *Jaborosa caulescens* and *Trechonaetes* spp. a feature is met with which so far has not been encountered in other pollen types, viz. two nexinous lists, protruding about 3·5–4 μ into the lumen of the grains, forming internal lines of demarcation between two rounded triangular areas around the poles and an intermediate equatorial belt.

STERCULIACEAE. This is a heterogeneous family in need of revision. According to a recent monograph on *Theobroma* and related genera it is correct to raise at least one of the former subgenera of *Theobroma* to generic rank. This is supported by features of pollen morphology.

STYLIDIACEAE. According to pollen morphology the Stylidiaceae are a heterogeneous family: the small colporate pollen grains in the Donatioideae are very different from those in the Stylidioideae. The latter are of much the same type as those in some campanulaceous genera (*Codonopsis, Cyananthus, Leptocodon, Platycodon*). There is also a faint similarity—not to be taken as an indication of relationship—with the grains in some species of *Nothofagus*.

SYMPLOCACEAE. This family is in need of monographic study for taxonomical as well as palaeobotanical reasons.

THYMELAEACEAE. Pollen morphology seems to be decidedly against the idea of a relationship between the Thymelaeaceae and the Proteaceae. Nor is there any similarity with the pollen grains in the Myrtales. On the other hand there is a fairly apparent similarity with the "crotonoid" pollen grains in the Euphorbiaceae, Buxaceae etc. Taxonomists may not agree that there is a real relationship between the Thymelaeaceae and the crotonoid Euphorbiaceae. Be that as it may, further studies of the pollen morphology—including comparisons between fine structure and exine stratification of the pollen grains concerned—may add more weight to the palynological arguments so far presented.

TILLIACEAE. A heterogeneous family. Besides more or less deviating types, there are two main pollen types, viz. the *Tilia*- and *Grewia*-type. The *Tilia*-type, familiar to every pollen analyst, is easy to identify but often difficult to describe as far as certain fine sexine details are concerned (Praglowski, 1962). These details make it sometimes possible to identify the species by means of pollen grain characters only, e.g. *Tilia cordata* and *T. platyphylla* as well as the hybrid between these.

TROPAEOLACEAE. The pollen grains in *Tropaeolum majus* are 3-colporate, of a rather common habit. This fact must not, as shown by Ricardi *et al.* (1957), lead to the belief that the genus is a stenopalynous one. On the contrary, there seems to be a great array of different sizes and shapes etc., as well as of chromosome numbers.

VELLOZIACEAE. Pollen grains 1-sulcate, of two distinct types: single or united in tetrads. In the former the grains are smaller and have a slightly thicker exine than the latter. To the former belong *Barbacenia*, *Barbaceniopsis* and the following "Vellozias": *dasylirioides, elegans, equisetoides, hereroensis, retinervis, splendens,* and *sessiflora* var. *villosa;* to the latter the true Vellozias, e.g. *V. aloaefolia circinans, compacta, dawsonii, phalocarpa, sellovii* and three undetermined specimens in the Herbarium, Rio de Janeiro (Duarte n. 2114 and Perreira n. 1290 and 2808).

The pollen morphology supports the new classification by Smith (1962).

VERBENACEAE. The Verbenaceae are, in contradistinction to the Labiatae, a pronouncedly eurypalynous family with among other things many large pollen types (longest axis in *Bouchea* e.g. up to 160 μ, in *Stachytarpheta* up to 150 μ). In his monographical surveys of the family (published in Phytologia, etc.), Moldenke has not taken pollen morphology into account. It would, however, be worth doing so in order to find out whether pollen morphology fits with his systematic treatment or not. Examples of an interesting exine stratification are found in e.g. *Bouchea* (Pl. III, 5–7).

XYRIDACEAE. See under Abolbodaceae.

ZINGIBERACEAE. See under Musaceae.

ZYGOPHYLLACEAE. A heterogeneous, eurypalynous family. The pollen grains in *Kallstroemia* and *Tribulus* are very similar to those in *Cobaea* and *Viviania*. A broad survey of the morphogenetics, fine structure etc. of the maybe more or less disparate elements belonging to this morphological group would probably yield interesting results.

B. GYMNOSPERMS

Attention will here only be drawn to a few facts.

Cedrus brevifolia should probably, for reasons including pollen mor-

PLATE XI

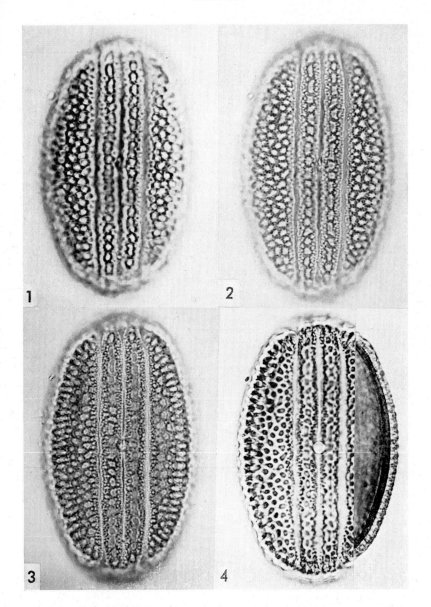

PLATE XI. *Hypoëstes antennifera* (acetolyzed pollen grain at different foci from high to low). (×1000.)

PLATE XII

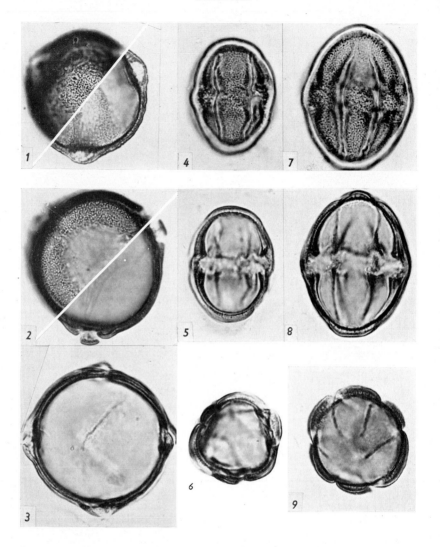

PLATE XII. Pollen grains of *Sanguisorba*.—Fig. 1, *Sanguisorba minor* ssp. *muricata* (2 n=28; Nordborg 0768, Gotland, Sweden).—Figs. 2, 3, *S. minor* ssp. *muricata* (2 n =56; Nordborg 0769, Gotland).—Figs. 4–6, *S. officinalis* (2 n=28; Nordborg 8035, Gotland).—Figs. 7–9. *S. officinalis* (2 n =56; Nordborg 0758, Norway). (×1000.)

phological ones, be regarded as a proper species, not as a taxon of sub-specific rank.

Larix and *Pseudotsuga* must be closely related, i.a. since pollen grains as well as megaspore membranes are of much the same type in both genera. *Cathaya* has small pinoid pollen grains and cannot be referred to *Pseudotsuga* as suggested by Greguss on anatomical grounds.

The genus *Dacrydium* is heterogeneous and should be subdivided. In some species the pollen grains are very podocarpoid, in other species they look much more "ancient" (Erdtman, 1957; Ueno, 1960).

VI. Palynology and Palaeobotany

The principles of modern pollen analysis have been dealt with at length in several text-books and will not be discussed here. A valuable account will be found in the second edition of the text-book by Faegri and Iversen which is scheduled to appear in 1962. The chief concern will be to examine some of the findings of pollen analyses as they affect palaeobotany. We will begin with the quaternary deposits, in which consideration will mainly be restricted to two items, surface samples and general trends.

A. QUATERNARY DEPOSITS

"The present provides the key for unlocking the past." In view of this axiom it is strange that only relatively feeble efforts have been made to show how present-day vegetation is reflected by pollen grains and spores in surface samples. If recent sediments, e.g. ooze under formation or peat *in statu nascendi*, are not available, sporomorphs (pollen grains and spores) in moss cushions, lichen thalli etc. can be studied. Here, the sporomorphs do not, in fact, reflect the vegetation exactly in the same way as do the sporomorphs in peat and ooze under formation, but they are, nevertheless, of interest as indicating for example how far, and in what quantities, sporomorphs can be carried by the wind. Such facts are of interest from a purely biological point of view as well as in connection with the theory of pollen statistics.

About 20 years ago an appeal was made in Sweden to follow, by means of pollen statistics, the history, beside the Swedish, of soils in general. That appeal, to which was added a chart showing the average pollen flora of various plant communities and soil profiles from southern Sweden to Lapland, was, however, largely in vain. It is only in the last few years that similar points of view have been taken up, e.g. in England by Dimbleby (1961), in Sweden by M. Fries (1960). The latter has i.a., studied pollen spectra of acid soils and been able to demonstrate that some of these soils were cultivated at an earlier date.

8

Interesting glimpses of plant life in times long past have also been obtained from studies of the pollen flora in soil samples from the arid zone, from the débris in caves, e.g. the famous Shanidar cave in northern Iraq. It should be remarked, however, that the pollen flora in such samples is often very scanty and difficult to extract. It is thus impossible to drawn conclusions from it in exactly the same way as from the pollen flora in peat, acid mineral soil, or permanently frozen ground.

The ^{14}C-method is an important, exact complement to pollen statistics as a method of dating. It is gratifying to know that many pollen analytical as well as geochronological datings, made 30 or 40 years or more ago, have been corroborated by ^{14}C-determination (Godwin, 1956).

The history of agriculture (the history of cereals and weeds, etc.) is a very fascinating topic which has contributed to the popularity of pollen statistics among archaeologists. It is in keeping with the great archaeological traditions in Denmark that most of the pioneer work in this line of pollen statistics has been made there—Iversen (1941), Jessen (1920), Troels-Smith (1955). Von Post et al. (1939) in Sweden, also recommended close collaboration with archaeologists and several scientists, notably Firbas (1937) in Germany, have called attention to the possibility of identifying the pollen grains of cultivated grasses.

Lennart von Post, Iversen, Firbas and others have tried to substitute the climatological subdivisions of Late Quaternary times put forward by Blytt and Sernander ("Boreal", "Atlantic", Subboreal", etc.) by other climatological-vegetational subdivisions. Thus we may speak of *cryocrats*, i.e. plants living during an Ice Age and along the border of a retreating continental ice (Gr. *kryos* = ice). They were replaced, in succeeding times with better climate, by *protocrats*; these in their turn, around the postglacial climatic optimum, by *mesocrats*. The following climatic deterioration, in which we actually live, is the time of the *telocrats*, which, if there were to be another Ice Age, would be followed by cryocrats etc. This sequence—cryo-, proto-, meso-, and telocrats—apparently also holds good for earlier (Interglacial) epochs.

From a gross ecological point of view there seems to be reason also to speak of "apocrats" and "statocrats" (or "topocrats"). The apocrats are "pioneers", not influenced, except perhaps to a minor extent, by climate. They can, therefore, appear practically any time, be it during an Ice Age or a climatic optimum. Their foremost requirement is freedom from competition, such as in areas uncovered during the swift retreat of glaciers, laid bare, e.g. at the sudden drainage of large, ice-dammed lakes or emerging from the sea in connection with eustatic sinking of the ocean level or with secular land upheaval. A temporary freedom from competition also accompanies the upsetting of the natural balance by axes, ploughs, bull-dozers and other machines in forests and fields, along highways and railroads, in the outskirts of towns and other settlements. In the stepp-like "Alvar" district in the south of the island of Öland in the Baltic the shallow soil is broken up each year by the action of the frost, which thus in some way plays the same role here as the bull-dozers outside the Palynological laboratory in Stockholm-Solna. At all events, apocrats thrive in both places: in the Alvar there are even relicts from Late Glacial times, such as *Helianthemum oelandicum* (other relicts, e.g. *Ephedra*, have disappeared), in Solna *Artemisia*, *Melilotus*, and many other weeds. There is a remarkable connection between Late Glacial vegetational elements (e.g. *Centaurea cyanus*) and "modern weeds", a connection which should be kept in mind when discussing the history of weeds. The first identifications of fossil *Artemisia* pollen were, by the way, made in the early thirties in banded glacial clays from Solna and from Late Glacial sediments on Storna Karlsö, near the island of Gotland. At that time the identifications were regarded with much suspicion. The situation has changed in the last two decades: now a prize will be awarded the palynologist who is the first to discover *Melilotus* pollen in Swedish Late Glacial deposits.

Pioneers have often been said to be heliophilous. I do not quite agree with this. Freedom from trees and the shadow produced by them is an essential condition, but still more essential is, no doubt, freedom from any competition and the presence of virgin, unleached soils. Heliophilous apocrats abound in the surroundings of Aranjuez, Budapest, Baghdad and other towns, but apocrats are also found in Prince Rupert Island, Bergen and Stornoway and other wet places with a more restricted amount of sunshine.

Apocrats are at present spreading and increasing at a rapid rate in consequence of the cultivation of the soil and the rapid devastation of natural vegetation. A surface sample from the Seychelles Islands in 1962 would tell a very different story if compared with a sample taken only a few years ago. By not working out the way in which the present-day natural vegetation in different parts of the world is reflected by the "pollen rain", palynologists may lose a number of keys for unlocking the secrets of the fossiliferous deposits.

One may ask why palynologists, as a rule, have not paid much attention to the pioneers, "opportunists", or apocrats (the last term was introduced in 1960). Probably the main reason is that apocrats usually do not grow in bogs and lakes or in the vicinity of places where microfossiliferous deposits are usually laid down. Instead they are often present in places where there is only a slight chance of their pollen grains being incorporated in fossiliferous deposits, e.g., on the steep slopes of weathering mountains or along declivities where eroded material is continually sliding down and being carried away.

This should not prevent palynologists, however, particularly if they have good luck, from finding polliniferous deposits with a fair admixture of pollen grains of apocrats. I recall in this connection a deposit not far from Kiel, Germany, where Schmitz, among others found stray pollen grains of *Centaurea cyanus*, from Late Glacial strata right up to more recent layers. I once visited a place called Schwedeneck, north of Kiel. Here the Baltic end moraine is continually being broken down by the sea. In the loose, steep slopes are found the moss *Dicranella varia* and many other apocrats. Along the upper end of the slope is a small path, only a foot or two from the declivity. In this narrow, marginal zone there grows, in places, a characteristic assemblage of plants (*Centaurea, Artemisia* etc.). There they are safe, neither man nor beast walks along the edge of an abyss. On the other side of the path are wide cultivated fields. I instantly thought of the possibility—theoretical at least—that this assemblage, lingering on a narrow edge, might be a relict of Late Glacial vegetation, that the instability of the soil was a prerequisite for the stability, for the remarkable conservatism with which the assemblage had piloted itself throughout proto-, meso- and telocratic times and all the vicissitudes induced by climatic and vegetational changes etc.

We turn now to earlier deposits.

B. PREQUATERNARY DEPOSITS

Not even a brief picture of the interesting palaeobotanical, Pre-Quaternary aspects of palynology can be attempted here. For one thing, there will presumably be a great many partially unexpected additions to this part of palynology in the near future, additions that may profoundly influence some of our present ideas. The discovery of restionaceous and proteaceous pollen grains in European Tertiary deposits and as it seems, of pollen grains of the *Nothofagus* type in the London Eocene Clay (Sein, 1961) provides a foretaste of what is about to happen. Reference is, instead, given to the following pollen and spore atlases and collocations, from which an idea of the palaeobotanical-palynological possibilities may be obtained. The first part (in stratigraphical order from Quaternary to Cambrian) is exclusively from the Soviet Union and is a token of the great interest in geopalynology taken in the U.S.S.R. The second part contains publications from outside the U.S.S.R. They are chosen more or less at random.

1. *U.S.S.R.* Quaternary: Dombrovskaja *et al.*, 1959; Miocene: Pokrovskaja, 1956a; Oligocene: Pokrovskaja, 1956b; Tertiary: Tchiguriaeva, 1956; Eocene, Palaeocene, Upper Cretaceous: Pokrovskaja *et al.*, 1960; Lower Cretaceous: Maljavkina, 1958; Upper Cretaceous:

Hlonova, 1960; Cretaceous, Jurassic: Maljavkina, 1949; Lower Creta-
ceous, Jurassic: Bolchovitina, 1959; Mesozoic: Bolchovitina, 1959;
Palaeozoic: Luber, 1955; Middle Carboniferous: Istchenko, 1952; Upper
Devon: Naumova, 1953; Lower Cambrium: Naumova, 1949; cf. also
Samoilovitch and Mtchedlishvili, 1961.

2. *Outside U.S.S.R.* (authors in alphabetical order): Artüz, 1957
(Carboniferous); Balme, 1957 (Mesozoic); Couper, 1960 (Mesozoic);
Góczán, 1956 (Liassic); Imgrund, 1960 (Carboniferous); Klaus, 1960
(Triassic); Kosanke, 1950 (Carboniferous); Krutzsch, 1959 (Eocene);
Pacltová, 1961 (Upper Cretaceous); Potonié and Kremp, 1955, 1956a, b
(Carboniferous); Rogalska, 1956 (Liassic); Schopf *et al.*, 1944 (Carboni-
ferous); Thiergart, 1940; Thomson and Pflug, 1953 (Tertiary).

It is stimulating indeed that the authors of palaeobotanical text-
books now gradually seem to be paying more attention to palynology
(Mägdefrau, 1953; Gothan and Weyland, 1954; Remy, 1959; Andrews,
1961). The attitude is still often somewhat critical. But at least it is
better to be too critical than to be the opposite: to non-specialists it
must be particularly difficult to try to separate the sheep from the
goats among the palynological-micropalaeobotanical papers now crowd-
ing the shelves of palaeobotanical institutions.

The best way of establishing a better understanding and closer con-
tacts between specialists in macro-palaeobotany and specialists in
micro-palaeobotany is to study pollen grains and spores *in situ*, still
within or attached to the sporangia. In this way the painstaking work
of the macro-palaeobotanists will be brought into contact with the
findings of the palynologists to their mutual benefit.

Pollen grains and spores are often useful as guide fossils. Spore slides,
from, for example supposed Upper Carboniferous, Lower Triassic, or
Upper Oligocene layers destitute of other recognizable fossils, often
immediately reveal to the trained palynologist whether a supposed age
is correct or not.

In coal or lignite-bearing areas sporology often affords a simple, safe,
and quick means of identifying parts of seams dismembered by thrusts
or dislocated by folding. A geologist may ask: Is this seam eight—or
is it more likely part of some seam from the group fifty-nine to sixty-
four? The palynologist in charge should know the answer. If it be seam
eight, borings would reveal the presence of other seams further down. If
it be the sixty-fourth (and last) seam, further borings would be out of
place. Pollen grains and spores are used in the same way in the quest
for oil and salt etc.

The way in which the fossil pollen grains and spores have been described is often discouraging. I really admire only one description of a
fossil pollen grain, viz. that given by Klaus (1954, pp. 114–131). There

are, of course, also many other good descriptions and diagnoses, but far too many follow the following pattern: "Genus *Parvisaccites* n. gen. Diagnosis. Body of grain generally broader than long; bladders small in comparison with body of grain, thickenings of bladders tending to a radial arrangement; bladders attached distally". What is meant, e.g., by "thickenings of bladders"? Another example: "*Decussatisporites* n. gen.: Furche schmal, sie führt von Pol zo Pol. An den Enden nicht erweitert. Form oval bis spindelartig. Die Streifen verlaufen longitudinal und transversal". This genus (Schweiz. Paläont. Abhandlungen, Bd. 72, Basel 1955, p. 43) has as genotype *Decussatisporites delineatus* n. sp., of which it is said that the colpus is open at its ends (which it should not be according to the generic description) but closed, or almost so, in its central part. And what does it mean that the colpus goes from pole to pole, etc.?

The present situation with regard to nomenclature and diagnoses are, in spite of the good offices by Potonié and others, quite unsettled. In some palynologists' opinions, chaos is just around the corner. Too many new "pollen and spore species" with insufficient diagnoses and photomicrographs showing practically nothing but a dark, diffuse fleck (cf. e.g. Palaeontographica, vol. 94 B, 1–4, Pl. 9, Figs. 126–132, 1954), are being published. About 90% or perhaps more, of these "species" have no scientific significance and should be amended or discarded altogether.

In christening these species there is often a tendency to choose as long names as possible. Names with fifteen letters are common, and twenty letters or more are not quite uncommon. Such names—referred to as *nomina sesqui-pedalia* (names one foot and a half long)—are often truly monstrous as the linguistical difficulties have a tendency to rise with the length of the names: *Proteacidites isopogiformis* should, I think, more correctly be *Proteaceidites* (or *Proteaceoidites*, i.e. Proteaceoid-ites) *isopogoniformis*, and *Abietineaepollenites* with its twenty letters may also be in need of a make-up. In each of the following three names, which are formed in the same way, there seems to be at least one error: *Tsugaepollenites, Lycopodiumsporites, Marattisporites.* These are names of "genera". Is it necessary to learn and remember them in order to be able to make correct references to the genera in question?

Most of the International Rules of Nomenclature were proposed a long time before botanists had any idea of the possibility of describing "new species" based solely on isolated pollen grains or spores. By placing their initials or names after the "n. sp." the authors, i.a. inform the palynologists of tomorrow where the responsibility for the happenings in this field in our days can be sought. It would no doubt be better not to give our successors that chance but to conceive, instead, a more rational system for a registration and description of fossil pollen grains and spores. The present nomenclature system will, I hope, fall to pieces long before the next century.

In order to give beginners in the study of palynology an idea of fossil pollen grains special slide collections have been established at the Palynological laboratory: here are thus slides made from Scandinavian and Extra-Scandinavian surface samples, from Late Quaternary material (with a special sub-division for the Tropics), from Interglacial material, Tertiary material (several sub-divisions: Pliocene, Oligocene, etc.), Mesozoic material (with several sub-divisions), as well as from Palaeozoic and older material (equally with several sub-divisions). Among these are some slides, made in 1935, of *"Eucommiidites"* and *Classopollis*, and some other spores which also may, or may not, have some bearing on the question of the origin of the angiosperms. They come from Jurassic and Cretaceous material collected in the Isle of Wight and the mainland opposite. I hastened to show them to some palaeobotanists in 1935 only to hear that they had more important things to do. The situation was not much better at the International Botanical Congress in Amsterdam 1937. The sporomorphs mentioned are now well known and much discussed. Times have changed.

Still that abominable mystery, the origin and early history of the angiosperms, has not been solved. Some palynologists argue, almost *modo monomaniaco*, that no indisputable pollen grains (and other remains) of angiosperms have been found before Mid-Cretaceous times (Aptian), cf., e.g., Harris (1960). They may be right. They may also be wrong. In weighing the pros and cons, answers to, i.a. the following questions are looked for:

1. What plants have produced the clumsy, early Jurassic spore tetrads (found i.a., in Eastern Greenland and Sweden) described as *Riccisporites tuberculatus* Lundblad? Can they possibly have been produced by predecessors to, e.g., *Dasymaschalon* (Annonaceae)?

2. Interesting arguments in favour of assuming gymnospermous affinities of *Eucommiidites* have recently been put forward. It should be observed, however, that pollen grains of *Cercidiphyllum*, if compressed, can assume much the same shape as *Eucommiidites* and/or similar sporomorphs.

3. *"Classopollis"* is supposed to be a product of gymnospermous (araucariaceous?) plants (cf. e.g. Couper, 1955). The pollen grains (which often are united in tetrads) are however so strange—entirely different from those in the modern gymnosperms—that there may be suspicion that Mesozoic strata still contain secrets, which, if unveiled, would help to clarify matters that are now very controversial.

4. Do we yet really know where to draw a safe line of demarcation between a primitive monocolpate (anacolpate) pollen grains of angiospermous type and certain pollen types of extinct gymnosperms? In this field electron micrographs of replicas should be particularly valuable (Yamazaki and Takeoka, 1962). This technique has, however, so far, only sporadically been applied on fossil material (Yamazaki and Takeoka, 1959a).

5. "Apocrats" (p. 196) may also be referred to as "opportunists". It is generally agreed that the unstable conditions under which they live may provide developmental stimuli. They grow in widely different places: in ravines, at the narrow margin of abysses, on volcanic soil, along tidal coasts, etc. Their pollen grains, however, may never, or only occasionally, be trapped (and later encountered) in polliniferous deposits. *Centaurea* has perhaps lived at Schwedeneck (p. 198) for about 15 000 years without leaving any records in the lake sediments and peat deposits in the country at large. Did the early angiosperms live as "progressive apocrats" in the vegetation of Early Cretaceous, Jurassic, and maybe still earlier times? Were they forerunners to plants of annonaceous, myristicaceous, lauraceous, magnoliaceous, liliaceous, eupomatiaceous, nymphaeaceous, cercidiphyllaceous, or amentaceous type? Were they entomophilous? Were their exines as a rule resistant to decay or not?

Apocrats sometimes take, and can take, a chance, invading a country at large, suppressing other plants, and establishing themselves as "topocrats". Birds began to take possession of the air in Jurassic times. Can the angiospermous outburst in Late Cretaceous times be explained along similar lines—as a sudden offensive with radical changes in a comparatively short time as a result?

VII. Pollen Grains in Honey, Drugs, Excrements etc.

The plants visited by bees can be traced by pollen grains in honey in much the same way as the composition of the anemophilous part of the vegetation is reflected by pollen grains in surface samples. Ancient honeys from Egyptian graves provided information on the plants visited by bees in the Nile valley thousands of years ago (Zander, 1941). Pollen analytical investigations of honey may give hints as to the time of year the honey was produced and also help to reveal dishonest practices: a supposed heather honey from Yorkshire with an abundance of pollen grains of Californian or South African plants cannot readily come from the Moors of Cleveland Hills, etc.

At present a very active interest is being taken in honey pollen research in many countries, e.g., Switzerland, France, U.S.S.R., and

India. This affords another example of how important it is to have an economic background to scientific investigations.

Pollen grains are sometimes of interest in qualitative analyses of drug powders (Knell, 1914) as showing up adulterations and substitutions, etc. In the last few years much has been written—particularly in newspapers and magazines—about the remarkable properties of certain pollen tablets, tonics, and creams, etc. Pollen tablets have been said to be useful in the treatment of prostatitis (Ask-Upmark, 1960). Further experiments have confirmed this statement.

By means of copropalynological studies, i.e., the study of pollen grains in excrements, information can be obtained about the types of plants eaten by herbivorous animals, grouse, hares, sheep, goats, donkeys, camels, etc. Guano deposits in caves inhabited by bats may— particularly in tropical and sub-tropical countries—contain a record of entomophilous plants visited, century after century, or millennium after millennium, by moths that were subsequently devoured by bats. An entomologist may, furthermore, perhaps be able to follow—from the identification of the tiny scales from the wings of the moths contained in the guano—the changes, if any, of the moth fauna through the ages.

The palynological possibilities buried in bat guano were revealed during an investigation in the Palynological laboratory of samples from a cave near Harir, in northern Iraq. Similar guano deposits also occur in Morocco, Japan, and other countries. Palynology in connection with hay-fever and asthma research is outside the scope of the present paper (for references see, e.g., Wodehouse, 1945 and Andrup, 1945).

VIII. Palynology and Criminology

Characteristic, easily identifiable pollen grains or spores from plants with a restricted distribution may, if found in dirt on shoes, clothes etc., indicate more or less clearly where the contamination occurred. This must be particularly true if the plants are hypogeic and at the same time rare (e.g. certain fungi), entomophilous (e.g. Leguminosae) or, if anemophilous, nevertheless possess pollen grains which are usually not carried very far by the wind (*Larix, Pseudotsuga, Zea* etc.).

In Sweden pollen analysis was resorted to not long ago in an effort to clarify a case of murder. Four specialists, a palynologist, a diatomologist, a pedologist, and a mineralogist, were called upon by the prosecution. They all, independently, arrived at the same conclusion, viz. that the dirt on the clothes of the murdered person probably did not come from the place where the corpse was found (June 1959, about a month after the crime had been committed). This statement was, for certain reasons, not to the advantage of the person accused of the murder!

The defence, in its turn, asked another palynologist to undertake an examination. He could testify to the correctness of the statements of the first palynologist but, nevertheless, he did not entirely agree with his conclusions. The main reasons were as follows. It was found that the pollen flora in the most superficial soil layers in May—the crime was committed that month—had not the same composition as the pollen flora in samples taken in the same places one month later (these were, e.g., more rich in pollen grains of grasses, *Rumex* and *Plantago*). This furnished a new aspect to the interpretation of the pollen grains in the dirt samples. In some of the latter, as well as in two samples from near the place where the corpse was found, several pollen grains of *Trifolium pratense* as well as spores of *Endogone* (det. J. A. Nannfeldt), a hypogeic phycomycete, were found. Neither pollen grains of *Trifolium pratense* nor spores of *Endogone* have previously been reported from Swedish soil or peat samples. *Endogone* has so far only been sporadically collected in Sweden: in the herbarium of the Museum of Natural History in Stockholm there is only one specimen, viz. from the Botanical Garden, Uppsala. Some of the *Endogone* spores, both from the soil and from the dirt on the clothes, are Zygo-spores, i.e., not of the more common azygospore type (addition, October 1962).

The above is a complicated example of the application of palynology in criminology. Other, more simple, examples can be added. They testify to the fact that the fate of a criminal can sometimes depend upon a few pollen grains, perhaps a single grain only.

REFERENCES

Afzelius, B. M. (1955). *Bot. Notiser* **108**, 141–143.
Afzelius, B. M. (1956). *Grana palynol.* **1** (2), 22–37.
Afzelius, B. M., Erdtman, G. and Sjöstrand, F. S. (1954). *Svensk bot. Tidskr.* **48**, 155–161. Reprinted in *Grana palynol.* **1** (1).
Agardth, J. G. (1858). "Theoria systematis plantarum". Gleerup, Lundae.
Andrews, H. N. Jr. (1961). "Studies in Palaeobotany". Wiley, New York and London.
Andrup, O. (1945). *Skr. VidenskSelsk., Christ.* **5**, 1–127.
Artüz, S. (1957). *Rev. Fac. Sci. Univ. Istanbul*, B **22** (4), 239–263.
Ask-Upmark, E. (1960). *Grana palynol.* **2** (2), 115–118.
Bailey, I. W. (1960). *J. Arnold Arbor.* **51**, 141–152.
Balme, B. E. (1957). *Commonwealth Sci. Ind. Res. Org., Coal Res. Sect.*, T.C. **25**, 1–48.
Beug, H.-J. (1961). "Leitfaden der Pollenbestimmung", Lief. 1, XIV. Fischer, Stuttgart.
Bischoff, G. W. (1842). "Handbuch der botanischen Terminologie und System-kunde", 2. Schrag, Nuremberg.

Bolchovitina, N. A. (1956). "Atlas of Pollen Grains and Spores in the Jurassic and Lower Cretaceous Deposits of the Vilioni Depression", Trud. geol. Inst. 2, Moscow.

Bolchovitina, N. A. (1959). "Pollen Grains from the Mesozoic Deposits of the Vilioni Depression and their Stratigraphical Importance", Trud. geol. Inst. 24, Moscow.

Bremekamp, C. E. D. (1953). Proc. Acad. Sci., Amst., Ser. C 56, 533–546.

Campo, M. van and Guinet, Ph. (1961). Pollen et Spores 3, 189–200.

Carlquist, S. (1961). Aliso 5, 39–66.

Couper, R. A. (1955). Geol. Mag. 92, 471–475.

Couper, R. A. (1958). Palaeontographica 103, 75–179.

Couper, R. A. (1960). Bull. geol. Surv. N.Z. 32, 1–88.

Cronquist, A. (1957). Bull. Jard. bot. Brux. 37, 13–40.

Dimbleby, G. W. (1961). J. Soil. Sci. 12, 1–11.

Dombrovskaja, A. V., Koreneva, M. M. and Turemnov, S. N. (1959). "Atlas of Plant Remains in Peat", Kalinin Peat Inst., Moscow and Leningrad.

Dyakowska, J. (1959). "Podręcznik palynologii. Metody i problemy". Kraków. Zake. Graf., Kraków.

Engler, A. and Diels, L. (1936). "Syllabus der Pflanzenfamilien". Borntraeger, Berlin.

Erdtman, G. (1943, 1956). "An Introduction to Pollen Analysis". Chronica Botanica Reprints, Waltham, Mass.

Erdtman, G. (1948). Bull. Mus. Hist. nat., Paris 20, 387–394.

Erdtman, G. (1952). "Pollen Morphology and Plant Taxonomy. Angiosperms". Almqvist and Wiksell, Stockholm, and Chronica Botanica Reprints, Waltham, Mass.

Erdtman, G. (1957). "Pollen and Spore Morphology and Plant Taxonomy. Gymnospermae, Pteridophyta, Bryophyta (Illustrations)". Almqvist and Wiksell, Stockholm, and Ronald Press Co., New York.

Erdtman, G. (1958). Flora 146, 408–411.

Erdtman, G. (1960). Bot. Notiser 113, 285–288.

Erdtman, G., Berglund, B. and Praglowski, J. (1961). Grana palynol. 2 (3), 3–92.

Erdtman, G. and Nordborg, G. (1961). Bot. Notiser 114, 19–21.

Erdtman, G. and Straka, H. (1961). Geol. Fören. Stockh. Förh. 83, 65–78.

Faegri, K. and Iversen, J. (1950). "Text-book of Modern Pollen-analysis". Munksgaard, Copenhagen.

Firbas, F. (1937). Z. Bot. 31, 447–478.

Firbas, F. (1949–52). "Spät- und nacheiszeitliche Waldgeschichte Mitteleuropas nördlich der Alpen", I (1949), II (1952). Fischer, Jena.

Fischer, H. (1890). "Beiträge zu vergleichenden Morphologie der Pollenkörner". Kern, Breslau.

Fitting, H. (1900). Bot. Ztg. 85, 107–165.

Forman, L. L. (1954). Kew Bull., 555–564.

Friedrich, H.-C. (1956). Phyton 6, 220–263.

Fries, M. (1960). Ann. Acad. Sci., Uppsala 4, 39–52.

Fritzsche, C. J. (1837). Mém. Sav. Etrang. Acad. Sci. Pétersb. 3, 1–122.

Gagnepain, F. (1903). C.R. Socs. sav. Paris & Dép. 150–159 (also in Bull. Soc. Hist. nat. Autun 16, 1–15).

Góczán, F. (1956). Évk. Magyar áll. föld. int. 45, 135–212.

Godwin, H. (1956). "The History of the British Flora". University Press, Cambridge.

Gothan, W. and Weyland, H. (1954). "Lehrbuch der Paläobotanik". Akademie-Verlag, Berlin.

Greguss, P. (1929). *Math. naturw. Ber. Ung.* **46**, 621–624.

Hallier, H. (1910). *Meded. Rijks-Herb.* **1**, 1–41.

Hallier, H. (1921). *Beih. bot. Zbl.* **39**, 1–178.

Harling, G. (1958). *Acta Hort. berg.* **18**, 1–428.

Harris, T. M. (1960). *Advanc. Sci.* **67**.

Hedberg, O. (1947). *Svensk. bot. Tidskr.* **40**, 371–404.

Hlonova, A. F. (1960). "Upper Cretaceous Spores and Pollen Grains from the Chulymo-Jenisei Lowland". Novosibirsk. (In Russian.)

Hutchinson, J. (1959). "The Families of Flowering Plants", Vol. I. Dicotyledons. Clarendon Press, Oxford.

Hyde, H. A. and Adams, K. F. (1958). "An Atlas of Airborne Pollen Grains". Macmillan, London.

Hyde, H. A. and Williams, D. A. (1945). *Nature, Lond.* **155**, 265.

Ikuse, M. (1956). "Pollen Grains of Japan". Hirokawa, Tokyo.

Imgrund, R. (1960). *Geol. Jb.* **77**, 143–204.

Istchenko, A. M. (1952). "Atlas of Microspores and Pollen Grains from the Middle Carboniferous of the Western Part of the Don Valley". Kiev. (In Ukranian.)

Iversen, J. (1941). *Danm. geol. Unders* **2**, 66.

Iversen, J. and Troels-Smith, J. (1950). *Danm. geol. Unders.* **4** R. 3, 8, 1–54.

Jeffrey, C. (1962). *Kew Bull.* **15**, 337–371.

Jessen, K. (1920). *Danm. geol. Unders.* **2**, 34.

Johansson, L. and Afzelius, B. (1956). *Nature, Lond.* **178**, 137.

Jussieu, A. de (1789). "Genera Plantarum". Herissant and Barrois, Parisiis.

Klaus, W. (1954). *Bot. Notiser* ——, 114–131.

Klaus, W. (1960). *Jber. geol. Bundesanst., Wien* **5**, 107–183. (Special Edn.)

Knell, A. K. (1914). "Die Pollenkörner als Diagnostikum in Drogenpulvern". Thesis, Würzburg.

Kosanke, R. M. (1950). *Bull. Ill. geol. Surv.* **74**, 1–128.

Krutzsch, W. (1959). *Palaeontographica* **105**, 126–157.

Kuprianova, L. A. (1961). *Bot. Ž.* **46**, 803–813.

Kuyl, O. S., Muller, J. and Waterbolk, H. Th. (1955). *Geol. en Mijnb.* **3**, 49–76.

Löve, D. (1960). *Rhodora* **62**, 265–292.

Luber, A. A. (1955). "Atlas of Pollen Grains and Spores from the Palaeozoic of Kazakstan". Alma-Ata.

MacLeod, J. (1887). *Bot. Zbl.* **29**, 116–121, 150–154, 182–185, 213–216.

Mägdefrau, K. (1953). "Paläobiologie der Pflanzen". Fischer, Jena.

Maljavkina, V. S. (1949). "Key to the Determination of Pollen Grains and Spores from Jurassic and Cretaceous Deposits", UNIGRI 33, Leningrad and Moscow. (In Russian.)

Maljavkina, V. S. (1953). "Sporen- und Pollen-Komplexe der Obertrias und des Unter- und Mitteljura aus dem Ost- und West-Vorural", UNIGRI 75, Leningrad and Moscow. (In Russian.)

Maljavkina, V. S. (1958). "Pollen Grains and Spores from the Lower Cretaceous of the Eastern Gobi Depression", UNIGRI 119, Leningrad. (In Russian.)

Maurizio, A. (1956). *Grana palynol.* **1**, 59–69.

Metcalfe, C. R. (1961). "Recent Advances in Botany", pp. 146–152. Univ. Toronto Press.

Mohl, H. (1834). "Beiträge zur Anatomie und Physiologie der Gewächse. Erstes Heft. Über den Bau und Formen der Pollenkörner". Bern.

Martius, C. Fr. Ph. de (1856). Flora Brasiliensis, Vol. 7, Monachii.

Mühlethaler K. (1953). Mikroskopie 8, 103–110.

Mühlethaler, K. (1955). Planta 46, 1–13.

Naumova, S. N. (1949). Bull. Acad. Sci. U.S.S.R. (Geol.) 4, 49–56.

Naumova, S. N. (1953). "Spores and Pollen Grains from the Upper Devonian of the Russian Plateau and their Stratigraphical Significance", Trud. geol. Inst. 143, Moscow. (In Russian.)

Nilsson, T. (1958). K. fysiogr. Sällsk. Handl., N.F. 69 (10), 1–112.

Pacltová, B. (1961). Sborn. ústřed. úst. geol., 26, 47–102.

Philipson, W. R. (1948). Ann. Bot. N.S. 12, 147–156.

Pla Dalmau, J. M. (1961). "Polen. Estructura y características de los granos de polen". Gerona.

Pohl, F. (1937). Beih. bot. Zbl. 56, 366–470.

Pokrovskaja, I. M. (1958). "Analyse pollinique". Annu. Serv. Inform. géol. B.R.G.G.M., 24 (1950).

Pokrovskaja, I. M., ed. (1956a). "Atlas of Miocene Pollen Grains and Spores from the U.S.S.R.", VSEGEI 13, Moscow. (In Russian.)

Pokrovskaja, I. M., ed. (1956b). "Atlas of Oligocene Pollen Grains and Spores from the U.S.S.R.", VSEGEI 16, Moscow. (In Russian.)

Pokrovskaja, I. M. and Stelmak, N. K., eds. (1960). "Atlas of Upper Cretaceous, Paleocene and Eocene Pollen Grains from Different Parts of the U.S.S.R.", Trud. VSEGEI, Leningrad. (In Russian.)

Post, L. von, Oldeberg, A. and Fröman, I. (1939). K. Vitterh. Hist. Antikv. Akad. Handl. 46, 7–98.

Potonié, R. (1934). Arb. Inst. Paläobot., Berl. 4, 5–24.

Potonié, R. (1956). Beih. geol. Jber. 23, 1–103.

Potonié, R. (1958). Beih. geol. Jber. 31, 1–114.

Potonié, R. and Klaus, W. (1954). Geol. Jber. 68, 517–546.

Potonié, R. and Kremp, G. (1955). Palaeontographica 98, 1–136.

Potonié, R. and Kremp, G. (1956a). Palaeontographica 99, 85–191.

Potonié, R. and Kremp, G. (1956b). Palaeontographica 100, 65–121.

Praglowski, J. (1962). Grana palynol. 3, 2, 45.

Raj, B. (1961). Grana palynol. 3, 1, 1–108.

Rauh, W. and Reznik, H. (1961). Bot. Jb. 81, 94–105.

Remy, W. and Remy, R. (1959). "Pflanzenfossilien. Ein Führer durch die Flora des limnisch entwickelten Paläozoikums". Akademie-Verlag, Berlin.

Ricardi, M., Marticorena, C. and Torres, C. (1957). Bol. Soc. Biol. Concepción 32, 17–19.

Rogalska, M. (1956). Inst. Geol. Warszawa Biul. 104, 1–89.

Rowley, J. R. (1959). Grana palynol. 2, 1, 3–31.

Rowley, J. R. (1960). Grana palynol. 2, 2, 9–15.

Saad, S. (1961). Grana palynol. 3, 1, 110–126.

Samojlovitch, S. (1953). Pap. Inst. Naphta Res. U.R.R.S., N.S. 75, 1–93.

Samojlovitch, S. and Mtchedlischvili, N. (1961). "Jurassic-Palaeocene Pollen Grains and Spores from Western Siberia", VSEGEI 177, Leningrad.

Schopf, J. M., Wilson, L. R. and Bentall, R. (1944). Rep. Invest. Ill. geol. Surv. 91, 1–93.

Sein, M. K. (1961). Nature, Lond. 190, 1030–1031.

Selling, O. (1946–48). Bernice P. Bishop Mus. Spec. publ. 37–39.

Sitte (1957). *In* "Die Chemie der Pflanzenzellwand" (ed. Treiber, C.), Springer Verlag, Berlin.

Sitte, P. (1960). *Grana palynol.* **2**, 2, 16–37.

Smith, L. B. (1962). *Contr. U.S. nat. Herb.* **35**, 251–292.

Smith-White, S. (1948). *Heredity* **2**.

Stix, E. (1960). *Grana palynol.* **2**, 2, 39–114.

Tchiguriaeva, A. A. (1956). "Atlas of the Microspores in Tertiary Strata in the U.S.S.R.". Harkow.

Thiergart, F. (1940). *Schr. Brennstoffgeol.* **13**, Enke, Stuttgart.

Thomson, P. W. and Pflug, H. (1953). *Palaeontographica* **94** B (1–4), 1–138.

Troels-Smith, J. (1955). "Pollen-analytische Untersuchungen zu einigen schweizerischen Pfahlbauproblemen", pp. 11–64 in "Das Pfaulbauproblem". Schaffhausen, 1954.

Ueno, J. (1960). *J. Inst. Polyt., Osaka* **11**, 109–136.

Vachey, G. (1961). *Pollen et Spores* **3**, 373–383.

Wagenitz, G. (1955). *Flora* **142**, 213–279.

Wang (1960). "Material towards a Chinese Pollen Flora". (In Chinese.) (Several hundred pages, many photomicrographs; seen by author in the Library of the Geol. Surv. of Hungary, Budapest.)

Wodehouse, R. P. (1933). *Bull. Torrey bot. Cl.* **60**, 479–524.

Wodehouse, R. P. (1935). "Pollen Grains. Their Structure, Identification and Significance in Science and Medicine". McGraw-Hill, New York and London (reprinted 1960).

Wodehouse, R. P. (1945). "Hay-fever plants". Verdoorn, Waltham, Mass.

Yamazaki, T. and Takeoka, M. (1959a). *Sci. Rep. Kyoto Univ.* **11**, 91–94.

Yamazaki, T. and Takeoka, M. (1959b). *J. Jap. For. Soc.* **41**, 125–129.

Yamazaki, T. and Takeoka, M. (1962). *Grana palynol.* **3**, 2, 3–12.

Zander, E. (1935–51). "Beiträge zur Herkunftbestimmung bei Blütenhonig", I, Berlin (1935). II, Leipzig (1937). III, Leipzig (1941). IV, Munich (1949). V, Leipzig (1951).

PERIODICALS DEVOTED TO PALYNOLOGY

Catalog of Fossil Spores and Pollen. College of Mineral Industries, The Pennsylvania State University.

Grana Palynologica. An International Journal of Palynology. Almqvist and Wiksell, Gamla Brogatan 26, Stockholm.

Palynologie Bibliographie. Service d'information Geologique du Bureau de Recherches Geologiques, Geophisiques et Minières, Paris.

Pollen et Spores. Museum Nation d'Histoire Naturelle, Laboratoire de Palynologie, Paris.

Metabolism and the Transport of Organic Substances in the Phloem

A. L. KURSANOV

Academy of Sciences, Moscow, U.S.S.R.

I. Introduction

The life of a plant is characterized by a continuous stream of reactions in which matter and energy undergo diverse transformations.

The basic principle underlying the self-regulation of metabolic processes consists in the fact that the final products of a given chain of reactions can serve as inhibitors of the initial-stages of this process. This self-inhibition is due either to the reversibility of individual links in the process, which (obeying the law of mass action) come to a state of reversible equilibrium upon accumulation of a definite amount of the final product, or is due to a more or less specific blocking of the initial reactions by an excess of the final products. Regulation of the rates of the individual processes (i.e. organization of metabolism in time) is thus achieved.

At the molecular level the structure of molecules influences as a guiding factor the chemical relationships that regulate the metabolism. This is especially true of the structure of such molecules as nucleic acids or proteins, with the configuration of which is associated the capacity for self-reproduction of molecules or the appearance of a specific enzymic action of certain proteins.

Of still greater significance in the control of metabolism is the organization of the protoplasm itself, its heterogeneity and diversity of structure, on the basis of which is achieved a localization and spatial division of substances and functions. This may be regarded as a second type of regulation of metabolic processes accomplished at a higher (cellular) level and providing for the organization of processes in space.

The deeper we penetrate into cellular organization the more obvious becomes the complexity of its structure. In the minute ribosomes (bearers of RNA) we find proteins being synthesized, as on a matrix, with a definite seriation of amino acids. The cellular nucleus with its double filaments of DNA is responsible for reproduction; in the semi-conductor-like chloroplasts is concentrated the system of reactions that lead to the transformation of radiant energy into the energy of stable chemical compounds; operating in the mitochondria is a complex cycle of di- and tricarbonic acids serving as the basis of oxygen respiration, etc.

In the unicellular and simplest organisms, in which all the metabolic activities are concentrated in a single cell, it is the organization of the cell that is the principal regulating system. In the more highly organized biological entities such as higher plants, with their well-defined differentiation of tissues and specialization of organs for co-ordination of metabolism at the molecular and cellular level, we have another feature: complex co-ordination that consists in control of the activities of the various tissues and organs in the plant as a whole.

There can be no doubt that even in the case of extreme specialization some of the functions, for instance respiration, protein synthesis, and certain others, remain common to all the living cells of a plant with differentiated structure. However, a number of other functions are found to be spatially separated and are realized in specialized organs that are frequently separated by a considerable distance. For example, photosynthesis occurs mainly in the leaves; absorption of nutrients and their primary assimilation, in the roots; the storing of nutrients, in specially adapted storing tissues, etc.

Specialization of functions was an important step in plant evolution. However, since the products of metabolism produced by each of the specialized organs are necessary for the whole plant, the process of specialization could arise and develop only in parallel with the development of the transport mechanism. This ensured an exchange of products of metabolism between the specialized parts of the plant as a whole.

Originally, this was apparently the transport of substances from cell to cell in non-specialized tissues. We find such similar movements of substances in existing plants, as for example in the thallus of *Marchantia* and also in the parenchyma cells of higher plants situated at a distance from the conducting tissues. This type of transport is difficult and therefore unsuited for moving substances over long distances. Nevertheless, it plays an important part even in the higher plants, ensuring the delivery of mineral substances from absorbing cells of the root to the vascular system, the export of assimilates from the green cells of a leaf into the phloem, and the export of sugars from sieve cells into the cells of the storing tissues. At first, the movement of substances over small distances was regarded as the result of diffusion. However, of late there is more and more evidence that the processes of free diffusion of organic and mineral substances are greatly restricted in living cells and that the penetration of metabolites from one cell into another is a complex physiological process involving intermediate enzymatic reactions and metabolic energy.

More economical (as regards expenditure of energy per unit length of path covered) is of course the transport of substances in a specialized conducting system, which in modern plants is a combination of phloem and xylem.

The structure of bark and wood is in general rather similar in all vascular plants, though the representatives of different groups do exhibit differences in the details of anatomical structure. For this reason, there are grounds for seeing a single principle in the organization of the transport of substances over long distances. In the most general form, this lies in the fact that the movement of organic substances from the leaves and the transport of mineral substances arriving from the roots are accomplished in different tissues and, apparently, require different sources of energy. The fact that the transport of water and of substances dissolved in it occurs in the xylem, along the trachea and tracheids that have no living contents, while organic substances are transported in the sieve cells of the bast that contain living protoplasm, compels us to search, in the former case, for more mechanical causes of the flow of solutions in a system of hollow tubes and, in the latter case, for phenomena more closely associated with metabolism and the semipermeable properties of protoplasm. Although the transport of substances in wood takes place in the hollow cells of the vessels, it is, however, on the whole excited and controlled by the metabolic activity of living tissues.

This is true, in a larger measure, for the transport of organic substances in the phloem, for in this case the conducting tissue itself, i.e., the sieve cells with the functionally related companion cells, is alive. For this reason, the transport of organic substances, for instance in the movement of assimilates from leaf to root, lies entirely within the sphere of metabolism. The process begins with the formation of sugars and other products of photosynthesis in the cells of the mesophyll, then successively passes through the stages of activation of assimilates, their transport over short distances through a series of parenchyma cells to the nearest bundle of phloem, active absorption (of the substance being transported) by the conducting cells of the phloem, the movement of this substance in a strictly defined direction through the sieve cells and, at a later stage in the movement, entrance into the parenchyma cells of the consuming or storing tissues. The latter stage is in turn probably associated with activation of the molecules of the substance undergoing transport by means of carriers through the protoplasmic barriers, and, finally, with secondary transformations in accordance with the nature of the metabolism of the receiving tissue. All these processes are intimately associated so that a disturbance in one of them immediately or gradually affects the functioning of the system as a whole.

Specialists in this field have not yet reached agreement on the mechanism of transport of organic matter in sieve cells. Some are inclined to consider that transportation occurs by means of a passive entrainment of the dissolved substance by the liquid flowing in the internal cavity of the sieve cells that form a continuous vacuolar system (the mass flow theory). Others think the molecules can be transported independently of a flow of liquid. This transfer is conceived of as being in contact with the protoplasm of sieve cells and, perhaps, the companion cells, the metabolism of which provides the process with the requisite energy (theory of active transport).

In any event, when studying this problem one should bear in mind that the process of evolution has reached in the xylem a point at which the conducting cells have lost all their living contents completely, while in the phloem this process has achieved only a partial simplification of the protoplasm as exemplified by the loss of the sieve cells' nuclei at an early stage. Still, the fact that the protoplasmic membranes of the sieve cells retain semipermeable properties and have many metabolically active structures compels us to regard the transfer of substances in sieve cells otherwise than in vessels devoid of living matter. We believe this conclusion to hold both in the case of mass flow and in the case of active transfer of molecules. For this reason, in any case of transport, the metabolism of sieve cells and adjacent and related cells must be the subject of a very detailed study. It is precisely from this point of view that we shall attempt to consider, in this survey, some of the problems of the transport of organic substances.

Studies of the transport of organic substances in plants began to be studied broadly only at the end of the 1940's when researchers could use ^{14}C and other tagged atoms to follow the movement and localization of substances. Previously, fundamental investigations into the movement of organic substances had been carried out in a number of countries. Of particular importance were the studies of the English workers Mason, Maskell and Phillis in 1926–36 and Münch's fundamental work published in Germany in 1930.

There are several reasons for the present increasing interest in the problem of transport of substances in plants. One of them is that this process appears to unite and co-ordinate the activities of all plant organs; for this reason, facts about the forces and the mechanism of transfer, the composition of the moving products and the regularities of their distribution open the way to a transition from the physiology of individual functions to the physiology of the whole plant. This transition would signal a new advance of plant physiology to a level at which it could more profoundly evaluate the state of vegetative organisms and could more boldly attempt to control their activities.

A profound understanding of the laws of movement of substances is also of direct importance in agriculture, for it is precisely as a result of the transport of substances that growing tissues are provided with food and there is an accumulation of reserve materials in seeds, fruits, roots, and other parts of plants that are of prime importance in crop yields.

II. Transport of Assimilates from the Mesophyll into the Phloem

The transport of organic substances in the phloem begins in the leaves. It is therefore best to begin a consideration of the problem of the functioning of this conducting system with the delivery to it of assimilates from the mesophyll.

The development of isotope techniques has made it possible to establish that in the very earliest stages of photosynthesis many organic substances are formed in assimilating cells. The substances formed during the very first seconds include not only triose phosphates but phosphoric esters of hexoses, many organic acids, and amino acids (see Bassham and Calvin, 1961). At the same time, it has been demonstrated that the first free (non-phosphorylated) sugar formed in photosynthesis is not hexose, as formerly thought, but sucrose (Calvin and Benson, 1949, Turkina, 1954, Norris et al., 1955, and others) or starch (Vittorio et al., 1954). The reason for this is that hexose phosphates, which appear at an earlier stage in photosynthesis, are already the activated residues of sugars that more readily undergo enzymatic transfer with the formation of a free disaccharide (sucrose) or polysaccharide (starch) than hydrolytic splitting into hexose and inorganic phosphate. Incidentally, the relative rates of these transformations, which determine the evolution of the primary products of photosynthesis, may be different, varying with the species but also undergoing substantial changes depending upon the age of the leaf, conditions of light, mineral nutrition, and other factors. That is why the range of assimilates accumulating in photosynthesizing leaves may be different in different circumstances (Nichiporovich et al., 1957).

The movement of assimilates through the cells of the mesophyll to the conducting tissues of the phloem cannot be regarded as simple diffusion along a falling gradient of concentration. Ordinarily, in the path of this movement are several tens and even hundreds of living cells that form a system of semipermeable barriers and which, by virtue of their metabolism, actively interact with the substances undergoing transportation towards the sieve cells. The studies of Mason and Maskell (1928) and also of Kursanov and Turkina (1954) and of certain others have shown that the yield of assimilates from the mesophyll into the conducting cells of the phloem can occur in a direction *counter* to the concentration gradient and, consequently, takes place as a result of some kind of active processes involving an expenditure of energy.

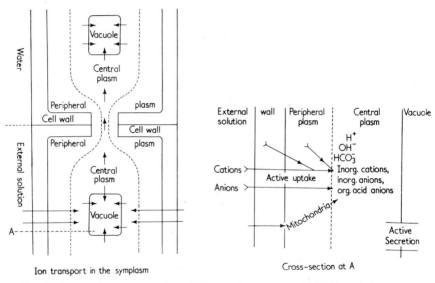

Ion transport in the symplasm   Cross-section at A

FIG. 1. Schematic representation of the uptake processes in *Vallisneria* leaves. Two adjoining cells are interconnected by a plasmodesma. The ions are absorbed in the peripheral plasm by a passive and in the central plasm by an active process. From the central plasm the ions are actively secreted into the vacuoles or translocated through the plasmodesmata to contiguous cells. (After Arisz, 1960.)

Detailed investigations by Arisz (1958, 1960) into the absorption and transport of inorganic ions in the leaves of *Vallisneria spiralis* led to the conclusion that the movement of ions from one parenchyma cell to another is accomplished within the central part of the protoplasm and through plasmadesmata and is therefore not involved in the overcoming of semipermeable barriers (Fig. 1). It may be that in the case of the movement of assimilates through the parenchyma tissue of the leaf the overcoming of protoplasmic barriers is not obligatory either, since organic substances form inside the cytoplasm in photosynthesis and can be transported from cell to cell via the plasmodesmata without having to pass the semipermeable surfaces each time.

Along the path of movement a portion of these substances is actively secreted through the tonoplast into the vacuoles of the parenchymata cells themselves. Another, larger, portion is drawn to the conducting bundles and ultimately concentrated in the sieve cells where, in turn, the incoming substances are distributed between the protoplasm and the vacuolar sap.

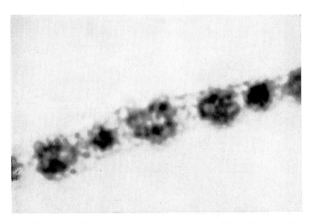

Fig. 2. Autoradiograph of a cross-section of maize leaf showing mainly concentration of ^{14}C labelled substances in the cells. surrounding vascular tissue (10 min after the leaf was fed with $^{14}CO_2$). (After Pristupa, 1962.)

The experiments of Kursanov *et al.* (1959a) showed that in a rhubarb leaf radioactive assimilates may already be detected in the conducting system 2 to 3 min after feeding an adjacent section of the mesophyll with $^{14}CO_2$. In the same laboratory Pristupa found by means of micro-autoradiography that, in maize leaves, assimilates formed in 10 min of photosynthesis are mainly concentrated in green cells, surrounding the bundles (see Fig. 2). If this is a result of movement of assimilates the rate must be much greater than the speed of free diffusion. Moreover, as Canny (1961) has recently shown in experiments with willow cuttings, the rate of influx of assimilates into the conducting system is, broadly speaking, independent of their concentration in the blade. All this compels us to regard the uptake of assimilates by the phloem as an active transfer of molecules by means of some kind of metabolic mechanism. The carrier theory, which at present is being developed mainly with respect to absorption of mineral elements by cells (Epstein, 1960, and others), is probably applicable also to the case of the transport of sugars and other products of photosynthesis. However, the factual material to support such concepts is still greatly inadequate.

A certain analogy to the movement of assimilates towards the sieve cells is found in the process of secretion of nectar, which has been studied in far greater detail; and although these two processes take place in opposite directions as it were, we can find certain elements of a common organization in them. At the present time there is a unanimous opinion that nectar is an exudate from sieve cells, the secretion of which is associated with the metabolism of nectary tissues (see, for example, Frey-Wyssling and Häusermann, 1960). Ziegler (1955) was first to note the respiration intensity peculiar to nectaries, while Shuel (1959) and others demonstrated that the secretion of nectar is reduced by the action of respiratory poisons. It is also known that the composition of the secreted nectar is not identical with that of the phloem sap. For instance, nectar contains much fewer nitrogenous substances and has a relatively higher percentage of sugars. This indicates that the composition of the secreted nectar is controlled by the tissues of the nectary. Lüttge (1961), who made a comparative study of the nectaries of many plants, came to the conclusion that the capacity for selective retention of N-substances and the secretion of sugars increases with the structural complexity of the nectaries so that, for example, in the highly organized nectaries of *Robinia pseudoacacia* the quantity of N-substances is 5000 times less than in the phloem sap. Apparently this selection is accomplished by the cells of the nectaries which are capable of actively reabsorbing the N-substances secreted from the phloem such as glutamic acid (Ziegler and Lüttge, 1959), phosphate (Lüttge, 1961), and probably many other compounds. Nevertheless nectar contains, in addition to sugars, also a diversity of amino acids, amides, ureides (allantoic acid and allantoin in *Acer*, for example), ascorbic acid, many acids of the Krebs cycle, and in certain cases (for instance, in *Musa*) also proteins (Lüttge, 1961). In addition, the cells of the nectaries frequently contain a very active acid phosphatase (Vis, 1958); this points to the phosphoric exchange metabolism of these tissues. Lüttge's data (1961) for *Musa sapientum* indicate that the phosphorus secreted with the nectar is largely associated with organic substances; the same author found that UDP contained in the nectary tissues of *Abutilon striatum* rapidly exchanges its phosphorus for ^{32}P, which indicates that phosphate is very actively utilized in the processes occurring in these tissues. Most interesting in this respect was a study by Matile (1956) which showed that the nectaries of *Euphorbia pulcherrima* contain a considerable quantity of sucrose phosphate. These same nectaries exhibited glucose 6-phosphate and fructose 1-phosphate.

Apparently sucrose phosphate plays an important part in the carbohydrate metabolism of plants, though it is not always detectable due to its intermediate position. Nevertheless, sucrose phosphate has repeatedly been found in small quantities among the early products of photosynthesis, while Buchanan (1953) has even demonstrated that there can be more sucrose phosphate than hexose phosphate in the leaves of sugar beet. The fact of considerable quantities of phosphorylated sugars in the tissues of nectaries led to the widely accepted view that the secretion of sugars, and maybe also their transport through the cells of the nectaries, is associated with their temporary phosphorylation, which involves the expenditure of respiratory energy conveyed via ATP. It is assumed, for one thing, that sucrose itself in the form of its phosphoric ester is transported through the nectary tissue (Lüttge, 1961, and others).

Let us now once again come back to this initial stage in the transport of assimilates, i.e., to the movement of the products of photosynthesis from the green cells into the conducting tissue of the phloem. Using ^{14}C, Kursanov et al. (1959a) showed in the case of the leaves of *Rheum rhaponticum* that labelled assimilates can be detected, several minutes after the onset of photosynthesis, in the fine branches of conducting bundles at a distance of 1 to 2 cm from the zone where $^{14}CO_2$ had been assimilated. Since during such a short time the secondary transformations of substances that have entered the phloem cannot be considerable it appeared possible, by direct analyses, to compare the composition of substances that had entered the phloem with that of the products forming in the assimilating cells.

These experiments showed that, for a brief assimilation of $^{14}CO_2$, the major portion of radioactivity in the mesophyll is found in sucrose (84–90% of the total radioactivity of the sugars). At the same time, the total quantity of sugars in the leaf is characterized by a lower relative quantity of sucrose (50–60%); this may be due either to a secondary inversion of sucrose or to its more rapid utilization in other processes, and above all to transport from the leaf (see, for example, Jones et al., 1959). A comparison of the sugar composition in the nearest conducting bundles of rhubarb has shown that the relative quantity of sucrose in them, both that labelled with ^{14}C and that not labelled, amounts to 84–85%, which is close to the proportion that we have during the first few minutes of photosynthesis in the mesophyll. These data suggest that not all of the sugars participate in equal measure in their outflow from the cells of the mesophyll. Apparently, that part of the sugars that has just formed in photosynthesis, and consequently is in the protoplasm, has the advantage, while the reserve of sugars in the vacuole is probably less amenable to movement and is drawn into the transport process only gradually (see also Nelson, 1962).

Since movement is regulated by cell metabolism, some substances (for instance, sugars) are rapidly conveyed to the phloem and soon appear in the conducting cells, while others remain stationary or, after attaining the conducting tissues, accumulate in the surrounding cells and do not get into the flow path. In this same study (Kursanov et al., 1959a) it was shown that organic acids formed in the rhubarb leaf during photosynthesis exhibit a tendency to accumulate in the vein zone and that only a small quantity of malic and citric acids enter the flow, at least in the case of short exposures, whereas the rest remain on the boundary with the conducting tissues.

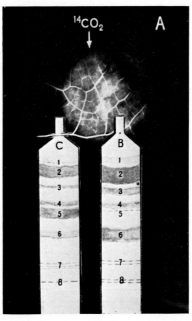

FIG. 3. The selective entry of amino acids from the mesophyll into the conducting bundles of *Rheum rhaponticum*. A. Autoradiograph of a section of the leaf which has photosynthesized $^{14}CO_2$ for 2 min. B, C. Diagram of autoradiochromatograms of amino acids from the mesophyll and from the closest conducting bundles respectively.

1. aspartic acid, 2. serine, 3. glycine, 4. glutamic acid, 5. threonine, 6. alanine, 7. proline, 8, γ-aminobutyric acid.

The selective nature of export of assimilates into the phloem is still more marked in the case of amino acids. In the case of rhubarb, for example, threonine is particularly mobile; it accounts for 24–35% of the total radioactivity of amino acids in the conducting systems, whereas in the assimilating cells the radioactivity of threonine did not exceed 3–4%. Serine and alanine also very actively enter the phloem from the mesophyll, whereas aspartic acid and proline, for example, are appreciably less mobile in this plant (Fig. 3). These differences can change with the age of the plant and due to other factors (Nelson, 1962).

Subsequent experiments showed that the movement of assimilates from the mesophyll to the phloem is accelerated when the leaves are enriched with ATP. Experiments of this kind were conducted by Kursanov and Brovchenko (1961) with sugar beet the leaves of which, after 5 min of photosynthesis in the presence of $^{14}CO_2$, were infiltrated with a 0·006 m. solution of ATP (experiment) or water (control). In

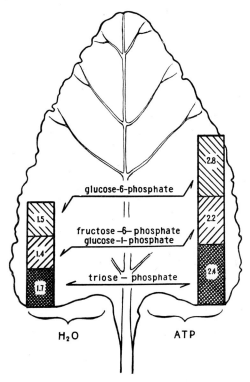

FIG. 4. The accumulation of ^{14}C-phosphoric esters in a sugar beet leaf with infiltration of 0·006 mole ATP (right half): the control is infiltrated with H_2O (left half). Numbers denote radioactivity of P-esters after 5 min of photosynthesis in presence of $^{14}CO_2$ and after 20 min in the dark (in thousands of counts per min/g dry weight).

order to allow the movement of the assimilates to take place and, at the same time, to avoid a photosynthetic formation of ATP in the blades themselves, both plants (after photosynthesis) were placed in the dark for 20 min. It was then demonstrated that enrichment of the leaves with ATP increases the amount of phosphoric esters, particularly hexose phosphates (Fig. 4) in them; in other words, the ATP introduced from without promotes carbohydrate-phosphoric meta-bolism.

At the same time, there is an enhanced outflow of assimilates from these leaves, thus compelling us to seek a connexion between the phosphorylation of sugars and their movement from the mesophyll into the conducting system (Fig. 5). Normally, in first-year sugar beet the

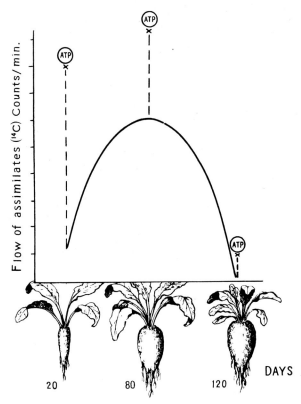

Fig. 5. Outflow of assimilates into the root of sugar beet of various ages. Solid line is export under natural conditions; broken line show the activation of export by means of ATP.

transport of sugars from the leaves is of varying intensity at different periods of vegetative growth. For instance, in 50-day plants in the phase of leaf formation, the transport of assimilates to the roots is several-fold less intense than in 80- to 90-day plants in the phase of intense accumulation of sucrose in the roots. Later (110–120 days) the outflow of assimilates from the leaves again exhibits a sharp reduction, indicating the completion of the process of sugar accumulation. Using ATP it is possilbe to increase the outflow of assimilates during all phases of growth. This effect, however, is more pronounced in young plants where the outflow can be accelerated 7- to 8-fold, and is weaker in mature plants, particularly at the end of vegetation (see Fig. 5).

The impression we get is that in young plants the transport of assimilates from the leaves is more restricted by the content of ATP than in mature plants. At first glance this result appears to be unexpected, since the photosynthesis and particularly the respiration of young leaves—both suppliers of ATP—are very intense. However, one has to bear in mind that the young leaves expend a great deal of ATP in their own metabolism (in the continuing processes of growth, for one thing); this should create a competition between the various processes for the energy-rich phosphate bonds.

The decisive part in the activation of transport of assimilates is apparently played by the third phosphate group in the ATP molecule, that is, by the energy-rich phosphate residue that activates the sugar *via* the hexokinase reaction. This follows from the fact that in these experiments enrichment of leaves with ADP or AMP did not result in a substantial stimulation of transport of assimilates in the case of sugar beet.

The problem of the immediate mechanism of action of ATP on the export of assimilates into the phloem requires further investigation. This effect may prove to be both the result of general stimulation involving also processes of transport, and a more direct result based on the better transportability of phosphorylated products. Kursanov and Brovchenko (1961) found that ATP accelerates transport to the phloem mainly of sugars and, to a lesser degree, of amino acids. On the other hand, organic acids, which are still less mobile in the leaves of sugar beet than in those of rhubarb, are not activated by ATP to emerge into the conducting system.

Thus, one of the prerequisites for transport of sugars in the parenchyma cells or for crossing protoplasm barriers is phosphorylation. It may be thought that the role of phosphate as an activator or carrier ends with the entrance of sugar into the sieve cells, because the amount of hexose phosphates in the phloem and particularly in the phloem exudate is disproportionately low as compared with the quantity of sugars undergoing transportation (Ziegler, 1956). In the experiments of Kursanov and Brovchenko (1961), enrichment of the leaves with ATP led to the accumulation within them of hexosephosphoric esters; however, very small quantities of phosphoric esters permeated into the conducting bundles extracted from these same leaves, despite the fact that there was a perceptible acceleration of the transport of assimilates. It may thus be considered that at the beginning of the transport of assimilates through the cells of the mesophyll the phosphate is carried by hexokinase with ATP to the sugars and is again broken off at some stage in the entry of the sugars into the conducting system. In this process, respiration or photosynthesis should play the role of suppliers of ATP.

Barrier and Loomis (1957) found that ^{32}P-labelled phosphate deposited on soya blades in the form of a solution penetrates into the veins six to seven times as fast if 5% sucrose is added to the phosphate. These findings may indicate that a part of the phosphate participating in activation of sugars during passage through the protoplasma barrier is carried into the conducting cells of the phloem after dephosphorylation, but this occurs separately from the sugars.

Since among the products of photosynthesis there is always a predominance of sucrose in the blade and since the same disaccharide is dominant also in the phloem, it is natural to assume that it is sucrose phosphate which can move about in the protoplasm and overcome protoplasmic barriers. It would appear that this presumption is corroborated also by Buchanan's data (1953) which demonstrated the presence (in the leaves of sugar beet) of noticeable quantities of sucrose phosphate. True, in most cases it has not been possible to detect sucrose phosphate in leaves containing various hexose phosphates. However, this may be due to its rapid dephosphorylation immediately following photosynthetic formation or, if the sucrose phosphate is indeed a highly transportable compound, it may be due to its transport into the phloem, where it likewise loses its phosphate residue. Another possibility is that phosphorylation is only a means for activation of molecules for their transfer to more specialized substances (carriers) which accomplish transport in the cells of the parenchyma (see Fig. 6). There are not sufficient grounds to consider one or the other mechanism as proved.

Kursanov and Brovchenko (1961) enriched sugar-beet leaves with ATP and did not find sucrose phosphate in them. However, they did find that ATP accelerated the uptake of monosaccharides by the phloem more than that of sucrose. These data rather seem to favour the view that the transport of sucrose from the mesophyll into the conducting paths is preceded by its transformation into two molecules of hexose monophosphate, which enter the conducting cells of the phloem directly or by means of special carriers and are rapidly resynthesized there into sucrose. From the work of Pavlinova (1955) we know that the resynthesis of sucrose in the conducting bundles of sugar beet can indeed take place very rapidly (see Fig. 6). That is why the phloem always has an abundance of sucrose. It may be that, in the accelerated outflow of assimilates due to ATP, the rate of uptake of hexose phosphates by the phloem begins to exceed the rate of resynthesis of sucrose and this leads to an accumulation, in the phloem, either of hexose phosphates or already dephosphorylated hexoxes, which was what we observed in these experiments.

True, allowing for such a mechanism of transport of assimilates from the green cells into the phloem, we in this way appear to preclude the possibility of continuous transport of sucrose although such a continuity could be extremely advantageous as regards retaining the chemical-bond energy of this disaccharide in the cells.

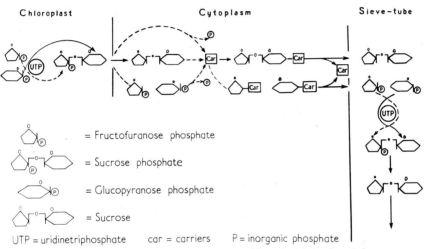

FIG. 6. Transport scheme of sugars from chloroplast into sieve tube.

However, in the system of reactions assumed in this case, all transformations should occur at the level of phosphorylated sugars in possession of an additional reserve of free energy. Such mutual transformations are catalyzed by transferase-type enzyme and proceed in either direction without appreciable alterations in the free-energy reserve in the system. For one thing, the hexose residues of phosphoric esters, too, can participate in many exchange reactions without supplementary activation. The foregoing can be represented by the diagram in Fig. 6.

In animal tissue studies, the problem of the necessity of phosphorylation of sugars for their transport was posed even earlier (Verzar, 1936). However, this problem is still unsolved, for along with facts in favour of the necessity of phosphorylation there are those which indicate that the poisoning of oxidative phosphorylation can even accelerate absorption of monosaccharides by animal membranes (Randle and Smith, 1958 a, b; Battaglia and Randle, 1959). These facts must also be borne in mind when studying the transport of sugars in vegetable tissues, though the character of these processes may be different.

Thus, the only thing that we know more or less definitely in regard to the export of assimilates from the mesophyll into the conducting cells is that this process is selective and requires ATP.

III. Uptake of Assimilates from the Phloem
by Growing and Storing Tissues

Processes very much related to those that take place when assimilates leave the mesophyll and enter the phloem occur when organic substances from the conducting cells of the phloem enter the growing and storing tissues, though in these cases the transport is in the reverse direction.

Bieleski (1960a) has shown that stalks of sugar cane cut into small pieces are capable of absorbing sucrose, glucose, and fructose from an external solution. Under these conditions, the absorption can continue for 2 or 3 days and lead to a rather considerable enrichment of the tissues with sugars. As Bieleski found, this ability is exhibited only by young tissues that have not yet reached the limit of sugar accumulation. Particularly intense sugar absorption is exhibited by the internodal tissue, that is, those sections of the stalks that accumulate most of the sugar under natural conditions.

The process of absorption takes place in two stages. The first lasts about an hour and, judging by the rate of disappearance of the sugars from the external solution during this time, it is the most productive. However, the sugars absorbed during this period are very weakly retained by the tissues so that when the stalks are placed in clean water they readily give up these sugars to the external medium. During this first period, the accumulation of sugars in the tissues takes place in proportion to their concentration in the external solution and is independent of the conditions of aeration of the cells. The initial absorption seems therefore to be determined by such phenomena as diffusion and a weak binding of the sugar on the cell surface and in the outer zone of the protoplasm. This stage is apparently not controlled by metabolism. However, it is important in the general process of biological absorption since it leads to a concentration of sugars in the immediate vicinity of the zone in which the metabolic transport of molecules begins. To summarize, then, by analogy with the concepts developed by Epstein (1960) with regard to absorption of the ions of mineral substances, it may be considered that the first stage of absorption of sugars saturates the so-called outer or free space from which the sugar molecules are subsequently conveyed metabolically into the inner spaces.

The second stage of absorption proceeds more slowly and for a longer time; it requires constant aeration of the tissues and is accompanied by enhanced respiration (by 30–40%). Bieleski (1960b) found also that this second stage of sugar absorption is inhibited by respiratory poisons (KCN, Na-azide, monoiodoacetate), and also by chloramphenicol, an inhibitor of protein synthesis. The conclusion from the foregoing is that at this stage the absorption of sugars is actively metabolic in character. According to Bieleski's data, the rate of this process in the tissues of young sugar-cane stalks gives an uptake of 1–5 mgm/g of fresh weight per day, continuing for up to 60 h in isolated tissues. The sugars absorbed are not washed out of the cells by water and, within a broad range, their absorption is independent of the concentration of the external solution. Here, the accumulation of sugars can occur against the concentration gradient, for example from solutions that are 100 to 200 times less concentrated than the contents of the cells themselves (Bieleski, 1960a).

Different sugars are absorbed by the tissues of sugar cane differently. For instance, according to the observations of Glasziou (1960), glucose is absorbed from an external solution ten times faster than sucrose; yet it is sucrose that is the "inner", i.e. non-leachable, sugar. This suggests that the sugars entering the inner spaces from without are rapidly reduced (via enzymatic transformations) to the proportion that is at the given period peculiar to the cells absorbing them.

In a later study, Glasziou (1961) demonstrated that the ratio of sucrose to fructose in the inner spaces of the cells does not change even if the proportion of one of these sugars in the external solution is increased 700 times. This stability of cells, discovered by Glasziou in sugar-cane stalks, as regards maintenance of a peculiar relationship of sugars was still earlier described for the leaves of many plants that had been infiltrated by the solutions of various sugars (Kursanov, 1941).

According to the data of Glasziou (1960), ^{14}C-sucrose forming in the tissues of sugar cane from ^{14}C-glucose proves to be equally labelled in both monose residues in the inner space, although these cells themselves contain an appreciable quantity of non-radioactive fructose. This result allows the conclusion that the synthesis of sucrose from glucose entering the cell is spatially separated from the sugar supply proper concentrated in the deeper lying layers of the protoplasm and in the vacuoles and that, consequently, it takes place in the early stages of the transport of glucose from the outer spaces into the inner spaces. A synthesis of this kind should be preceded by phosphorylation of glucose and its isomerization into the phosphoric ester of fructose, after which (via UDPG) there can occur an enzymatic synthesis of sucrose (1) or sucrose phosphate (2) that yields sucrose after dephosphorylation:

UDPG + fructose → sucrose + UDP (1)

UDPG + fructose 6-phosphate → sucrose phosphate + UDP (2)

A similar picture was observed in our laboratory by Pavlinova (1955) who enriched the fibro-vascular bundles of the leaf petiole of sugar beet with ^{14}C-fructose. To do this, she put the petioles of cut leaves into a solution of uniformly labelled fructose and by means of the transpiration current brought about a rapid penetration of this sugar into the xylem. From here, as from the free space, fructose was energetically absorbed by the living cells and primarily by the phloem, where it was transformed into sucrose that was equally labelled in the fructose and glucose residues.

Thus, in this instance also the additional hexose necessary for the synthesis of sucrose (glucose, in this case) was not taken from the inner reserve of the cells but was formed from sugar entering from without. And no free ^{14}C-glucose was found; this indicated that conversions of sugars associated with their penetration into the cells do not reach free hexoses but are executed at the level of their phosphoric esters. However, Pavlinova explains this, not as the result of spatial separation of sugar being absorbed and of the sugars of the cells themselves (see Glasziou, 1960, 1961), but as a consequence of phosphorylation of hexose that has crossed the protoplasmic membrane. In this process, the hexose phosphate can, via isomerization, readily convert into the phosphoric ester of the second component, which ester is necessary for the synthesis of sucrose. This mechanism is more advantageous energetically and for this reason proceeds faster than direct phosphorylation of the inert molecules of hexose that makes up the reserve proper of the cell. Despite certain differences in interpretations of this fact, we can already draw the conclusion that the active transport of sugars through the protoplasm is accompanied by their rapid mutual transformations that can occur only in the condition of phosphorylation of the sugars undergoing translocation.

Though the phosphorylation of sugars penetrating through the protoplasm of vegetative cells is at present extremely probable, the relation of this process to the transport, as such, of the molecules has still been inadequately studied. We do not know whether the phosphoric esters of the sugars are in themselves compounds that directly pass through the protoplasmic barrier or whether the role of phosphorylation consists only in activation of molecules that later participate in a more complicated process.

In general form, the uptake of sugars by the cell may be illustrated by the following scheme (after Glasziou, 1960):

$$\text{sugar} + [\text{R} - \text{X}] \to [\text{sugar} - \text{X}] + \text{R} \tag{1}$$

$$[\text{sugar} - \text{X}] + \text{carrier} \to [\text{sugar} - \text{carrier}] + \text{X} \tag{2}$$

$$[\text{sugar} - \text{carrier}] \to \text{sugar} + \text{carrier} \tag{3}$$
$$\text{(outer space)} \qquad \text{(inner space)}$$

Here, $\text{R} - \text{X}$ may be ATP by means of which the sugar is activated, converting into a phosphoric ester (sugar $-$ X). Further, it is assumed that the sugar residue is transferred to another, more specific, acceptor by means of which it penetrates the protoplasm; for this reason, the second acceptor is called the carrier. Finally, when the sugar enters the inner space of the cell, it releases the carrier, as indicated by the Glasziou scheme (3), or maybe is again transferred to the phosphate to form phosphorylated sugars capable of further transformations without supplementary activation. To take one instance, if these are hexose phosphates, they condense into sucrose, given appropriate enzymes.

We are still lacking information about the nature of the carriers, and even their very existence has not been proved. Nevertheless, a series of facts indicate indirectly that there are substances in the protoplasm with which sugars penetrating through the protoplasmic barrier have to combine temporarily. It is believed that the cell has several different carriers, each of which is peculiar to a definite substance or group of substances; this may be the underlying cause of the cell's capacity for selective absorption of substances. To illustrate, according to the data of Bieleski (1960a), the absorption of sucrose by the tissues of sugar-cane stalks suppresses the absorption of glucose. On the other hand, Glasziou (1960 a,b) found that glucose (when present jointly with sucrose) inhibits the absorption of the latter. This leads one to think that in sugar cane both sugars require one and the same carrier, for which there is competition in the case of joint entry into the cell.

A more detailed study of the problem of the competing action of monosaccharides for carriers was made by Battaglia and Randle (1959) in the penetration of sugars through the diaphragm of white rats. Although these results were obtained with animals, they are of great interest also for cases of the penetration of assimilates into the storing cells of plants. In this study sugars were again found to compete for passage into the cells. However, it was also found that not all sugars compete for the same carriers. By this feature they may be divided into three groups. The first group includes, for example, glucose, xylose, and arabinose, which inhibit the absorption of each other, and it may therefore be supposed that they are transported by the same carrier. The second group includes fructose and galactose which mutually compete for their carrier but do not influence the absorption of sugars of the first group. Finally, the third group of sugars, which includes maltose, penetrates into cells non-metabolically.

The different nature of the transport mechanisms of sugars of the first and second groups is evident from the fact that each of them is associated with different aspects of metabolism. For example, absorption of sugars of the first type require SH groups and ceases when they are blocked. Apparently, the transport of such sugars is closely bound up with substances of a protein nature, which is also indicated by Bieleski's experiments (1960b) in which it was demonstrated that the absorption of sucrose and glucose by sugar-cane tissues is suppressed by chloramphenicol that inhibits protein synthesis. Ensgraber (1958) points to the possibility of transporting mono- and disaccharides in conjunction with substances of a protein nature. He extracted from the roots and bark of plants protein substances (which he called Phythämagglutinine) that readily form labile bonds with sugars. On the other hand, the absorption of galactose and fructose by animal membranes is highly influenced by the level of oxidative phosphorylation, a fall in which causes a rise in the uptake of these sugars by the cells (Randle and Smith, 1958 a,b; Battaglia and Randle, 1959). Apparently, ATP is either not at all needed for transport of galactose and fructose or in excess it prevents these sugars from reacting with the carrier and directs them to other transformations.

To summarize, though the general picture of absorption of sugars by living cells is still largely obscure, available data suggest that sugars accumulate in storing and growing tissues with the help of various mechanisms of a metabolic nature.

IV. ABSORBING CAPACITY OF CONDUCTING TISSUES

As already demonstrated above, in a leaf the transport of assimilates towards the phloem involves the means of metabolism of the parenchyma cells, while in the growing and storing tissues transport is accomplished by the active absorption of substances from the phloem. This does not however mean that the conducting system itself remains passive to the accumulation or release of substances.

Turkina (1961) has investigated, in our laboratory, the ability of conducting tissues to absorb sugars. She used the fibro-vascular bundles of sugar beet and *Heracleum Sosnovskyi* after their extraction from the leaf petioles. Experiments show that such bundles were capable of absorbing sucrose, glucose, and certain other substances from an external solution. However, the conducting tissues have a large affinity for sucrose, i.e. the substance that is always predominant in them and which is the chief organic compound transported in plants over long distances.

On the average, this absorption is of the order of 45–65 μmoles of sucrose per g of dry weight or 6·8–9·8 μmoles per g of wet weight of bundles, per h, i.e. from four to seven times in excess of the absorption of sucrose by the parenchyma tissues of leaf petioles, which in the whole plant are close to the vascular-fibrous bundles (Fig. 7).

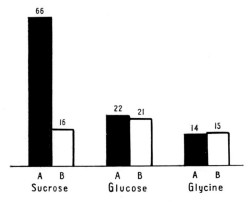

FIG. 7. Abosrption of some substances by the conducting bundles (A), and the parenchyma of the petioles (B) in sugar beet (μ mole/g dry weight/h). (After Turkina, 1961.)

The absorption of sucrose by the conducting tissues is but slightly dependent upon the concentration of the external solution and readily takes place against the gradient. This indicates that the process requires a constant influx of energy. Indeed, in Turkina's experiments suppression of respiration of isolated fibro-vascular bundles of sugar beet by means of 10^{-3} moles KCN retarded the absorption of sucrose by 60%. Absorption was just as strongly inhibited in the case of poisoning of oxidative phosphorylation by means of DNP.

By splitting the bundles extracted from the leaf petioles of *Heracleum Sosnovskyi* into the phloem and xylem, Turkina found that a high sucrose affinity was peculiar not only to the phloem but also, in about the same degree, to the xylem; this suggests that the xylem has extremely active cells (Fig. 8). At the same time, this allows the entire fibro-vascular bundle to be regarded as a system capable of selectively accumulating considerable quantities of sucrose from the external solution, on the condition that energy is expended. The selective nature of this absorption is confirmed by the fact that glucose and certain other substances are not absorbed by the conducting bundles more strongly than by the parenchyma tissues of the petioles (see Fig. 7).

The significance of selective sucrose-affinity of the conducting tissues may lie in the fact that, due to this property, conditions are created in the zone of transition of assimilates from the mesophyll to the phloem for accumulation and preferential penetration (into the conducting cells) precisely of this disaccharide. The high sucrose affinity of conducting tissues is, apparently, also one of the causes that confine the spread (diffusion) of this disaccharide into surrounding tissues during its translocation over long distances.

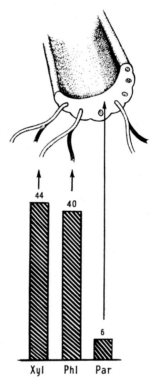

Fig. 8. Absorption of sucrose by various tissues of the petiole of *Heracleum Sosnovskyi* in μmole/g dry weight/h (after Turkina, 1961). Xyl=xylem, Phl=phloem Par=parenchyma.

Not only sugars enter the conducting tissues from the leaves, there are also amino acids that make up 7–10% of the total mass of assimilates entering the stream (Kursanov *et al.*, 1959a). Experiments conducted by Brovchenko (1962) demonstrated that the fibro-vascular bundles do not remain passive with respect to this group of substances either.

Brovchenko was also able to observe active absorption of these compounds by the fibro-vascular bundles extracted from sugar beet and rhubarb leaf blades placed (under good aeration) into solutions of various amino acids (Fig. 9). The process is more active in the first 45 min, during which time from 0·4 to 1·5 μmoles of a given amino acid were absorbed by 0·5 g of fresh bundles. However, as may be seen from Fig. 9, not all amino acids are absorbed equally: the bundles of beet absorb glutamic acid, glycine and serine most strongly, while

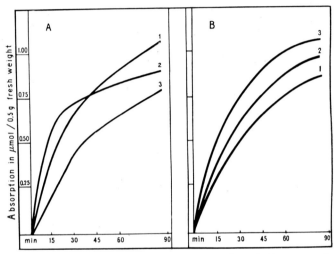

FIG. 9. Absorption of amino acids by conducting bundles (A) from sugar beet leaves, and (B) from the leaves of *Rheum rhaponticum*. 1=glutamic acid, 2=serine, 3=glycine.

alanine accumulation is least active. The other amino acids occupy intermediate positions. Thus, the conducting tissues exhibit selectivity in respect of the absorption of amino acids as well. This selectivity probably determines also the selective character of uptake of amino acids into the phloem. As shown by Kursanov *et al.* (1959a) and also Nelson and Gorham (1959b) and Nelson *et al.* (1961) the composition of amino acids released from the mesophyll and transported to the phloem may differ with the species of the plant and the age.

The concept of cellular carriers is applicable also to cases of the absorption of amino acids. To illustrate, Ratner and Böszörmenyi (1959) showed that the absorption, by the roots of wheat, of one of the amino acids can be suppressed by the addition of others, the mutual inhibiting action being the stronger the more similar are the competing amino acids structurally. All this is in favour of the transport of amino acids through the protoplasm by means of a common carrier. Unfortunately, insufficient study has been made of the competing

relations among amino acids in absorption in conducting tissues. However, Brovchenko (1962) discovered in the conducting bundles of sugar
beet and rhubarb a phenomenon of the mutual displacement of amino
acids, which consists in the following: when some one amino acid is
undergoing intense absorption, the other amino acid comes out of the
tissues into the external solution. The most readily supplanted is
γ-aminobutyric acid, which, as we know, is not a component of proteins
and is in the free state in cells. Significant quantities of it are released
into the external solution particularly during the absorption of glutamic
acid. The appearance of γ-aminobutyric acid in the external solution
cannot be accounted for by the decarboxylation of glutamic acid,

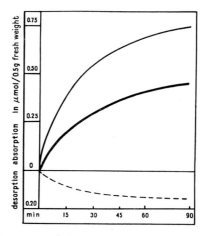

Fig. 10. Enhanced absorption of alanine by the conducting bundles of sugar beet due
to ATP and suppression of absorption due to 2,4-dinitrophenol (DNP). Thick solid line
=alanine; thin solid line=alanine+ATP; broken line=alanine+DNP.

because in a number of cases the quantity of displaced γ-aminobutyric
acid exceeds that of the absorbed glutamic acid. Besides this, γ-amino-
butyric acid is also displaced from the conducting tissues of sugar beet
during the absorption of threonine and, under the influence of glutamic
acid, serine may be released. These results may be interpreted as an
expression of competition for common acceptors, with respect to which
the individual amino acids have varying degrees of affinity.

The absorption of amino acids by the conducting tissues (like the
absorption of sugar) is regulated by metabolism. This follows from the
fact that activation of phosphoric metabolism by means of ATP increases the absorption (by the conducting tissues) of amino acids (Fig.
10).

On the other hand, by suppressing oxidative phosphorylation by means of DNP (10^{-4} mole) it is possible to stop the absorption of amino acids and even cause them to be released from the cells (Fig. 10).

A more detailed study has been made of the accumulation of amino acids by animal tissues, mainly by tumour cells and erythrocytes (see Christensen, 1955). It was shown that their absorption is largely dependent upon the structure of the amino acids; for instance, it increases with the molecular weight and with the presence of the oxy-group in the molecule. It was likewise shown that the absorption of amino acids by the intestine walls involves phosphorylation. The process may be inhibited by DNP and again activated by the addition of ATP (Shishova, 1956). Thus, there seems to be a certain similarity between the absorption of amino acids by vegetable and animal cells.

The foregoing allows the conclusion that the conducting bundles do not remain passive with respect to assimilates that accumulate around them, and can absorb them actively.

V. On the Transport Mechanism of Substances in the Phloem

The problem of the mechanism of transport in the sieve cells of the phloem has been most heatedly debated. Adherents of the mass flow theory accept one or another of Münch's schemes (1930) and assume that organic substances are transported by virtue of the flow of solutions in the inner cavities (vacuoles) of the sieve cells. This flow should take place under the influence of a difference in turgor pressure that is set up at opposite ends of the conducting system. In leaves, the sieve tubes are constantly being supplied with assimilates from photosynthesizing cells and this causes an influx of water to them from the xylem to create a high turgor pressure. At the same time, at the opposite end, for instance in the growth zone or at sites of reserve deposits where the substances undergoing translocation enter the surrounding tissues, the osmotic pressure in the sieve cells, and therefore the turgor pressure, is low. To summarize, then, conditions are postulated in the longitudinal row of sieve cells which would lead to a flow of solutions of organic substances through a system of communicating vacuoles towards falling turgor pressure. In this form, the transport of organic substances in sieve cells should be similar to the transport of substances in the xylem, with one difference, however; the inner wall surface of the conducting cells of the phloem covered with a layer of protoplasm, can ensure the turgor of these cells and control (by virtue of its semi-permeability) the entry and exit of assimilates.

The most difficult aspect of the mass flow theory is the assumption of a continuous vacuolar cavity that would permit a free flow of liquid.

Weatherley *et al.* (1959) found the radius of the willow sieve tube to be 11·5 μ. Taking into account the number of pores in one plate and also the fact that there are about 60 sieve plates per cm each being 5 μ thick, these authors calculated that the additional resistance due to the sieve plates (with open pores) acting against the solution would come to about 1 atm/m of path length. In the opinion of Weatherley and others, this resistance does not exclude the possibility of transporting organic substances in the phloem via the mass flow of solutions. However, these calculations were made on the assumption that the pores are completely open, which is not the case in reality. The sieve plates are, as a rule, covered with a layer of protoplasm and this creates the impression that the central vacuole of each sieve cell is isolated from its neighbours and that its contents can communicate only by diffusion or by means of metabolic processes. Search for channels straight through the sieve plates have not been successful. What is more, it was found that part of the opening of the sieve plate pores may be closed with a polysaccharide callose (β-1,3 glucan) which is readily formed in sieve cells and can completely close the pores (Kessler, 1958). For this reason, even if we assume that the plasmodesmata passing through the pores of sieve plates are hollow tubes with a fine opening the possibility of a through flow of solutions should be so restricted that this transport mechanism could only be of secondary importance.

Nevertheless, the mass flow theory continues to attract the attention of many workers, who strive to prove (either by direct or indirect observations) the fact of a mass flow of solutions from one sieve cell to another. For a long time, sufficiently cogent facts that would compel one to accept or reject the idea of a mass through flow were not forthcoming. It was only just recently (and almost simultaneously) that results were obtained which, on the one hand convincingly corroborate the existence of a mass flow in the phloem, and, on the other hand, equally convincingly exclude the possibility of such movement of solutions through series of sieve cells. This only aggravates the problem of the mechanism of transport of substances in the phloem. However, since these facts and the contradiction itself will have to find a place in any other theory that aspires to a complete picture of the transport of substances in the phloem, we shall dwell on these studies in a little more detail.

In 1953, Mittler and also Kennedy and Mittler (1953) found that after a sieve cell in the bark of *Salix* is pierced by an aphid, a sucrose solution is secreted for a long time from the damaged cell. This is very conveniently observed if one cuts off the proboscis of the insect after it has penetrated into the sieve cell. The emerging solution very much resembles (in both composition and concentration) the phloem exudate obtained from cut sieve cells. In most plants it contains mainly sucrose and a certain quantity of hexoses and amino acids. Mothes and Engel-brecht (1957), who made a detailed study of this phenomenon in *Symphytum officinale*, demonstrated that the outflow of solution from the punctured sieve cell may continue in this plant for several weeks, and that the general concentration of organic substances in the out-flowing solution diminishes only very gradually. This makes highly improbable the idea of a simple "washing out", with water (entering from the xylem), of the contents of the damaged cells. Similar results were obtained by Weatherley *et al.* (1959) in experiments with willow. True, Zimmermann (1960) found that when the bark of *Fraxinus americana* is punctured with a sharp instrument (which to some extent is similar to the bite of an aphid), the phloem exudate diminishes its concentration by 10 to 30% in several minutes. However, on the whole, the quantity of organic substances released from the damaged phloem is so great even in this case as to suggest an influx of nutrients to the damaged cells from a large number of adjacent cells. The more or less rapid decline in the original concentration of the outflowing sap only suggests intimate communication of the phloem with the supply of water of the xylem, and is in no way in contradiction with the con-clusion of the mass influx of organic material into the damaged cell from other cells. From this it may be assumed that the sucrose solution flowing out of the punctured cell is precisely the liquid which, according to the mass flow theory, moves through a system of communicating vacuoles in a long series of sieve cells. Canny's experiments (1961) with willow, in which the leaf above the aphid bite was fed with $^{14}CO_2$, give added support to the conclusion that the liquid released from the punctured sieve cell contains precisely those substances (which are radioactive in Canny's experiments) that are transported from the leaves via the conducting cells of the phloem.

These studies and also a series of other investigations carried out by the Mittler's method are weighty arguments in favour of the mass flow theory. True, there still remains the question of whether such a pro-longed flow of solution from the punctured sieve cell is the result of pathological changes caused by substances contained in the saliva of the insect. However, these experiments show that, at least under certain conditions, the sieve cells can be accessible to the mass flow of liquid. The rate of flow, as computed by Weatherley *et al.* (1959) on the basis of aphid experiments, comes out to about 100 cm/h, which is close to the established rate of movement of organic substances in the phloem.

A still more convincing demonstration of the movement of water in the phloem was made by Ziegler and Vieweg (1961). The underlying idea of these experiments was that the water (if it moves through the conducting cells of the phloem) may be detected as a heat carrier in the local heating of a small section of the phloem. To this end, these workers carefully extracted from a *Heracleum mantegazzianum* petiole a small section of the conducting bundle in such a manner that its ends retained a normal connection with the plant. They then divided the free portion of the bundle into xylem and phloem and illuminated a narrow section of the phloem with a fine pencil of light to raise its temperature.

It was shown that the heat from the heated zone is propagated (in the phloem) faster downwards than upwards. For example, when the illuminated zone was heated $0.9°C$ (this precluded any danger of over-heating) the authors recorded a temperature rise of $0.1-0.2°C$ in 2–3 sec at a distance of $2-2\frac{1}{2}$ cm downwards from the heated section.

At the same time, when the phloem was severed above the heating site the unilateral propagation of heat down the plant ceased. This is easily explained by a cessation of the flow of water across the cut portion. Since such a rate of heat propagation (and in one direction, too) could not occur by simple diffusion, the authors rightly concluded that the conducting cells of the phloem contained a solution that carried the heat in the direction of its movement. Approximate calculations and experiments with models showed that the rate of motion of the heat carrier in the phloem came to 35–70 cm/h, which is rather close to the speed of transport of organic substances measured by means of ^{14}C assimilates (60–120 cm/h).

Thus the mass flow theory has been experimentally confirmed during recent years by facts of fundamental significance. It would appear that the translocation mechanism in the phloem has been solved by these investigations definitively and that it is not worth while searching for other mechanisms of transport. However, in actuality the role of the sieve cells as conductors of organic substances proved considerably more complicated.

Whereas physiological experiments were more and more in favour of the movement of solutions in the sieve cells, microscopic investigations of the structure of the protoplast of sieve cells and particularly the pictures obtained by the electron microscope focused attention on the internal organization of these cells pointing to the possibility of metabolic forces for the transport of substances. Very interesting in this respect are the studies of Kollmann (1960 a, b) of the organization of the protoplast of sieve cells of *Passiflora coerulea*. By means of optical and electron microscopy it was shown that the inner surface of mature functioning sieve cells is covered with a thin layer of protoplasm ($0.1-0.3$ μ) which completely covers also the sieve plates thus blocking the way to continuous communication between vacuoles. The nucleus in sieve cells is known to degrade at a rather early stage in cell development, and this is invoked by the followers of the mass flow theory as proof of the metabolic inadequacy of mature sieve cells. However, Kollmann (1960b) demonstrated that even after the disappearance of the nucleus there remains in such cells a large clear-cut nucleolus and also plastids and mitochondria which are bearers of active metabolic processes. It has also been noted that the nucleolus and mitochondria in sieve cells are situated in the zones adjacent to the sieve plates, which (due to transport of substances) is apparently the site of the most intense metabolic processes. A characteristic peculiarity of the protoplasm and nucleolus of sieve cells is their filamentous and, in some cases, also reticular structure. These filaments are from 70–130 Å in diameter. As the filaments of the protoplasm approach the sieve plates they fold into a dense strand oriented in the elongated direction of the cell and in this form they penetrate the pore (Fig. 11). Similar structure was also described for sieve cells of *Metasequoia glyptostroboides* by Kollmann and Schumacher (1961).

On the whole, this is evidence that the protoplasm is continuous in the lengthwise rows of sieve cells and at the same time makes highly improbable direct communication between the vacuoles of neighbouring cells. True, a careful study of the protoplasmic strands penetrating the pores of the sieve cells permits of the assumption that the filaments have fine canals filled with a colourless substance of a different refraction. However the diameter of these canals does not exceed several tens of Å. For this reason, the resistance to a flow of solution through them would be so great as to make highly improbable direct flow between vacuoles via these canals. The organization of the protoplasts of sieve cells and especially the organization of the filaments that permeate the pores of the sieve plates should probably be regarded as a characteristic feature of the so-called endoplasmic reticulum. On the whole, it resembles a system with a highly developed surface adapted to absorption and conduction of substances in a strictly definite direction. These results likewise compel us to regard the sieve plate zones as sites of the most intense metabolism, for it is precisely here that active transport of substances should attain greatest density and tension. It is not by accident, therefore, that precisely here we find the nucleolus and the greater portion of mitochondria (Kollmann, 1960b). Hepton and Preston (1960) who have made a study of the structure of this zone in the sieve cells of different plants have also come to the conclusion that the protoplasmic strands that penetrate the pores of the sieve plates are metabolically extremely active and, apparently, capable (like mitochondria) of producing large quantities of energy. Spanner (1958) has also pointed to the active role of the sieve plates in the

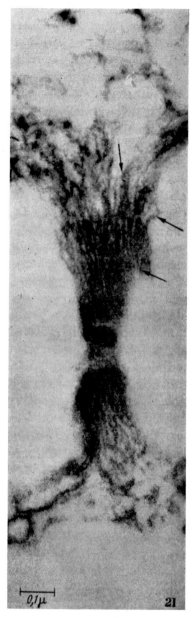

Fig. 11. Electron micrograph of longitudinal section of sieve pore, filled with proto-plasmic strands. In some places, marked by arrows, one can see the tubular structure of the strands or their double-membrane structure. (Passiflora coerulea).

The photograph also shows the transition of protoplasmic strands into a fine network of cytoplasm in contact with the wall. (Magnification 96 000.) (After Kollmann, 1960b.)

transport of organic substances in sieve cells. He proposed an electro-osmotic theory that suggests the appearance of a difference of potential precisely in the zones of the sieve plates. Finally, according to the findings of Esau and Cheadle (1955), a number of plants exhibit a tendency, during the formation of the secondary phloem, to form shorter sieve cells or (and this is the same thing) to an increase in the number of sieve plates per unit path-length.

From the foregoing we may draw the conclusion that the zones of sieve plates which, in terms of mass flow, must be regarded as formations hampering transport may in reality be something in the nature of metabolic "pumps" that accelerate the transport of organic substances in the phloem. (See also Weatherley, 1962.) This makes a further study of the structure and properties of sieve plates and adjacent zones highly desirable.

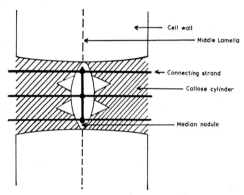

FIG. 12. Schematic structure of a pore in the sieve area of *Vitis* as seen in section (after Hepton and Preston, 1960).

An advance in this direction was made by Hepton and Preston (1960) who obtained a number of good electron micrographs that reveal the fine structure of sieve plates. They demonstrated that in plants of different systematic groups (*Pinus sylvestris, Sorbus aucuparia, Vitis vinefera, Cucurbita pepo*) the pores of sieve plates are structurally similar. In all the plants, these pores are sealed on both sides by callose plugs. However, these plugs do not meet, but leave a free space in the middle that communicates, via canaliculi in the callose plugs, with the cavity of each of the cells (see Fig. 12).

Such canals are filled with substances of a protein and lipoid nature and are an unbroken continuation of the protoplasts. In the opinion of the authors, these strands are not hollow inside so as to be able to permit the direct flow of cell sap. Thus, these data again offer convincing proof that the pores of sieve plates are of an extremely complex structure and (this is particularly important) they are tightly sealed against the direct flow of solutions through them from one vacuole to another. True, cytologic pictures of the fine organization of protoplasm can sometimes lead to erroneous conclusions due to the ready denaturation of labile structures. However, in the works of Kollmann (1960 a,b) and Hepton and Preston (1960), the internal organization of the pores of sieve plates and protoplasmic strands were so peculiar and complex and, what is more, they were so consistently repeated in different plant species that the possibility of an artefact is highly improbable here.

We thus arrive at a rather paradoxical conclusion that the sieve cells are impermeable to a direct flow of vacuolar sap from one conducting cell to another, and yet in a number of cases the flow of solutions in the phloem was demonstrated to a high degree of probability (Mittler, 1953; Ziegler and Vieweg, 1961). Further investigations will help to resolve this contradiction. However, I believe we can already presume that the mass flow theory, just like the theory of metabolic transport of substances in the phloem, each of which originally aspired to an independent resolution of the problem, are in reality only different aspects of a single phenomenon. For example, we can give up the idea of a "pushing" of threads of water through sieve cells by means of a difference in turgor pressure at the ends of the conducting system and we can visualize the movement of water from one conducting cell to another as the result of its metabolic "pumping" through the protoplasmic strands that penetrate the sieve plates.

The action of this mechanism may be visualized in analogy with what is known regarding the swelling of the colloids of mitochondria. Price and Davies (1954) found that the mitochondria taken from rat liver, or from the breast muscle of a pigeon readily swell in pure water, and in the presence of ATP give up their water. Sabato (1959) pointed to the possibility of regulating the swelling of mitochondria by altering the conditions for oxidative phosphorylation in the medium. For instance, the presence of succinate or α-ketoglutarate, as the substrates undergoing oxidation, restricted the swelling of the colloids. Similar was the action of DPNH. Likewise, the experiments of Chappell and Greville (1959) point to a dependence of mitochondrial swelling on processes of oxidation.

It might very well be, therefore, that also in the sieve cells of plants there is a similar "metabolic pump" functioning in the sieve plate zones that makes the plasma colloids periodically swell on one side of the plate and release water on the other side. One should also bear in mind the exceeding ease with which water is exchanged in living cells. This was demonstrated by Vartapetian (1960) and by Vartapetian and Kursanov (1959, 1961) using $H_2^{18}O$ and also by Lebedyev (1960),

Ducet and Vandewalle (1960), Hübner (1960), and Ordin and Gairon (1961) using deuterated water.

However, no matter what the transport mechanism of water in the phloem may be, it cannot be identified with the translocation of organic substances occurring in the same cells. True, the vacuolar sap of sieve cells contains a great deal of sucrose and other organic substances the movement of which in the phloem may be readily proved by means of ^{14}C-labelled assimilates. However, there can scarcely be any doubt that the protoplasm of sieve cells surrounding the vacuole and piercing it with a network of fine filaments (see Kollmann, 1960 a,b) does not remain passive in the distribution of substances between the cell sap and the protoplasm itself.

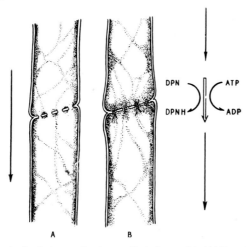

Fig. 13. Transport of substances in sieve cells (schematic). (A) Free flow of liquid from one cell to another via open pores in sieve plates (mass flow). (B) Combined transport when sieve pores filled with protoplasm are impermeable to mass flow. In the zone of the sieve plate, in this case, free flow is replaceable by metabolic transport.

The role of the protoplasm in the sieve plate zones, where the water molecules have to pass through a layer of protoplasm and its connecting strands should be still more active. At these sites the selective character of absorption should be particularly marked and, therefore, the composition of substances passing from one cell into another should be most rigidly controlled precisely at these sites. Thus, the transport of organic substance in the phloem cannot under any conditions be likened to the movement of water in the xylem. One has to bear in mind that throughout its pathway in the phloem the solution of organic substances washes living protoplasm which, due to its own metabolism, exhibits a selective affinity for certain compounds.

In the case of a through-flow of vacuolar sap from cell to cell via the open pores of the sieve plates, as is envisaged by the mass flow theory in its pure form, the role of protoplasm would be still more restricted and would reduce mainly to regulation of the composition of the substances undergoing transport by fixing it or releasing it into the moving solution (Fig. 13A).

However, if the path between the sieve cells is obstructed by protoplasmic bridges (and this at present seems most probable), the role of the living contents of sieve cells as a factor controlling the composition of substances undergoing transport and their rate of movement must be much more marked. In this case, each sieve plate zone must function like a glandular cell, absorbing substances from the vacuolar solution, fixing them with appropriate carriers or particles (Thaine, 1961), transporting these complexes through the connecting strands of the protoplasm via sieve plates and, finally, again releasing them into the vacuolar solution of the next sieve cell (Fig. 13B). The significance of the simultaneous transport of water by the very same system of sieve cell might probably consist in the accelerated transfer of substances from one zone of metabolic transport to another.

Finally, yet another case is possible if the transport of organic substances in sieve cells takes place throughout in contact with the protoplasm and is caused by the metabolic activity of the latter. This type of transport should, essentially, be similar to the earlier considered cases of the transport of substances in parenchyma cells, for instance the export of assimilates from the mesophyll into the conducting tissues. Here, in such continuous metabolic transport, the role of the protoplasm would be greatest. The vacuolar liquid could, like a buffer solution, maintain the necessary equilibrium and concentration of substances in the translocation apparatus of the protoplasm. Recently, with the help of phase-contrast microscopy, Thaine (1961, 1962) in Preston's laboratory, has discovered fine threads, 1–7 μ in diameter on which occur granules (0·5 μ diameter) recalling the mitochondria-like particles and small plastids in the living sieve tube elements of some plants. These granules move at a rate of 3–5 cm/h and easily penetrate from one cell to another. In Thaine's opinion these corpuscles are bearers of organic substances and are involved in translocation. But not a single type of transport in sieve cells has as yet been proved.

Let it be noted that the fibrous structure of the protoplasm of the sieve cells, which is their characteristic peculiarity (Kollmann, 1960a, b), is very reminiscent of the construction of a conductor extended in the direction of the flow of transport.

All the foregoing cases are no more than attempts of the most general kind to map out possible forms of communication between the movement of substances by means of a "water transporting agent" and metabolic transport. The range of possiblities here extends from the free flow of solutions through a series of sieve cells to the continuous metabolic transport of substances in contact with the protoplasm. A compromise between the two extreme viewpoints may be the concept of a "combined" system of transport in the phloem, in which organic substances are carried from one end of a sieve cell to the other by a moving solution and are then forced through the sieve plate zones metabolically. To a certain extent, this view permits correlating the results of investigations carried out by Mittler and Ziegler and Vieweg, who demonstrated the flow of solutions, with the data of Kollmann and Hepton and Preston, who demonstrated that the vacuolar sap could not get through the sieve plates. However, no matter what the solution of the problem of the mechanism of transport in the phloem may turn out to be, metabolism of the conducting cells remains an essential aspect of this process. For this reason, investigations into the metabolic potentials of conducting tissues is of very great interest to the problem of transport at large.

VI. Substances Transported in the Phloem

The phloem, a complex tissue consisting of a variety of cell types, contains a large diversity of compounds, among which the substances undergoing translocation cannot be distinguished from the products of tissue metabolism proper without special techniques. True, Mason and Maskell 1928 a,b) even on the basis of an overall analysis of the phloem of cotton, which proved rich in sucrose, drew the correct conclusion that sucrose is the principal substance transported in this plant from the leaves. This conclusion, however, is always risky and proved true in this case only because up to 90% of the entire mass of assimilates undergoing translocation is sucrose. Obviously, the other substances contained in smaller quantities in the flow cannot be convincingly detected by means of an overall analysis of the phloem.

More definite information about the substances transported in the phloem is obtainable from analyses of the phloem exudate secreted when sieve cells are punctured. This solution certainly permits the determination of only a part of the contents of the sieve cells, because substances associated with the protoplasm do not, in this case, fall within the scope of the technique. But in these instances the investigator actually studies substances from the conducting cells themselves and this makes such data more interesting, especially if we view the transport of substances in sieve cells as a mass flow of solutions.

Isotope techniques open up a much wider field for the study of transported substances. They permit detailed determinations of the composition of compounds undergoing transport and also measurements

10

of the rate and direction of their movements in the plant. For this reason, we shall rely mostly on data obtained by isotope techniques in the analysis of the phloem sap when considering the composition and the rate of movement of substances transported in the phloem.

In 1934, Votchal noted that the conducting bundles of sugar beet can be shown histochemically to contain a considerable quantity of sucrose. Later, Okanenko (1936) isolated the fibro-vascular bundles from the petioles of sugar beet and demonstrated by direct analyses that the conducting tissues contain principally sucrose whereas the surrounding tissues of the parenchyma contain monosaccharides. Nevertheless, it was still not clear whether the sucrose in the conducting bundles of this plant is a transported compound or whether it accumulates here as reserve product. This problem was resolved through experiments with $^{14}CO_2$. Turkina (1954) fed the leaves of sugar beet with $^{14}CO_2$ and a few minutes later extracted the conducting bundles from the petioles. In this way it was demonstrated that the first radioactive substance entering the phloem is sucrose and that precisely this disaccharide is rapidly transported to the root where it can be detected within 15 min. At the same time, Pavlinova (1954) using ^{14}C-labelled hexoses demonstrated that the root tissues of sugar beet are capable of synthesizing sucrose from glucose and fructose only at the very earliest phases of development. Later, for instance during the period of intense sugar accumulation, this synthesis has not been observed. In the leaves and in the conducting bundles the synthesis of sucrose from hexoses continues at a high rate during the entire period of vegetation. It was also shown that sucrose was mainly transported from the leaves and that, consequently, the root itself receives the product it stores in ready made form. These findings shook the earlier prevalent views of Colin (1916) that the movement of assimilates from the leaves into the roots of sugar beet occurs in the form of hexoses.

Recently, the problem of the ability of sugar beet roots to synthesize sucrose was again raised in the work of Dutton et al. (1961). These authors found in mature sugar beet roots a series of nucleotides and enzymes required in the synthesis of sucrose from hexoses. What is more, they accomplished the synthesis itself of sucrose from UDPG and fructose by means of a preparation taken from the roots. Thus, the biochemical apparatus for synthesis of sucrose is retained also in mature roots (Dutton et al.); however, the synthesis itself (when glucose and fructose are introduced into the tissue) proceeds weakly or does not take place at all (Pavlinova). This contradiction may probably be accounted for by the fact that in a rapidly growing root crop activated hexoses are directed to other processes. For instance, the presence of UDPG in the roots makes it probable that during this period the glucose undergoes a rapid transformation into galactose, and this latter oxidizes into galacturonic acid and is used for the synthesis of pectins. It is likewise probable that UDPG is directly utilized in the synthesis of polysaccharides of the cell walls. This is in accord with the observations of Pavlinova (1954) that showed that when ^{14}C-glucose is introduced into the root it is rapidly consumed in some other direction and does not convert into sucrose.

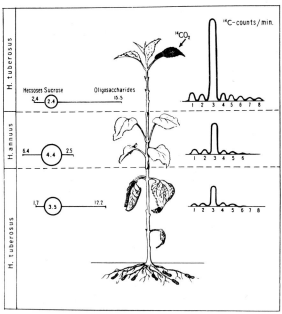

Fig. 14. Diagram of double graft (Jerusalem artichoke—sunflower—Jerusalem artichoke) Left: relative amount of sugars in the bark of grafted plants. Right: relative radioactivity of different sugars in bark (4h after feeding the leaf with $^{14}CO_2$). 1=fructose; 2=glucose; 3=sucrose; 4–8=oligosaccharides in the order of increasing molecular weight.

At the present time, the predominant role of sucrose in the transport of assimilates has been demonstrated by many authors for plants belonging to the most diverse systematic groups and having unlike types of metabolism. This was, for example, demonstrated in the experiments of Kursanov et al. (1958) with grafted plants in which sucrose invariably remained the dominant substance transported in whichever of the grafted components it was formed and through whichever it was transported (see Fig. 14). In addition to sucrose there were certain quantities of monosaccharides that are transported in the phloem but their proportion is small ranging between 2 and 10%. In a number of cases, there were also hexose phosphates in the assimilates transported in the phloem. For example, Biddulph and Cory (1957) found in the conducting tissues of 12-day bean plants, small quantities of fructose 1,6-diphosphate and glucose 6-phosphate.

Kursanov and Brovchenko (1961) in experiments restricting the possibility of a secondary formation of hexose phosphates in the conducting cells themselves, found glucose 6-phosphate and fructose 6-phosphate among other substances entering the phloem of sugar beet. Nevertheless, the quantity of phosphoric esters in the general mixture of sugars undergoing translocation is so small (about 1%) that it is hardly possible to assign to them the role of the principal form of sugar transport along the whole path length.

In one of his studies, Zimmerman (1958b) gives the results of an analysis of the exudates of the sieve cells of sixteen species of trees. In most of them the dominant substance was sucrose. However, in the case of three species (*Fraxinus americana*, *Ulmus americana* and *Tilia americana*) considerable quantities of raffinose, stachyose and a little verbascose in addition to sucrose were found. To illustrate, in *F. americana*, with a total concentration of substances in the exudate equal to 0·5–0·6 mole, stachyose makes up 0·3–0·5 mole, raffinose 0·1 mole and sucrose 0·15–0·2 mole.

Almost at the same time, Pristupa (1959) using the isotope method in our laboratory, found that up to 80% of the assimilates in *Cucurbita pepo* is transported in the form of oligosaccharides of the stachyose type. Peel and Weatherley (1959) also established the presence of raffinose (about 15% of the total quantity of sugars) in the phloem exudate of *Salix viminalis*. It thus appears that in some plants the main portion of the assimilates is transported not in the form of sucrose itself but in the form of oligosaccharides constructed on the basis of sucrose. These oligosaccharides include those in which one, two, or three galactose residues are attached to the glucose moiety of sucrose. Oligosaccharides of this type appear in the stream even in the very first minutes after the onset of photosynthesis, and rapidly move along the conducting cells of the phloem.

According to the findings of Pristupa (1959), the removal of a pumpkin leaf which has been fed with $^{14}CO_2$ leads to a rapid decline of tagged oligosaccharides in the conducting tissues of the stem; this is accounted for by their movement into other parts of the plant. Similar results were obtained by Zimmermann (1958a) who in July removed all the leaves from a specimen of *F. americana* and found a rapid decline of stachyose in the phloem sap; yet the stachyose reappeared when the leaves grew out again. At the same time it was demonstrated that the disappearance of stachyose in *F. americana* is accompanied by the accumulation of sucrose. This enabled Zimmermann (1958 a,b) to express the view that as the stachyose and other similar oligosaccharides move along in the sieve cells they are subjected to the action of the enzyme α-d-galactosidase, as a result of which the galactose residues separate one after the other and are utilized in the metabolism of the surrounding tissues. Due to the small quantity of invertase in the sieve cells, sucrose does not break down in transit but plays the part of a conductor that remains after the separation of the galactose residues in the sieve cells. In our laboratory, Pavlinova (1959) showed that the phloem sap of *Fraxinus* sp. does actually contain α-galactosidase, which, it would appear, is in agreement with Zimmerman's hypothesis.

However, Zimmerman himself calls attention to the fact that neither in the sieve cells nor in the surrounding tissues has it been possible to detect free galactose. It therefore appears to us improbable that the use of the galactose residues of stachyose should begin with its hydrolytic splitting, in the course of which the galactose would undoubtedly accumulate, all the more so since, according to the data of Ziegler (1956), the galactose introduced into the sieve cells of *Robinia pseudoacacia* does not metabolize. It is far more probable that the use of galactose residues begins with their enzymatic transfer to UTP. UDPGal is then formed, which passes into glucose under the action of galactowaldenase (Caputto *et al.*, 1950; Leloir, 1951). The elevated content of uridine phosphates and UDPG in the nucleotides of the conducting tissues of sugar beet as reported recently by Pavlinova and Afanasieva (1962) makes this transformation pathway particularly probable. What is more, Ziegler (1960) isolated UDPG from the phloem of *Heracleum mantegazzianum*.

Nevertheless, the basic conclusion that sucrose is the main transport substance is not invalidated even for plants that transport their assimilates mainly in the form of galactose-containing oligosaccharides. On the contrary, this shows that sucrose is not only by itself the main transport substance but that other substances, such as galactose, can be transported with it.

In addition to sugars that make up as much as 90% of the assimilates that are transported, many other compounds are carried in the phloem. Among them, of great importance are amino acids and amides, the amount of which in different plants and during different periods ranges from 0·2 to 12% and, in some cases (after Nelson, 1962), can even reach 50% of the total quantity of assimilates undergoing translocation (Ziegler, 1956; Zimmermann, 1958b; Kursanov *et al.*, 1959 a,b). The amino acids transported in the phloem are extremely diverse, and usually number about ten to twelve in all. The composition and relative proportion of amino acids varies significantly depending upon the plant species and its general state. For example, according to the data of Kursanov *et al.* (1959a), at the beginning of vegetative growth there is an enhanced export from the leaves of *Rheum rhaponticum*, of serine, alanine, and threonine, and also perceptible quantities of aspartic acid and proline. During the latter half of the summer the export of aspartic acid and proline ceases, but γ-amino-butyric acid, which earlier had not entered the sieve cells, becomes mobile. According to observations by Nelson and Gorham (1959b), young soybean plants exhibit a particularly enhanced translocation of serine, whereas asparagine and glutamine are rather immobile. These relationships change with age, and the amides begin to be transported while the serine remains at its site.

A change in the composition of amino acids in the phloem exudate has likewise been observed by Ziegler (1956) in *Robinia pseudoacacia*. According to these data, during the summer *Robinia pseudoacacia* transports in the phloem mainly glutamic acid, leucine and valine, while during the autumn it is mainly proline and also leucine and valine. It was also demonstrated that in the autumn the concentration of amino acids in the phloem sap is appreciably higher than in summer; this is to be explained by the breakdown of proteins in yellowing leaves and the retreat of amino acids into wintering organs. Zimmermann (1958b) also observed a considerable rise in amino acid concentration in the phloem exudate in autumn in *Fraxinus americana*.

Since amino acids and other of the simplest nitrogenous compounds are transported not only from the leaves through the phloem but also from the roots through the xylem, it would be very important to compare the composition of these substances moving simultaneously downwards and upwards.

Less typical for transport in the phloem are organic acids. As has been shown by Kursanov *et al.* (1959a), in *Rheum rhaponticum* acids are weakly exported from the assimilating cells into the phloem, some of them entering more readily (for instance, malic and citric acid) and others remaining in the leaf. Nevertheless it is always possible to find organic acids among the moving substances in the conducting tissues. A portion of them, together with other assimilates, is exported from the mesophyll, while another portion is formed from sugars as a result of the metabolism of the conducting cells themselves (see Turkina, 1959). However, in nitrogen-deficient plants the organic acid content in mobile assimilates may significantly increase due to concurrent decrease of the amino acid content. In particular, Nelson *et al.* (1961) found that in the soy bean plant a large part of mobile assimilates which usually consist of serine is replaced by malate under conditions of nitrogen-deficiency. In addition to substances of a more universal character, the phloem can carry also compounds of a rather specific nature. For instance, Warren Wilson (1959) showed by means of grafting alkaloid plants on the roots of non-alkaloid plants that, in representatives of Solanaceae, alkaloids can be transported through the phloem in the descending direction. Barrier and Loomis (1957) showed that a solution of 2,4-dichlorophenoxyacetic acid deposited on soybean leaves gradually penetrates into the sieve cells and via them gets into other parts of the plant. Zimmerman (1957) found (in the phloem exudate from *Fraxinus americana*) in addition to sugars also mannite with a concentration ranging from 0·05 to 0·20 mole. Wanner (1953b) and other authors found in the phloem exudate various enzymes which should possibly be regarded as transported compounds. Unfortunately, information on the movement of substances of a specific nature in the phloem is still very insufficient. This aspect of the transport problem merits the most careful study, for it would clearly lead to an elucidation of still finer mechanisms of correlation between the different parts of the plant.

A characteristic peculiarity of phloem sap is its alkaline reaction. For instance, in *Robinia pseudoacacia*, according to the findings of Ziegler (1956), the contents of the sieve cells have a pH of 7·45–8·60. This is due to the fact that the conducting cells of the phloem are very rich in potassium. Peel and Weatherley (1959), for example, found that there is about 2% potassium in the phloem sap of *Salix viminalis*. Another peculiarity of the ash composition of sieve cells is the low percentage of Ca and Na. The transport of sugars in a complex with borate (see Skok, 1957) is highly improbable because there are negligible quantities of boron in the phloem sap. (See also Dyar and Webb, 1961.) The large quantity of potassium in the conducting cells of the phloem merits the greatest attention. Spanner (1958) expressed the view that between the sieve cells and companion cells in the sieve plate zone there is a constant circulation of water which moves K^+ with it. In the author's opinion, this should produce a difference of potential at the opposite sides of the sieve plates, thus accelerating the passage of sugars through the plasmodesma. This view has in no way been proved yet. It would seem more important that potassium is a necessary component of a number of enzymes such as phosphofructokinase, pyruvatekinase, aminopherase and others, upon the action of which in large measure depends the energy exchange of the cells. It is therefore possible that, due to the high potassium content in the protoplasm of sieve cells and maybe accompanying cells, intense enzymatic process may develop that are connected with the transport of sugars and other organic compounds. These processes should be particularly substantial in the sieve plate zones where the substances undergoing translocation pass through the plasmodesmata and the protoplasm. In particular, one may expect an intense phosphorylation of sugars by means of enzymes activated by potassium ions.

VII. Rate of Transport of Substances in the Phloem

The rate of transport of substances in the phloem was measured fairly exactly only in comparatively recent times since isotope techniques became available. In grass and arboreal plants this rate ranges mostly between 40 and 110 cm/h (Table I).

Table I

Rate of Transport of Substances in the Phloem of Certain Plants (Measured by Means of Assimilated ^{14}C)

Plant type	Plants	Rate of transport cm/h	Author and year
Herbaceous	Soybean	84	Vernon and Aronoff, 1952
	Red kidney bean	107	Biddulph and Cory, 1957
	Sugar beet	85–100	Kursanov, Turkina and Dubinina, 1953
	Pumpkin	40–60	Pristupa and Kursanov, 1957
	Potato	20–80	Mokronosov and Bubenschchikova, 1961
Arboreal	Concord Grape (*Vitis labruscana*)	60	Swanson and El-Shishiny, 1958
	Tuberolachnus salignus (willow)	100	Weatherley *et al.*, 1959

This high rate of movement of substances in comparatively narrow and structurally complex sieve cells is a severe test both to the mass flow theory and to the theory of active transport. Indeed, the pushing of solutions at this speed through long series of sieve cells over distances of several metres, sometimes tens of metres, requires a very great moving force which, according to the mass flow theory, results from the difference in turgor pressure at opposite ends of the conducting system. The problem is further complicated by the fact that each metre of pathway through the sieve cells has about 6000 sieve plates permeated by narrow pores (Weatherley *et al.*, 1959). Even if we assume that there is still some kind of opening inside these pores for a continuous flow of vacuolar sap, it is apparently so small as to offer a very considerable resistance on the whole. Apparently, to expect the transport of substances in the phloem at a sufficient speed by means of mass flow, one would have to search for some simplifying factors that would reduce the resistance to the passage of solutions through the sieve plate zones.

As has already been mentioned, such conditions may be visualized if we presume that the flow of solution in the sieve cells gives way to a forcing through the sieve plate zone by means of metabolic transport of substances through connecting strands of the protoplasm from one conducting cell to the next. In this process, as in the case of transport in parenchyma cells, there should be an absorption of all or some substances dissolved in the vacuolar sap, after which the transported substances pass (by means of carriers) through the protoplasm and, after being once again released, enter the next vacuole. Water, due to its greater mobility, probably passes through this barrier with less expenditure of energy; yet even its unidirectional movement through the protoplasm should be mediated by a mechanism similar to a metabolic pump (see above). Under such conditions, substances entrained in the phloem by the water flow must pass through zones of active transport about 6000 times over the course of 1m, and in these zones the composition and relative rate of passage of the individual components may be controlled by the metabolic activities of the protoplasm.

This suggests that the transport of assimilates in the phloem requires not only a difference in turgor pressure at the ends of the conducting system, a pressure sufficient to push through solutions, but also an intense metabolic activity on the part of the conducting cells themselves. Taking this view of the metabolic participation of the conducting cells in the transport of substances we can explain some phenomena that do not fit into the scheme of the mechanical flow of solutions.

This refers, in particular, to the possibility of the simultaneous transport of substances in the phloem at different speeds. Biddulph and Cory (1957) found that when a solution of phosphate (^{32}P), sucrose (^{14}C) and tritiated water (THO) was deposited on the leaves of 12-day-old bean plants, all these substances penetrated into the phloem and moved in it. But the speed of transport differs: for sucrose it is 107 cm/h, for THO and ^{32}P it is about 87 cm/h. Gage and Aronoff (1960) also demonstrated that sugar moves faster than water. In their experiments they deposited ^{14}C-fructose in a solution of tritium water on a cut petiole of a 3-week-old soybean plant and found that the sugar is transported 2 to 3 orders of magnitude faster than THO. In the opinion of the authors these results show that sugars and water move in the phloem largely independently of one another. There are also indications that different sugars are transported in the phloem at different speeds. Swanson and El-Shishiny (1958), who studied the movement of labelled assimilates in *Vitis labruscana*, found that at points along the phloem at increasing distance from a leaf that had received $^{14}CO_2$, the relative content of sucrose in the phloem increases, while that of monosaccharides diminishes. This might be interpreted as the result of a faster movement of sucrose in the phloem. Similar observations have been made by Kursanov *et al.* (1958) in their study of the transport of assimilates in grafted plants. (See also Burley, 1961.)

Some workers have noticed a different speed of transport of amino acids in the phloem. To illustrate, Nelson and Gorham (1959a) give rates of from 370–1370 cm/h in the case of 20 to 24-day-old soybean plants for different amino acids. The slower ones are asparagine, glutamine, glutamic acid, glycine, and urea, while the faster ones are alanine, serine, arginine, and aspartic acid. These relationships can probably change substantially with the age of the plants and under the influence of external agents. Nevertheless, this is an indication that the transport of N-substances in the phloem (at least in some part) is not associated with the mass flow of solutions. A similar conclusion was arrived at by Turner (1960 a,b) who studied the movement of N-substances in *Pelargonium* by means of annular cuts in the bark.

The problem of the rate of transport of assimilates has recently been re-examined. Nelson *et al.* (1958) have noticed that in soybean, in addition to the more dense flow of assimilates moving in the phloem with the usual velocity (100–120 cm/h), a small portion of labelled substances (after leaf feeding with $^{14}CO_2$) can during the first few minutes be moving towards the roots with a velocity 10 to 20 times the normal. Later, this phenomenon was studied in more detail (Nelson *et al.*, 1959) and it was found that the "express" rates can reach 50-70 m/h (!) (Nelson, 1962). The nature of the substances transported in plant tissues at such a speed is still obscure: however, it is known that it is not $^{14}CO_2$, but some kind of organic compounds. It is not quite clear either whether such transport takes place in the phloem or in other tissues. It is, however, obvious that in this case the transport of these substances cannot involve the mass flow of solutions.

Mokronosov and Bubenschchikova (1961) studied the transport of assimilates in potato and also noticed different rates and types of transport of organic substances. In some cases, 15–30 min after leaf feeding with $^{14}CO_2$ the autoradiographs showed a dense flow of radioactive substances along the petiole and stem towards the roots. The rate did not exceed 80 cm/h, i.e., it was in accord with the usual rate of transport of assimilates in the phloem (Fig. 15a).

In the opinion of Mokronosov, this type of transport is characteristic of very young potato plants or of plants put into short-day conditions. At the same time, one can frequently observe a very fast movement of small quantities of radioactive substances both upwards and downwards without appreciable accumulations in the conducting tissues themselves (Fig. 15b). The authors called this type of transport "diffuse". Finally, though more infrequently, we find another type of transport of assimilates which consists in the rapid transfer of a small quantity of substances to some part of the plant, for example, the terminal bud, without retention of radioactivity in the stems through which this transport should occur (Fig. 15c). Mokronosov called this

FIG. 15. Various types of transport in potato (30 min after leaf is fed with $^{14}CO_2$) Autoradiographs: (a) ordinary transport towards roots 20–80 cm/h), (b) diffuse transport and (c) impulse transport (after Mokronosov and Bubenschchikova, 1961).

third type of transport "impulse" transport. The velocity of the "diffuse" and "impulse" transport of substances greatly exceeds the ordinary rate and could not, therefore, be measured by the authors. It is however probable that the rapid transport observed by Nelson et al. in soybean and by Mokronosov and Bubenshchikova in potato is one and the same phenomenon.

VIII. Metabolism of the Conducting Tissues

The active participation of living cells in the transport of substances over large distances may be seen even from a comparison of the rates of simultaneous movement of water and the compounds dissolved in it. Thus, Kursanov and Zaprometov (1949) immersed the shoots of wheat (cut stems) into solutions of amino acids and found that the amino acids are transported to maturing seeds in far larger quantities (sometimes 50-fold and more) than could occur if their solution were passively sucked up with the transpiration stream. At the same time, the amino acids are more or less passively carried into the leaves whence the major portion of the water is directed, because the ratio of evaporated water to arrived amino acids is in general in accord with the concentration of amino acids in the external solution. The findings of Chen (1951), Nelson and Gorham (1959 a,b), Biddulph and Cory (1957), Turner (1960 a,b), Gage and Aronoff (1960), and some others also indicate a certain independence in the rate of transport of substances in the phloem of the transport of water.

Further confirmation of the concept of the metabolic nature of transport in the phloem has been given in experiments that show the

TABLE II

Respiration of Leaves, Fibro-vascular bundles, and Parenchyma of the Petioles of Various Plants

| Plant | Respiration (in μ liter O_2/g of fresh weight/h.) | | | | |
	Fibro-vascular bundles	Leaves	Petioles (without bundles)	Temperature	Authors
Sugar beet	572	416	100	30°	Kursanov and Turkina, 1952
Sugar beet	462	232	95	30°	Tsao Tsung Hsun and Liu Chih-yii, 1957
Plantago major	820	374	228	30°	Kursanov and Turkina, 1952
Plantago major	500	—	146	25°	Willenbrink, 1957
Primula leesiana	309	—	52	25°	Willenbrink, 1957
Heracleum mantegazzianum	230	—	32	25°	Ziegler, 1958
Cucurbita pepo	540 (phloem)	—	60	25°	Duloy and Mercer, 1961

dependence of this process upon the respiration of the conducting tissues. Kirsanov and Turkina (1952) and then Tsao Tsung Hsun and Liu Chih-yii (1957), Willenbrink (1957), Ziegler (1958), Duloy and Mercer (1961) demonstrated that the fibro-vascular bundles extracted from the petioles of various plants possess respiration that exceeds several-fold the intensity of respiration of the parenchyma tissues of the same petioles (Table II). This alone pointed to a high metabolic activity of the conducting tissues, and it was natural to link this up with the functions they perform.

Original fears that the high respiration of the bundles might be the result of damage during extraction from the petioles were removed by the experiments of Canny (1960), who succeeded in measuring the respiration of the phloem in the bundles of grapes without extracting them from the plant. To do this, Canny fed one of the plant leaves with $^{14}CO_2$ and then, allowing time for the labelled assimilate to get into the phloem of the petiole, he enclosed the latter (without detaching it from the plant) in a chamber in order to measure respiration rates. Since in this plant the radioactive assimilates strictly follow the phloem, the entire $^{14}CO_2$ released by the petiole should be attributed to the respiration of the phloem. By appropriate calculations Canny found that the respiration of intact phloem comes out to about 200 μl CO_2/g of fresh weight/h, which is rather close to that obtained, for example, by Ziegler with extracted bundles.

Incidentally, intense respiration is peculiar not only to the phloem. Ziegler (1958), using in his experiments the fibro-vascular bundles of *Heracleum mantegazzianum* and dividing them lengthwise into phloem and xylem, determined the respiration of each of the tissues separately. It was found that the xylem, despite the considerable quantity of cells in it devoid of living contents, respires just as intensely as the phloem. This result is an indication of the high activity of the living cells of the xylem as well, the participation of which in the transport of substances has not yet been studied to any great extent. At the same time, it permits of the conclusion that the conducting bundle on the whole possesses a considerable physiological activity.

It is natural to presume that the energy generated in the process of respiration is in some way utilized in the basic function of these tissues, i.e., in the transport of substances. Indeed, as early as 1929 Curtis noticed that localized cooling of a stem down to 2–5°C makes this section pass organic substances with difficulty; this is probably due to reduced respiration of the conducting cells.

The experiments of Willenbrink (1957) demonstrated that poisoning the respiration of the conducting bundles in *Pelargonium zonale* by DNP (which inhibits oxidative phosphorylation) by HCN (which blocks terminal oxidases) or by arsenite (which inhibits oxidation of succinic acid in the Krebs cycle), likewise stops the transport of assimilates. Similar results were obtained in our laboratory by Dubinina (see Kursanov, 1955) who poisoned (using CO) the terminal stages of respiration ($Fe^{+++} \rightleftarrows Fe^{++}$) of the conducting tissues of sugar beet and in this way stopped the export of ^{14}C of assimilates from the leaf.

Nelson and Gorham (1959a), using their method of depositing a solution of labelled substances on a petiole base, showed that in soybean, poisoning the conducting tissues with HCN stops the transport of arginine. In similar fashion one can stop the transport of sugars and amino acids by killing a small section of the stem with hot steam, as Nelson and Gorham (1957b) and Gage and Aronoff (1960), and others have done.

At the same time, certain observations show that the transport of different substances may be suppressed by one and the same respiratory poison used in different degrees. Willenbrink (1957) found that HCN inhibits more strongly the transport of N-compounds than of P-compounds in *Pelargonium zonale*; Nelson and Gorham (1957b) found that respiratory poisoning of the conducting bundles in 30-day-old soybean plants far more strongly inhibits the transport of sucrose than of hexoses. Later, the same authors (Nelson and Gorham, 1959a) reported that according to their observations HCN does not affect the transport, in soybean, of such amino acids as alanine and asparagine, but completely stops the transport of arginine.

Unfortunately, these data have as yet been obtained in a limited number of experiments and the problem, therefore, requires further study. However, judging by these first results the transport of various organic substances in the phloem could take place by means of different mechanisms, as has been demonstrated by Battaglia and Randle (1959) for the transport of sugars in the rat diaphragm or in the intestinal wall. At the same time, all this corroborates the view that the transport of organic substances in plants is associated with the metabolic activity of the conducting tissue.

As yet we do not know how the energy of respiration of the conducting tissues is transformed into the work performed in the transport of these substances. It may be that here, as in passage through the protoplasmic membrane, one of the necessary stages is the phosphorylation of sugars by means of the ATP produced in respiration. At any rate, the presence, in the conducting cells of the phloem, of glucose 6-phosphate, fructose 6-phosphate, fructose 1,6-diphosphate and certain other phosphorylated compounds (see Kursanov *et al.*, 1959b; Biddulph and Cory, 1957; Kursanov and Brovchenko, 1961, and others) makes possible this pathway of sugars in active transport. According to the data of Pavlinova and Afanasieva (1962), the fibro-vascular bundles of sugar beet contains (in 1000 g of fresh material) 173 μ moles glucose 6-phosphate and 58 μmoles fructose 6-phosphate. This is more than twice the amount of hexose phosphates in the parenchyma cells of the petiole adjacent to the bundles, thus indicating an enhanced hexokinase reaction in the conducting tissues.

However, the quantity of sugar phosphates in the phloem is still too small compared with sucrose to allow the phosphoric esters themselves to be the compounds undergoing direct transport. We may therefore take it that here phosphorylation is only the initial stage of activation of molecules, after which they are conveyed to more specific carriers that conduct them through the sieve plate zone and, maybe, to still greater distances. In this respect it would be of great interest to search, in the phloem, for labile bound sugars (something like protein-hexose complexes) that are formed on the basis of the phythämagglutinins isolated by Ensgraber (1958) from the bark and roots of certain plants.

A more detailed investigation of the respiration of conducting tissues made by Turkina and Dubinina (1954) showed that the respiratory coefficient (CO_2/O_2) in fibro-vascular bundles of sugar beet is close to unity, while in *Plantago* it is 1·2. Similar values were obtained by Willenbrink (1957) for the bundles of *Pelargonium zonale*, *Primula* sp., and certain other plants, and by Ziegler (1958) for the bundles of *Heracleum*. All this indicates that the respiratory process takes place in the conducting bundles mainly at the expense of aerobic oxidation of carbohydrates. To this we can, to a small extent, add alcoholic fermentation, which is mainly peculiar to meristematic tissues (Ramshorn, 1961) and probably to cambium. (See also Ullrich, 1961.)

In our laboratory, a more detailed study was made of respiratory transformations of sugars in the case of fibro-vascular bundles of sugar beet. Turkina (1959) placed bundles isolated from the petioles in a 0·05 mole solution of uniformly ^{14}C-labelled sucrose and found that the sucrose accumulates in the conducting tissues. Part of it is utilized in cellular metabolism so that within 60 min the fibro-vascular bundles exhibit a whole series of labelled organic acids characteristic of the Krebs cycle. Of particular interest is the formation of various keto acids, of which the following were identified: pyruvate, oxypyruvate, α-ketoglutarate, oxaloacetate, and glyoxylate. On their basis, if the conducting tissues have ammonia or an organic source of NH_2 groups, there is just as rapid a formation of amino acids, of which fourteen were identified. Thus, Turkina's experiments show that sucrose, which is the principal substance transported in the phloem, is partially utilized by the conducting tissues themselves. Judging by the data obtained, the main pathway of these transformations (that leads to pyruvate) passes through glycolysis and, following the Krebs cycle, leads to the formation of many acids and amino acids produced on their basis. This entire path of transformations is schematically shown in Fig. 16. From the diagram it will be seen that sucrose moving in sugar beet at a rate of

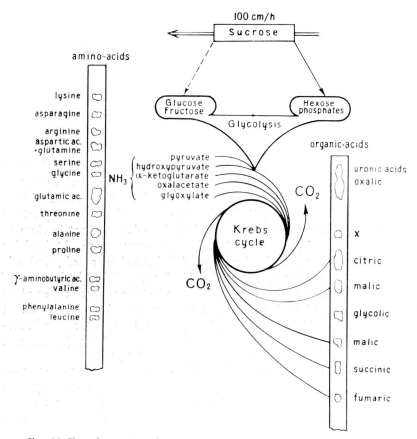

Fig. 16. Transformations of sucrose in the conducting bundles of sugar beet.

about 100 cm/h is partially utilized by the conducting tissues themselves for respiration and secondary transformations. Judging by the data obtained by Canny (1960) for grapes, the phloem in 7 h uses about 5% of the sucrose in the conducting cells for respiration which, taking into account constant transit, amounts to less than 1% of the transported sucrose. Thus, very little is spent on the metabolism of the conducting cells when compared with the amount of transport itself; however, this expenditure is apparently sufficient to provide the conducting cells with energy-rich phospho-organic compounds like adenosine phosphates and uridine phosphates.

Pavlinova and Afanasieva (1962) have made a detailed study of these compounds in the fibro-vascular bundles of sugar beet.

They investigated the nucleotides on a comparative basis: in the fibro-vascular bundles extracted from petioles, in blades and in the parenchyma of petioles. This made it possible to compare the energy level of the conducting bundles with such a metabolically active tissue as the mesophyll, on the one hand, and with the rather inactive parenchyma of petioles, on the other. It was demonstrated that, in relation to the total amount of acid-soluble nucleotides, the conducting bundles are not only not inferior to the blades, but even exceed them somewhat (115 and 108 μmoles per 1000 g of fresh weight, respectively). The petioles are considerably less rich in these compounds (30 μmoles), thus corresponding to their lower metabolic activity. Each of these tissues contains both adenosine phosphates and uridine phosphates, but their ratio is different in different organs, and this reflects the peculiarities of their metabolism (see Table III). The leaves contain more adenosine phosphate, while the conducting tissues are greatly dominated by uridine phosphates. A more detailed analysis of the substances that make up these groups showed that ATP, the reserves of which can readily be replenished in photosynthesis, is pecular to the leaves.

TABLE III

Ratio between adenosine phosphates and uridine phosphates in the various tissues of sugar beet (in μmoles per 1000 g fresh tissue weight, after Pavlinova and Afanasieva, 1962)

	Total nucleotides	Adenosine phosphates (A)	Uridine phosphates (U)	A/U
Leaves	107·7	50·6	54·7	0·93
Conducting tissues	115·3	33·9	77·2	0·44
Parenchyma of petiole	30·3	4·9	24·7	0·20

The conducting tissues contain $2\frac{1}{2}$ times less ATP than the leaves, but they are twice as rich in AMP, which may indicate a rapid utilization, by the conducting tissues, of energy-rich phosphate bonds of ATP and, at the same time, a more difficult replenishment at the expense of respiration (Fig. 17). Of the uridine phosphates, the fibro-vascular bundles exhibit a predominance of UDP and UDPG, that is, compounds closely related to the synthesis of di- and polysaccharides and to the mutual transformations of sugars. In this respect, the conducting tissues are apparently better equipped than the blades. Ziegler (1960) also refers to UDPG in the phloem of the conducting bundles of *Heracleum*.

In the case of the generally low amount of nucleotides, the parenchyma of petioles, like the fibro-vascular bundles, exhibits a predominance of uridine phosphates.

Thus, a comparative analysis of nucleotides that play an important part in the energy exchange of cells permits the conclusion that the conducting bundles are among the most active tissues of the plant and can be placed alongside the blades. The peculiarity of their nucleotide composition also indicates that they are capable of rapid synthesis of sucrose, which is known to proceed with the participation of UDPG

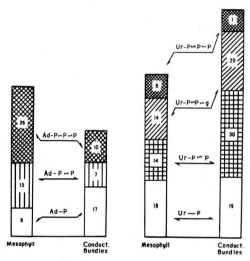

Fig. 17. Relative amount of nucleotides in the mesophyll and conducting bundles of sugar beet (in μmoles/1000 g fresh weight). (After Pavlinova and Afanasieva, 1962).

Ad—P∼P∼P=adenosine triphosphate, Ad—P∼P=adenosine diphosphate, Ad—P =adenosine monophosphate, Ur—P∼P∼P=uridine triphosphate, Ur—P∼P G=uridine diphosphateglucose Ur—P∼P=uridine diphosphate, Ur—P=uridine monophosphate.

and, maybe, of callose, the synthesis of which takes place (according to the data of Kessler, 1958) in the sieve cells at a high rate. This conclusion finds confirmation in the direct experiments of Kursanov and Kazakova (1933), William (1948, 1952) and Pavlinova (1955), who observed the rapid synthesis of sucrose in the conducting paths of sugar beet.

The conclusion concerning the metabolic activity of the conducting tissues is, at first glance, in contradiction with the data of Wanner (1953b) who studied enzymes in the phloem exudate of *Robinia pseudoacacia*. In this study, it was shown that the liquid released from punctured bark is relatively poor in enzymes. It was found to contain phosphoglucomutase and phosphatase, but did not contain such typical enzymes of glycolysis as aldolase or hexokinase. Invertase was also absent. These results led Wanner to the conclusion that sieve cells (from which most of the liquid comes) have a reduced enzymatic apparatus; for example, they have lost the function of glycolysis, which corresponds to a partial degeneration of the protoplasm in mature sieve cells. In the opinion of Wanner, such cells cannot participate metabolically in the transport of assimilates, but must be the passive conductors of a flowing solution. These data were made wide use of by many other workers as a weighty argument against the possibility of active transport in the phloem. It should, however, be borne in mind that the liquid in which Wanner determined the enzymes was actually the vacuolar sap of sieve cells, that is, the solution which, according to the mass flow theory, flows from cell to cell entraining the substances within it. Therefore, the enzyme composition of such a solution could at best suggest the transport of certain enzymes in the phloem, which is interesting in itself but could in no way characterize the enzymatic composition of the protoplasm. For this reason, the results of determinations of enzymes in the tissues themselves may be more significant for an appraisal of the metabolic activity of conducting tissues.

Kuprevich (1949) found in the young bark of a number of trees, amylase, invertase, thirozinase, catalase, and other enzymes. By means of cytochemical techniques it was also shown that peroxidase is present both in the accompanying cells and in the sieve elements themselves. Ullrich (1961) found that in the phloem of *Pelargonium zonale* the oxygen respiration necessary for transport of fluorescein involves peroxides instead of free oxygen. Again, Turkina and Dubinina (1954) demonstrated that there is in the fibro-vascular bundles of sugar beet and peculiar to them an exceptionally active cytochromoxidase, the suppression of which by HCN reduces 85% the uptake of O_2 by the conducting tissues. At the same time, the parenchyma cells of the petioles and other tissues of the same plant do not exhibit a high activity of cytochromoxidase. Ziegler (1958) found cytochromoxidase predominant over other oxidases in the bundles of *Heralceum mantegazzianum*.

Frey (1954) showed that a high activity of phosphatase is peculiar to the phloem and to the adjacent parenchyma cells and, as Bauer (1953) demonstrated, the action of phosphatase is strongest in the accompanying cells and sieve cells. The activity of phosphatases in the zone of the sieve cells is ordinarily associated with the passage of sugars through the protoplasmic barrier during their transport from the photosynthesizing cells into the conducting cells, which has already been mentioned (see also Wanner, 1952, 1953b). However, if this is so, there are no grounds for denying the possibility of subsequent participation of the processes of phosphorylation-dephosphorylation in the transport of sugars in the sieve cells themselves, or at least when they pass through sieve-plate zones.

A characteristic feature of the phloem is the low activity of invertase in it. Wanner (1953b) and Ziegler (1956) did not find this enzyme at all in the phloem exudate of *Robinia pseudoacacia* and other trees. Kursanov and Turkina (1954) found invertase in the conducting bundles of sugar beet, but noted its extremely low activity. Apparently, this peculiarity is very important for the transport of sugars because the absence of active invertase limits the possibility of an hydrolytic splitting of sucrose in the conducting cells. On the contrary, the presence, in the bundles, of uridine phosphates, adenosine phosphates, and hexose phosphates and also the presence of active phosphatases, phosphorylase (Kursanov and Turkina 1954) hexokinase and phosphohexoisomerase (Kursanov *et al.*, 1959b) indicates that the transformations of sucrose in the phloem are accomplished on a higher energy level, corresponding to phosphoric esters.

In the fibro-vascular bundles of sugar beet and in the phloem of *Fraxinus*, Pavlinova also found a rather active α-galactosidase, which is interesting in connection with the fact that galactose-bearing oligosaccharides of the stachyose type are found in the phloem of certain plants as the compounds undergoing translocation (Zimmermann, 1958a; Pristupa, 1959).

Finally, a special study of the enzymes of glycolysis in the fibro-vascular bundles of sugar beet that was undertaken in our laboratory (see Kursanov *et al.*, 1959b) enabled us to find extremely active aldolase and hexokinase and less active phosphohexoisomerase and apirase. In addition, Ziegler (1958) found in the phloem of *Heracleum mantegazzianum* active succinodehydrogenase, which is very important in the transformations of organic acids in the Krebs cycle. To summarize, these data confirm the conclusion of Turkina (1959) concerning the transformation of sugars in the conducting tissues via glycolysis and the cycle of di- and tricarbonic acids (see Fig. 16).

It will be recalled that these processes are related to the formation of energy-rich P-bonds, which can in any event be utilized in the transport of assimilates. From this viewpoint it is interesting to note that, judging by the data of our laboratory (Kursanov *et al.*, 1959b), the activity of hexokinase (the enzyme that carries the energy-rich phosphate residue with ATP to hexose) has a tendency, in sugar-beet bundles, to increase in the middle of August during the period of most active movement of sugars from the leaves to the roots, and diminishes at the end of September, at the end of the period of sugar accumulation.

Gerola and Barbesino (1956), who identified dehydrogenase in the bark of three species of tree, also found that in the periods of most intense movement of organic substances, namely in spring and autumn, there is an increase in the activity of dehydrogenase in the phloem.

The foregoing leaves no doubt that the conducting tissues have a well-developed enzymatic apparatus capable of accomplishing diverse transformations of substances and frequently with great intensity. True, a large part of the research does not as yet refer specially to the sieve cells but to the phloem as a whole or even to the fibro-vascular bundles that consist of cells of different types and designations. Nevertheless, even at this stage, by isolating the finest conducting paths from a plant we realize that they possess metabolism which exceeds the intensity of that of the surrounding tissues. The close interaction between phloem and xylem and also the indissoluble connection between sieve cells and companion cells (see, for instance, Zimmermann, 1960) permit us to regard the conducting bundle as a unified system, the metabolism of which may be not less important for the transport of substances than that of the sieve cells.

Still unresolved is the question of the mechanism of transport itself, i.e., the forces that move the substances contained in the sieve cells. Now that much more has become known about the transport of substances in the phloem we can no longer so rigidly put in opposition the mass flow theory and the concept of active transport of molecules. It is highly probable that further progress in the study of the moving forces of transport will bring these two viewpoints closer together, because the osmotic and metabolic forces here probably comprise two aspects of a single system.

IX. TRANSPORT OF ASSIMILATES IN THE WHOLE PLANT

At the present time, isotope techniques make it easy to observe the direction and rate of movement of organic substances in plants, while chromatographic methods of separation enable us to establish the composition of transported compounds in small samples. Thus, broad vistas open up for a study of the regularities of movement and distribution of assimilates in plants. More detailed research in this direction should elucidate the general picture of the interaction of organs and tissues. The end goal of these investigations is to learn (using appropriate techniques) to control the composition of products undergoing translocation and also the direction and rate of their movement. This would open up possibilities for regulating the processes of growth and storing of substances, which is one of the central tasks of plant raising. Unfortunately, work in this direction is just beginning and so we have as yet very incomplete information. Nevertheless, some of these studies are worth examining, though they are only fragments of a future complete picture.

As far back as 1953 we obtained a number of autoradiographs which showed that in 20-day-old pumpkin plants ^{14}C-labelled assimilates are transported from the leaves mainly in a descending direction (see Kursanov, 1955). This problem was studied in more detail by Pristupa and Kursanov (1957) who found that in 22-day-old pumpkin, 1 to 2 h after leaf feeding with $^{14}CO_2$, the roots exhibit from 18–50% labelled assimilates that have entered the conducting system. In addition, a rather considerable portion of assimilates is in the conducting bundles between the leaf that has received the $^{14}CO_2$ and the roots; which permits us to regard assimilates this part, too, as undergoing transport to the roots. A preferential movement of assimilates rootwards has also been observed by Nelson and Gorham (1957b; 1959b) in young soybean plants, by Jones et al. (1959) in tobacco at the beginning of vegetation; by Turner (1960 a,b) in *Pelargonium* cuttings, and by certain others. All this indicates that in young plants that have not yet begun to flower, the roots are particularly intensely supplied with assimilates. This corresponds to the period of active formation of the root system and maximum display of its absorbing activity.

Later, the transport of assimilates to the roots becomes less intense and more periodical, following the alternation of day and night. To illustrate, according to the data of Pristupa and Kursanov (1957), transport of assimilates to the roots occurs almost exclusively at night in the same pumpkin plants but at the age of 50 days when the stems are already elongated and the plants have begun to flower. During the day-time, on the contrary, the assimilates either do not enter the roots at all or in very small quantities (4 to 8% of the total entering the conducting system). At the same time, during the day the products of photosynthesis can be transported to the growing zones of the shoots and to the flowers. Besides, a substantial portion of them is accumulated in the stem in a section 15–20 cm long below the leaf that has received $^{14}CO_2$. The impression one gets is that at specific hours of the day, one of the nodes of the stalk of the phloem becomes almost impassable for the assimilates, while at night this restriction is removed and the products of photosynthesis that have accumulated in the stems reach the roots readily. Nelson and Gorham (1957a) have also pointed to the dependence of movement of assimilates upon the alternation of day and night. In their experiments it was shown that after 15 min of photosynthesis in the presence of $^{14}CO_2$, 30-day-old soybean plants (remaining in the light) transport in 3 h about 2% of their total radioactivity to the top of a shoot and 4·4% to the roots. In the dark, the shoots receive only 0·5% of the assimilates, while the roots get 16·5%.

Still more marked is the effect of light on the transport of labelled sugars when their solutions are deposited on the blades. The same authors showed that in this case only 1% of the radioactivity leaves the leaves in 14 h of illumination whereas at night 40% of the deposited sugars move in the direction of the roots.

The immediate causes of these differences are as yet unknown. However, judging by the available data the delay in transport of assimilates may be due to a number of reasons. One may be the formation, at definite sections of the stem, of callose plugs that are readily synthesized and again hydrolyzed in sieve cells (Kessler, 1958; Zimmermann, 1961). On the other hand, one may expect also a more active role of the protoplasm itself of the conducting tissues. For one thing, the experiments of Nelson and Gorham (1957b) demonstrated that by reducing the respiration (by means of poisons) of the portion of the stem in which sugars have accumulated, it is possible to bypass the "traffic congestion" of substances moving towards the roots. It might very likely be that the periodical halts in transport have to do with absorption phenomena, as a result of which the mobile components are stopped in definite zones of the conducting tissues or of the parenchyma around them, and then again acquire mobility. Finally, the nature of leaf activity must also affect the periodicity of transport of assimilates; this is evident at least from the fact that variations in volume of transport correspond to periods of light and dark. However, it is scarcely possible to relate directly the transport in the phloem to the quantity of assimilates in the leaves, since the yield of the products of photosynthesis from the mesophyll, as a metabolic process is, in a broad range, independent of their concentration in assimilating cells. This is also indicated by the fact that, in the cases mentioned, the movement of assimilates is enhanced in the dark and not in the light. For this reason, it is more probable that the regulating role of the leaf in the transport of substances in the phloem is determined primarily by that aspect of metabolism which causes the movement of substances from photosynthesizing cells to the conducting cells.

As the plants come to exhibit a predominance of reproductive processes, the original transport scheme of assimilates directed rootwards from the leaves (Fig. 18, 1) changes more and more, acquiring a zonal character. To illustrate, as Belikov (1955 a,b) showed in the case of soybean, with increasing number of leaves and elongation of the stem, the distribution of various strata of definite nutrition zones between the leaves becomes more marked. In general this finds expression in the fact that several strata of the lower leaves continue to send their assimilates to the root system, the leaves of the middle strata mainly feed the flowers closest to them and the fruits (in soybean the fruits receive the products of photosynthesis only from the leaves of their nodes (Belikov, 1955 a,b)), and the upper leaves—if they are mature enough—supply their assimilates to the growing shoots and young leaves, which even after the onset of their own photosynthetic activity still continue for a long time to obtain assimilates from mature leaves (Fig. 18, 2). Finally, during the final stage of vegetation, when the maturing fruits begin to accumulate more and more organic substances, the leaf zones that send their assimilates to each group of fruits become more and more isolated; this frequently leads to cessation of transport of nutrients to the roots and young shoots, which during this period reduce their activities (Fig. 18, 3).

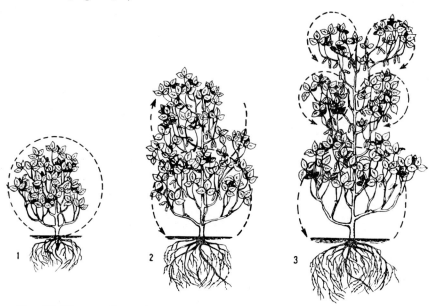

FIG. 18. Diagram illustrating the direction of movement of assimilates in soybean during ontogenesis. 1 shows the development phase of leaves, 2 the elongation phase of stem, and 3 the fruit-bearing phase (after Belikov, 1955 a,b).

In cereals, this period is characterized by mass movement of assimilates from all or the greater part of the functioning leaves in the ascending direction to the maturing seeds. The assimilates of the green parts of the stem and of the ear itself take part in this movement. To illustrate, according to the findings of Watson *et al.* (1958), in some varieties of barley up to 26% of all organic substances stored in the seeds are the products of photosynthesis of the ear itself. Similar results have been obtained for wheat by Asada *et al.* (1960) and for rice Yin Hung-chang *et al.* (1958). There are also indications that during the maturing period of the rye seeds not only the assimilates of the leaves are mobilized, but also the carbohydrate components of the cell walls of the stem (Paleyev, 1938). However, Wanner and Bachofen (1961) have found that flow of assimilates from illuminated pods of bean plants occurs when the rest of the plant is darkened.

It is sometimes thought that the turn in the transport of organic substances is determined by a sharp fall in the concentration gradient. However, in reality this phenomenon is of a more complex nature and is determined not so much by the "molecular vacuum" in the growing zone as by the active function of the conducting tissues that direct organic substances into such zones. Indeed, as far back as 1936 Okanenko demonstrated that the transport of sugars from the leaves of sugar beet is sometimes so intense that the parenchyma tissues of the root do not have time to absorb, from the sieve cells, all the sugar coming to them. During such periods, according to Okanenko's observations, a part of the sugars enter the parenchyma of the petioles and form a temporary store without reaching the root. Sveshnikova (1957) studied the cytological pattern of oil formation in maturing mustard seeds (*Sinapis alba*) and also noted a temporary storage of excess sugars being transported to the fruits in the tissues of the fruit stem in the form of a special half-soluble starch. Information on excess sucrose transported in rice to maturing seeds has also been reported by Tanaka (1961). He found that in this plant the upper part of the stem can temporarily store up to 10% of all the assimilates in transit for seed formation.

The foregoing shows that the export of organic substances to growing and storing tissues is not determined by direct consumption of these substances, but is accomplished by means of forces lying outside the consuming tissues, and oriented towards the stimulus spreading from the growing zone.

At the present time there is a widely current opinion that this stimulus may be heteroauxin (indole-3-acetic acid) or other growth stimulants both of biological and artificial origin. An examination of the extensive literature on this problem is beyond the scope of our review. We shall only say that the very fact of enhanced growth of tissues on sections treated with heteroauxin or the rapid increase in the size of fruits treated with 2,4-dichlorophenoxyacetic acid are, in themselves, proof that the transport of assimilates in plants may be oriented towards substances that stimulate growth.

Iakushkina *et al.* (1956) showed that in wheat seeds in the phase of milky ripeness, the quantity of its own substances of the auxin type greatly increases from stem to ear, which corresponds to the direction of transport of organic substances during this period. According to the data of the same authors, the movement of sugars towards the seeds may be stopped or the sugars may even be caused to move in the opposite direction if the opposite end of the stem is enriched with heteroauxin.

In another study, Iakushkina (1962) demonstrated that when trusses of tomatoes are treated with 2,4-dichlorophenoxyacetic acid there is not only an enhanced influx of substances from the adjacent leaves to the young fruits, but there also begins a transport of assimilates from the more distant leaves, which do not normally supply these fruits.

Povolotskaya et al. (1962) using ^{14}C glucose found that enrichment of the conducting tissues themselves with heteroauxin (by sucking in the solution through a cut petiole) leads to an accumulation of additional sugars in the conducting cells of rhubarb. Heteroauxin stimulates accumulation, in the tissues, not only of sugars but also N-substances. For one thing, Reinhold and Powell (1956) demonstrated that the stem cuttings of 6-day-old sunflower plants increase (due to hetero-auxin) the absorption of glutamic acid and glycine by 43 to 72%. At the same time there is an enhanced inflow of water into the tissues excited by the heteroauxin.

The mechanism of hormonal control of the physiological functions of plants is still rather obscure, but it may be expected that progress in the general development of this problem will help to find a specific answer also to the problem of how growth substances affect the transport of assimilates. However, it is now already probable that the system of such regulators is more complex than the action of a heteroauxin alone and that its individual components function on the basis of different principles. In particular, this follows from a comparison of the action of indoleacetic acid and gibberellin (see, for instance, Wareing, 1958). A specific effect on transport is also produced by kinetin (Mothes et al., 1959) and kinetin-like substances discovered in the bleeding sap of plants by Kulayeva (1962) that mainly activate the transport of N-substances in the parenchyma tissues of the leaf.

The experiments of De Stigter (1959) demonstrated the complex pattern of control of the transport of assimilates on the part of various parts of the whole plant. These experiments showed that grafts of Cucumis melo on Cucurbita ficifolia succeed only if the seedling stock retains its leaves (Fig. 19, 1). Experiments with radioactive $^{14}CO_2$ showed that the reason for this is that the assimilates of the graft (Cucumis melo) are not transported to the stock (Fig. 19, 2) as a result of which the leaves proper of Cucurbita ficifolia are necessary for nutrition of the roots. However, it is sufficient to graft even one leaf of the stock to the crown of the scion (as in shown in Fig. 19, 3) for the unhindered transport of assimilates to the roots from the grafted leaf and from the leaves proper of the scion. It may be that in this case we have to do with a hormonal or enzymatic effect (De Stigter, 1961), or an electrophysiological phenomenon.

Electro-physiological methods open up new horizons for studying the role of the conducting parts of a whole plant. They enable one to study its functional state without damaging the conducting system. Sinjuhin (1962), for example, demonstrated that when pumpkin roots come in contact with KCl they exhibit a rhythmical series of electric pulses that are propagated along the conducting bundles at the rate of 2000 to 3000 cm/h at intervals of 3–5 min. This velocity is 10–20 times that of the transport of the potassium itself and approaches the super-fast transport of substances described by Nelson *et al.* (1959) and

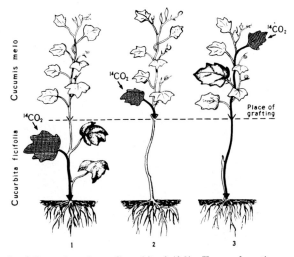

FIG. 19. Graft of *Cucumis melo* on *Cucurbita ficifolia*. For explanation, see text.

Mokronosov and Bubenshchikova (1961). It is also interesting to note that, according to Sinjuhin's data, when the bio-current pulse reaches the leaf it raises the photosynthesis in the latter within one minute. A similar type of bio-current, though with more infrequent pulses, is produced when KCl is deposited on the top of a shoot. Here, in all cases, poisoning of the conducting bundles with DNP or other respiratory poisons blocks the bundles to all such pulses. Thus, judging by these findings, the fibro-vascular bundles are not only conductors of sub-stances but can also serve for the propagation of bio-currents that arise in some part of the plant. It may be that the experiments of De Stigter (1959) will find an interesting continuation in precisely this direction, and at the same time problems of the mechanism of long-distance transport will be elucidated anew.

It has long since been noted that assimilates transported from one leaf almost never get into the other leaves. This is readily evident from autoradiographs obtained from plants in which one of the leaves has been fed with $^{14}CO_2$ (Fig. 20). This problem was studied in more detail by Belikov (1955 a,b), using soybean. According to his data, the assimilates transported in the phloem do not enter leaves even if these

FIG. 20. Autoradiograph of young pumpkin plant showing that assimilates weakly penetrate from one leaf into another.

are incapable of photosynthesis (see also Thaine *et al.*, 1959). This lack of mutual aid among mature leaves seems to be very important biologically, because it excludes the possibility of a plant retaining leaves ("hangers-on") which are at a disadvantage with regard to light. This severe law underlies the formation of foliage by the plant and at the same time of the crown itself. Incidentally, this does not include young leaves that are supplied by assimilates from lower-lying adult leaves until the former complete their growth. Jones *et al.* (1959) made a detailed study of the exchange of assimilates between the leaves of tobacco and found that the products of photosynthesis continued to be imported into young leaves even after they themselves have acquired the ability to photosynthesize. Thus, during the first stage of development the young leaves have mainly a heterotrophic type of nutrition and receive ready-made organic substances from adult leaves. Later, when their photosynthetic apparatus is already functioning they switch to a mixed type of nutrition since they begin photosynthesizing themselves but continue to receive ready-made assimilates from the mature leaves. It is interesting to note that during this period, according to Jones *et al.* (1959), the young leaves receive and release assimilates at one and the same time. With increasing age this ratio more and more is shifted towards the export of its own products of photosynthesis and, finally, the leaf completely ceases to receive ready-made assimilates. Apparently, at this age the young leaf is particularly active in the supported by the findings of Shiroya *et al.* (1961), according to which about 22% of ^{14}C assimilates is exported from a tobacco leaf in 5 h, whereas only 10 to 11% is exported from a more mature leaf. This difference is in good agreement with the high level of phosphorylation of sugars observed in young leaves. For example, in one of the experiments 28% radioactivity was found in sugar phosphates and 25% in sucrose after photosynthesis (in a $^{14}CO_2$ atmosphere) of young leaves that had completed their growth. In older leaves, only 13% of the assimilates was phosphorylated and 70% went to make up the sucrose. Shiroya *et al.* are inclined to regard these differences in relation to the necessity of phosphorylation of sugars for their exit from photosynthesizing cells into the phloem.

Incidentally, the original view on the total absence of exchange of assimilates between mature leaves now requires certain corrections. In particular, in the work of Jones *et al.* (1959) it was shown that, in tobacco, ^{14}C assimilates moving in a descending direction do not get into the leaves situated downwards along the stem, but partially get into the leaves above even if they are mature. Quantitatively, this export to other leaves is small, being from 0·5 – 3% of the mobile assimilates, and for this reason cannot be regarded as an essential source of supply of sugars to mature leaves. However, from the viewpoint of establishing a relationship between the separate leaves of a plant—this was demonstrated, for example, by the experiments of De Stigter (1959)—exchange of small quantities of substances between the leaves (an exchange which may be of a more specific nature) is probably very important.

Usually, the assimilates arriving from another leaf remain, in the mature leaf, in the conducting bundles or concentrate in the tissues adjacent to them. For this reason, autoradiographs exhibit this penetration mainly in the form of light sections (see Fig. 20). Jones *et al.* (1959) believe that the distribution of organic substances in these leaves is accomplished not in the phloem but in the vessels of the xylem, where a portion of the assimilates penetrate from the phloem and is moved by the transpiration stream into leaves situated higher up the stem. According to the findings of Shiroya et al. (1961), in tobacco, alien assimilates get into mature leaves mainly when these are situated close to the flowers; radioactivity is then found in the leaves above and below the leaf that has received $^{14}CO_2$. Other workers too have noted the penetration of small quantities of assimilates into mature leaves. (See also Wanner and Bachofen, 1961.)

The distribution of assimilates is hampered not only between different leaves on a plant but also between individual sections of one and the same mature leaf. This may be illustrated simply by darkening a part of the blade; the accumulation of assimilates is strictly restricted to the illuminated section. The transport of assimilates in any portion of the blade is so oriented towards the closest conducting cells of the phloem that this greatly restricts the possibility of transport of the products of photosynthesis along the leaf parenchyma in other directions. Apparently, the assimilates can remain in the leaf for a longer time only in the form of insoluble polymers like starch or, by penetrating through the tonoplast, they can accumulate in the vacuolar sap, whence the sugars must be subjected to appropriate activation in order to be transported. The experiments of Nelson (1962) with Pinus resinosa confirm this viewpoint. Thus, the ramifications of phloem bundles permeating the entire blade in the form of a dense network may be regarded as a collector into which the entire mobile portion of assimilates moves rapidly without having time to spread appreciably into the main tissue of the mesophyll. The experiments of Pristupa (1962) support this view. He demonstrated that in just a few minutes after completion of photosynthesis in $^{14}CO_2$, a large part of the labelled assimilates are attracted to the conducting bundles (see Fig. 2).

Somewhat apart is the problem of the transport of the products of photosynthesis from one part of the leaf to another in poecilophyllous plants which, besides green portions, also have colourless sections not capable of photosynthesis. Nevertheless, in these plants, too, the direct transport of assimilates from the green portion to adjacent colourless portions encounters serious obstacles. It may even be that the main supply of organic substances to such sections does not take place directly from the adjoining green zone but is accomplished by means of a cycle of matter moving in the phloem to the roots and rising again through the xylem to the above-ground organs. This is supported by the experiments of Mothes and Engelbrecht (1956), in which it is convincingly demonstrated that the albino sections on severed Pelargonium zonale leaves die quickly even when they are kept in conditions that ensure photosynthesis of adjacent green sections. At the same time, if measures are taken to form roots in such isolated leaves, the albino sections (and the green sections as well) remain alive indefinitely. It is possible, however, that the organic substances needed to maintain the albino cells do not always need to accomplish the entire cycle via the root system. In certain cases they can probably return to the blade without having reached the roots because part of the assimilates emerges into the xylem and is moved by the transpiration stream back into the leaves.

Investigations carried out by Belikov (1962) with the aid of $^{14}CO_2$ also showed that in poecilophyllous plants (*Pelargonium zonale*) there is an added possibility of direct transfer of assimilates within the confines of the leaf, although this process takes place slowly (from 9–24 h). The experiments of Mothes *et al.* (1959) and also Kulayeva (1962) showed, in addition, that the transport of amino acids from one part of a blade to another may be accelerated by kinetin, which is capable of causing the transport of nitrogenous compounds into those parts of the leaf on which it is deposited. But in these cases, too, the process is measured in hours and, in some plants, in days.

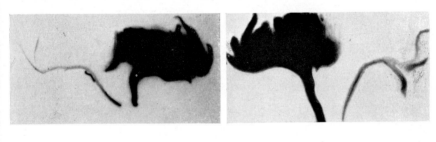

a   b

FIG. 21. Autoradiograph of capitulum of *Helianthus annuus*. (a) Plant received $^{14}CO_2$ via leaf on left side of stem. (b) Same on right side (after Prokofiev *et al.*, 1957).

The transport of substances in the phloem is characterized by a strict orientation in the longitudinal direction. As has already been mentioned, additional energy is required to activate substances so that the compounds undergoing transportation can leave the main stream and, moving in a crosswise direction, enter adjoining tissues such as the cambium, the storing parenchyma or the growth zones. The possibility of distribution of mobile substances between parallel rows of phloem that make up the separate fibro-vascular bundles or form a solid ring of bast is probably rather restricted. The result is that the assimilates from each leaf and, maybe, from its different sections follow a very definite bundle or strip of bast, without mixing with the contents of the parallel rows of such cells, even these latter suffer a lack of mobile compounds. As a result, each leaf on the stem, or group of leaves situated on one side of the stem, supplies its assimilates only to that portion of tissues that is directly connected with the longitudinal rows of conducting cells of the phloem of this side. Using the autoradiographic method, Prokofiev and Sobolyev (1957) and Prokofiev et al. (1957) vividly demonstrated this in the case of sunflower, mustard, and certain other plants. In their experiments, the movement of labelled substances always takes place only along the side of the stem subtending the leaf which received the radioactive substances. In some cases this was observed simply by wetting the blade surface with solutions of phosphate (^{32}P) or ^{14}C-acetate, in others by means of the photosynthesis of $^{14}CO_2$. The localization of labelled compounds was retained not only in transit but also in their distribution in the inflorescence or in the seeds (see Fig. 21).

Similar results were obtained by Jones et al. (1959) for tobacco, Asada et al. (1960) for rice and by others. True, the tendency of assimilates to spread, while moving crosswise in the phloem, can still be noticed with the aid of labelled atoms (see, for example, Zimmermann, 1960); however, as a rule this spread is very small and cannot supply with nutrients those parts of the plant which are connected with the reverse side of the stem. As a result, when leaves on one side are removed, the plant gradually becomes asymmetric, which is well known in practice and was recently demonstrated again by Prokofiev et al. (1957) in the case of the capitulum of sunflower, and by Caldwell (1961) in the development of stems in Coleus.

We thus arrive at the conclusion that the direction of transport of substances in the phloem is determined by the orientation of the longitudinal axis of the sieve tubes, the sieve plates of which, though closed by protoplasm, are readily amenable to such movement. At the same time, the composition of products undergoing transport, the rate of movement of each of the compounds and the direction of transport in the direction of a given part of the plant are determined and controlled by the metabolism of the entire plant and of its individual tissues.

REFERENCES

Arisz, W. H. (1958). Acta neerl. bot. 7, 1–32.
Arisz, W. H. (1960). Protoplasma 52, 310–343.
Asada, K., Konishi, Sh., Kawachima, J. and Kasai, Z. (1960). Mem. Res. Inst. Food Sci. Kyoto University 22, I–II.
Barrier, G. and Loomis, W. (1957). Plant Physiol. 32, 225–231.
Bassham, J. A. and Calvin, M. (1961). Preprint of a paper to be read at the Vth Intern. Congr. Biochem. Moscow, August, 1961. Symp. VI, Preprint 48.
Battaglia, F. and Randle, P. J. (1959). Nature, Lond. 184, 1713–1714.
Bauer, L. (1953). Planta 42, 367–451.

Belikov, I. F. (1955a). *Fiziol. Rastenii*, 2 354–357.
Belikov, I. F. (1955b). *C.R. Acad. Sci., U.R.S.S.* 102, 379–381.
Belikov, I. F. (1962). *Fiziol. Rastenii*, in press.
Biddulph, O. and Cory, R. (1957). *Plant Physiol.* 32, 608–619.
Bieleski, R. L. (1960a). *Aust. J. biol. Sci.* 13, 203–220.
Bieleski, R. L. (1960b). *Aust. J. biol. Sci.* 13, 221–231.
Brovchenko, M. I. (1962). *Fiziol. Rastenii*, in press.
Buchanan, J. (1953). *Arch. Biochem. Biophys.* 44, 140–149.
Burley, J. (1961). *Plant Physiol.* 36, 820–824.
Caldwell, J. (1961). *Nature, Lond.* 190, 1028–1029.
Calvin, M. and Benson, A. (1949). *Science* 109, 140.
Canny, M. (1960). *Ann. Bot. Lond.* 24, 330–344.
Canny, M. (1961). *Ann. Bot., Lond.* 25, 152–167.
Caputto, R., Leloir, L. F., Cardini, C. E. and Paladini, A. C. (1950). *J. biol. Chem.* 184, 333–350.
Chappell, J. and Greville, G. (1959). *Nature, Lond.* 183, 1737–1738.
Chen, S. L. (1951). *Amer. J. Bot.* 38, 203–210.
Christensen, H. N. (1955). "A Symposium on Amino Acid Metabolism." The Johns Hopkins Press, Baltimore, pp. 63–107.
Colin, H. (1916). *Rev. gén. Bot.* 28, 289–321.
Curtis, O. F. (1929). *Amer. J. Bot.* 16, 154–168.
De Stigter, H. C. M. (1959). *Proc. IXth. Intern. Bot. Congr.* 2, 90–91.
De Stigter, H. C. M. (1961). *Acta Botanica Nederlandica* 10, 466–473.
Ducet, G. and Vandewalle, G. (1960). *Fiziol. Rastenii* 7, 407–413.
Duloy, M. and Mercer, F. (1961). *Austral. J. Biol. Sci.* 14, 391–701.
Dutton, J. V., Curruthers, A. and Oldfield, F. T. (1961). *Biochem. J.* 81, 266–272.
Dyar, J. and Webb, K. (1961). *Plant Physiol.* 36, 672–676.
Ensgraber, A. (1958). *Ber. dtsch. bot. Ges.* 71, 349–361.
Epstein, E. (1960). *Amer. J. Bot.* 47, 393–399.
Esau, K. and Cheadle, V. I. (1955). *Acta. bot. neerl.* 4, 348–357.
Frey, G. (1954). *Ber. schweiz. bot. Ges.* 64, 390–453.
Frey-Wyssling, A. und Häusermann, E. (1960). *Ber. schweiz, bot. Ges.* 70, 150–162.
Gage, R. and Aronoff, S. (1960). *Plant Physiol.* 35, 53–64.
Gerola, F. and Barbesino, M. (1956). *Nuovo G. bot. ital.* 63, 37–45.
Glasziou, K. T. (1960). *Plant Physiol.* 35, 895–902.
Glasziou, K. T. (1961). *Plant Physiol.* 36, 175–179.
Hepton, C. E. and Preston, R. D. (1960). *J. exp. Bot.* 2, 381–393.
Hübner, G. (1960). *Flora* 148, 549–594.
Iakushkina, N. I. (1962). *Fiziol. Rastenii*, in press.
Iakushkina, N. I., Poroyskaya, S. M. and Filatova, T. G. (1956). *Fiziol. Rastenii* 3, 423–430.
Jones, H., Martin, R. and Porter, H. (1959). *Ann. Bot., Lond.* 23, 493–508.
Kennedy, J. S. and Mittler, T. E. (1953). *Nature, Lond.* 171, 528.
Kessler, G. (1958). *Ber. schweiz. bot. Ges.* 68, 5–43.
Kollmann, R. (1960a). *Planta* 54, 611–640.
Kollmann, R. (1960b). *Planta* 55, 67–107.
Kollmann, R. and Schumacher, W. (1961). *Planta* 57, 583–607.
Kulayeva, O. (1962). *Fiziol. Rastenii*, in press.
Kuprevich, V. F. (1949). *Bot. Zh.* 34, 613–617.
Kursanov, A. L. (1941). *Advanc. Enzymol.* 1, 329–370.
Kursanov, A. L. (1955). Peaceful Uses Atomic Energy. *Proc. Int. Confer. in Geneva* 12, 165–169.

11

Kursanov, A. L. and Brovchenko, M. I. (1961). *Fiziol. Rastenii* 8, 270–278.

Kursanov, A. L. and Kazakova, M. N. (1933). *Trud. Z.I.N.S.* 1, 3–13.

Kursanov, A. L. and Turkina, M. V. (1952). *C.R. Acad. Sci*, *U.R.S.S.* 84, 1073–1076.

Kursanov, A. L. and Turkina, M. V. (1954). *C.R. Acad. Sci.*, *U.R.S.S.* 95, 885–888.

Kursanov, A. L. and Zaprometov, M. N. (1949). *C.R. Acad. Sci.*, *U.R.S.S.* 68, 1113–1116.

Kursanov, A. L., Turkina, M. V. and Dubinina, I. M. (1953). *C.R. Acad. Sci.*, *U.R.S.S.* 93, 1115–1124.

Kursanov, A. L., Chailakhian, M.Kh., Pavlinova, O. A., Turkina, M. V. and Brovchenko, M. I. (1958). *Fiziol. Rastenii* 5, 3–15.

Kursanov, A. L., Brovchenko, M. I. and Pariskaya, A. N. (1959a). *Fiziol. Rastenii* 6, 6, 527–536.

Kursanov, A. L., Pavlinova, O. A. and Afanasieva, T. P. (1959b). *Fiziol. Rastenii* 6, 286–295.

Lebedyev, G. V. (1960). *Fiziol. Rastenii* 7, 398–400.

Leloir, L. F. (1951). *Arch. Biochem. Biophys.* 33, 186–190.

Lüttge, U. (1961). *Planta* 56, 189–212.

Mason, T. and Maskell, E. (1928a). *Ann. Bot.* 42, 189–253.

Mason, T. and Maskell, E. (1928b). *Ann. Bot.* 42, 571–636.

Mason, T. and Phillis, E. (1936). *Ann. Bot.* 50, 455–499.

Matile, Ph. (1956). *Ber. schweiz. bot. Ges.* 66, 237–266.

Mittler, T. E. (1953). *Nature, Lond.* 172, 207.

Mokronosov, A. T. and Bubenschchikova, N. K. (1961). *Fiziol. Rastenii* 8, 560–568.

Mothes, K. and Engelbrecht, L. (1956). *Flora* 143, 428–472.

Mothes, K. and Engelbrecht, L. (1957). *Flora* 145, 132–145.

Mothes, K., Engelbrecht, L. and Kulayeva, O. (1959). *Flora* 147, 445–464.

Münch, E. (1930). "Die Stoffbewegungen in der Pflanze," J. Fischer, Jena.

Nelson, C. (1962). *Canad. J. Res. (Botany)* 40, 757–770.

Nelson, C. and Gorham, P. (1957a). *Canad. J. Res. (Botany)* 35, 339–347.

Nelson, C. and Gorham, P. (1957b). *Canad. J. Res. (Botany)* 35, 703–713.

Nelson, C. and Gorham, P. (1959a). *Canad. J. Res. (Botany)* 37, 431–438.

Nelson, C. and Gorham, P. (1959b). *Canad. J. Res. (Botany)* 37, 439–447.

Nelson, C., Clauss, H., Mortimer, D. and Gorham, P. (1961). *Plant Physiol.* 36, 581–588.

Nelson, C., Perkins, H. and Gorham, P. (1958). *Can. J. Biochem. Physiol.* 36, 1277–1279.

Nelson, C., Perkins, H. and Gorham, P. (1959). *Canad. J. Res. (Botany)* 37, 1181–1189.

Nichiporovich, A. A., Andreyeva, T. F., Voskresenskaya, N. P., Nezgovorova, L. A. and Novitzkij, J. I. (1958). "Proceedings of the First (UNESCO) International Conference", Paris, 1957. *Radioisotopes in Scientific Research* 4, 411–431, London.

Norris, L., Norris, R. E. and Calvin, M. (1955). *J. exp. Bot.* 6, 64–70.

Okanenko, A. S. (1936). *Nauchnyie sapisky po saccharnoi promyichlennosti. Agronomicheski vyipusk. V.N.I..I.S.* 4, 46–65.

Ordin, L. and Gairon, S. (1961). *Plant Physiol.* 36, 331–335.

Paleyev, A. M. (1938). *Biokhimija* 3, 258–269.

Pavlinova, O. A. (1954). *Biokhimiya* 19, 364–372.

Pavlinova, O. A. (1955). *Fiziol. Rastenii* 2, 378–386.

Pavlinova, O. A. .(1959). Unpublished results.
Pavlinova, O, A. and Afanasieva, T. P. (1962). *Fiziol. Rastenii* **9**, in press.
Peel A., and Weatherley, P. E. (1959). *Nature, Lond.* **184**, 1955–1956.
Povolotskaya, K. L., Rakitin, J. V. and Khovanskaya, I. V. (1962). *Fiziol. Rastenii*, **9**, in press.
Price, C. and Davies, R. (1954). *Biochem. J.* **58**, xvii.
Pristupa, N. A. (1959). *Fiziol. Rastenii* **2**, 378–386.
Pristupa, N. A. (1962). *Fiziol. Rastenii*, in press.
Pristupa, N. A. and Kursanov, A. L. (1957). *Fiziol. Rastenii* **4**, 417–424.
Prokofiev, A. A. and Sobolyev, A. M. (1957). *Fiziol. Rastenii*, **4** 14–23.
Prokofiev, A. A., Zhdanova, L. P. and Sobolyev, A. M. (1957). *Fiziol. Rastenii* **4**, 425–431.
Ramshorn, K. (1961). *Fiziol. Rastenii* **8**, 29–41.
Randle, P. J. and Smith, G. H. (1958a). *Biochem. J.* **70**, 490–500.
Randle, P. J. and Smith, G. H. (1958b). *Biochem. J.* **70**, 501–504.
Ratner, E. I. and Böszörmenyi, L. (1959). *Fiziol. Rastenii* **6**, 537–543.
Reinhold, L. and Powell, R. (1956). *Nature, Lond.* **177**, 658–659.
Sabato, G. (1959). *Nature, Lond.* **183**, 997–998.
Shiroya, M., Lister, G., Nelson, C. and Krotkov, G. (1961). *Canad. J. Res. (Botany)* **39**, 855–864.
Shishova, O. A. (1956). *Biokhimija* **21**, 111–118.
Shuel, R. (1959). *Canad. J. Res. (Botany)* **37**, 1167–1180.
Sinjuhin, A. M. (1962). *Fiziol. Rastenii* **9**, in press.
Skok, J. (1957). *Plant Physiol.* **32**, 308–312.
Spanner, D. C. (1958). *J. Exp. Bot.* **9**, 332–342.
Sveshnikova, I. N. (1957). *Fiziol. Rastenii* **4**, 24–27.
Swanson, C. and El-Shishiny, E. (1958). *Plant Physiol.* **33**, 33–37.
Tanaka, A. (1961). *J. Fac. Agric. Hokkaido Univ.* **51**, 449–550.
Thaine, R. (1961). *Nature, Lond.* **192**, 772–773. Transcellular strands and particle movement in mature sieve tubes.
Thaine, R. (1962). *J. Exp. Bot.* **13**, 152-160. A translocation hypothesis based on the structure of plant cytoplasm.
Tsao Tsung Hsun and Liu Chih-yii (1957). *Acta bot. Sinica* **6**, 269–280.
Turkina, M. V. (1954). *Biokhimija* **19**, 357–363.
Turkina, M. V. (1959). *Fiziol. Rastenii* **6**, 709–718.
Turkina, M. V. (1961). *Fiziol. Rastenii* **8**, 649–657.
Turkina, M. V. and Dubinina, U. M. (1954). *C.R. Acad. Sci., U.R.S.S.* **95**, 199–202.
Turner, E. (1960a). *Ann. Bot.* **24**, 382–386.
Turner, E. (1960b). *Ann. Bot.* **24**, 387–396.
Ullrich, W. (1961). *Planta* **57**, 402–429.
Vartapetian, B. B. (1960). *Fiziol. Rastenii* **7**, 395–397.
Vartapetian, B. B. and Kursanov, A. L. (1959). *Fiziol. Rastenii* **6**, 144–150.
Vartapetian, B. B. and Kursanov, A. L. (1961). *Fiziol. Rastenii* **8**, 569–575.
Vernon, L. P. and Aronoff, S. (1952). *Arch. Biochem. Biophys.* **36**, 383–398.
Verzar, F. (1936). *In* "Absorption from the intestine," 142–144. Longmans Green, London, New York, Toronto.
Vis, J. H. (1958). *Acta bot. neerl.* **7**, 124–130.
Vittorio, P. V., Krotkov, G. and Reed, G. B. (1954). *Science* **119**, 906–908.
Votchal, E. F. (1934). *Nauchnyie sapisky po saccharnoi promyichlennosti V.N.I.I.S.* **XL**, 1–8.
Wanner, H. (1952). *Planta* **41**, 190–194.

Wanner, H. (1953a). *Ber. schweiz. bot. Ges.* **63**, 162–168.
Warnner, H. (1953b). *Ber. schweiz. bot. Ges.* **63**, 201–211.
Wanner, H. and Bachofen, R. (1961). *Planta* **57**, 531–542.
Wareing, P. (1958). *Nature, Lond.* **181**, 1744–1745.
Warren Wilson, P. M. (1959). *New Phytol.* **58**, 326–329.
Watson, D., Thorne, G. and French, S. (1958). *Ann. Bot.* **22**, 321–352.
Wheatherley, P. E. (1962). *Advancement of Science*, March, 571–577
Weatherley, P. E., Peel, A. and Hill, G. (1959). *J. Exp. Bot.* **10**, 1–16.
Willam, A. (1948). *Arch. Inst. bot. Univ. Liége* **18**, 1—104.
Willam, A. (1952). *Industr. agric.* **69**, 763–770.
Willenbrink, I. (1957). *Planta* **48**, 269–342.
Yin Hung-chang, Shen Yuen-kang and Shen Kung-mau. (1958). *Acta biol. Exp. Sinica* **6**, 105–110.
Ziegler, H. (1955). *Naturwissenschaften* **42**, 259–260.
Ziegler, H. (1956). *Planta* **47**, 447–500.
Ziegler, H. (1958). *Planta* **51**, 186–200.
Ziegler, H. (1960). *Naturwissenschaften* **47**, 140.
Ziegler, H. und Lüttge, U. (1959). *Naturwissenschaften* **46**, 176.
Ziegler, H. and Vieweg, G. H. (1961). *Planta* **56**, 402–408.
Zimmerman, M. (1957). *Plant Physiol.* **32**, 399–404.
Zimmerman, M. (1958a). *Plant Physiol.* **33**, 213–217.
Zimmerman, M. (1958b). *In* "The Physiology of forest trees", pp. 381–400 (1957). K. Thimann Ronald Press Company, New York.
Zimmerman, M. (1960). *Annu. Rev. Pl. Physiol.* **11**, 167–190.
Zimmerman, M. (1961). *Science* **133**, 73–79.
Zimmermann, M. (1960). *Beih. Z. schweiz. Forstver.* **30**, 289–300.

Water Relations of Plant Cells

JACK DAINTY

Biophysics Department, University of Edinburgh, Scotland

I. INTRODUCTION

This article on water relations of plant cells is necessarily limited in scope. A wide-ranging review over cells, tissues and whole plants at a satisfactory level of treatment would require a whole book and an amount of time which was just not available. Attention has therefore been confined to a discussion of the water relations of single vacuolated cells and I have emphasized the kinetic, transport, aspects rather than the equilibrium properties.

Even this restricted subject cannot be treated as fully as it deserves, but this is partly due to a great lack of precise, fundamental information. The vacuolated plant cell is a very complex object; from the simplest possible viewpoint it consists of a wet cell wall—a rather concentrated weakly acid, inhomogeneous, ion-exchange resin under considerable tension—surrounding the protoplast, which in turn encloses the vacuole. Most botanists would agree that the protoplast is bounded by two membranes, which I shall refer to as the plasmalemma and the tonoplast. The protoplast is likely to swell and shrink when gaining or losing water and its elastic and viscous properties will certainly affect the rates of swelling and shrinking. In both the cell wall and protoplast, with their high content of macromolecules, the matric potential may well be an appreciable part of the chemical potential of the water in these phases. [This term matric potential is another way—and a better one—of handling the older expressions such as imbibed and bound water (see Slatyer and Taylor, 1960, 1961; Taylor and Slatyer, 1962).] The above aspects of the plant cell have scarcely been considered by the majority of workers in this field; the protoplast as a whole is normally considered as a single membrane of negligible thickness and the cell wall simply as an elastic box and sometimes, even, its elastic properties are ignored.

I shall perforce have to take this oversimplified view of the vacuolated plant cell, but this oversimplification is not all loss for, because of it, the principles of water and solute transport can be clearly discussed and these principles apply at whatever level of sophistication one wishes to treat plant cells and tissues. Writing this article has impressed on me, however, the great scarcity of fundamental facts about water relations of plant cells and the urgency of the need to obtain simple quantitative information on water permeability, in absolute units, of the protoplast as a whole and particularly of the plasmalemma and tonoplast separately, on the swelling and shrinking of protoplasts, on the relationship between cell volume and turgor pressure, on the water relations of cell walls, and so on. In respect of the latter, Carr and Gaff (1962) (see also Gaff and Carr, 1961) have made some very interesting observations on the role of water in the cell walls in *Eucalyptus* leaves. Their contentions that the cell wall water is a large fraction (40%) of the total leaf water and that it acts both as a "buffering" system during periods of water stress and as the pathway of transpired water in the leaf raise some very interesting and fundamental problems. Weatherley (1962) also has evidence that the transpiration stream in leaves passes *via* the cell walls and not through the protoplasts and vacuoles of the cells. This kind of work needs to be taken up intensively, so that we can have some real data on the state of water in plant cell walls under various conditions; it also brings out a most important point that, so far as their water relations are concerned, plant cells and tissues may rarely be in a state of equilibrium.

One point which I do not discuss in the body of this article is the great range, apparently, of those very few water permeabilities that have been measured. First of all it is surprising to find so few actual values of water permeability in the literature. Bennet-Clark (1959) lists seven values and to these a few more can be added (Stadelmann, 1962). These "osmotic" permeability values range from about 2×10^{-8} cm sec^{-1} atm^{-1} for unplasmolyzed cells of beet (Myers, 1951) to 3×10^{-5} cm sec^{-1} atm^{-1} for the giant internodal cells of members of the Characeae (Kamiya and Tazawa, 1956; Dainty and Hope, 1959a). This range of values is quite large and it is somewhat disturbing that the most accurate determinations have been made on the large internodal cells of *Nitella* and its allies and have resulted in high values for the osmotic permeability coefficients of the order of 10^{-5} cm sec^{-1} atm^{-1}; and the method of transcellular osmosis used in these experiments causes minimum disturbance of the cells for they are still turgid. Are these Characean cells aberrant in their permeability properties? Or is there some doubt about the interpretation of the experimental results, either with the Characeae or with the other, mostly higher plant, cells? This whole question of the absolute magnitudes of water permeability coefficients needs re-examination both from the experimental and theoretical points of view. Perhaps this article, by pointing out the importance of unstirred layers and of the elastic properties of the cell wall and the protoplast, can help in this reappraisal.

As mentioned before, most of the article is devoted to transport processes but a brief discussion of the equilibrium water relations is given first. This is mostly concerned with an attempt to give a thorough thermodynamic approach and terminology along the lines of, though not quite identical with, the recent paper presented by Taylor and Slatyer (1962) at the Madrid UNESCO symposium on plant–water relationships in arid and semi-arid conditions. The main feature of this section is the stress laid upon the *chemical potential* of water as the basic concept and yet another plea to abandon the expressions *suction pressure, DPD*, and the like. Next the basic theory of water and solute transport across membranes is given. I am afraid that the only correct theory is that based on the theory of irreversible thermodynamics and this must be faced by all workers in this field.

Subsequent sections are concerned with special aspects of transport. The mechanism of osmosis has been the subject of some important recent papers, particularly one by Ray (1960). I outline the ideas which have developed about this mechanism and discuss the experimental tests which have been, or can be made to check whether osmotic flow through cell membranes is a bulk flow through water-filled pores. In the next section the swelling and shrinking of plant cells when they are transferred between solutions of different concentrations of either non-permeating or permeating solutes are discussed; the theoretical paper of Philip (1958) and the experimental papers of Collander (1949, 1950) and Werth (1961) are particularly relevant here.

One very important complicating feature in all transport phenomena, particularly with rapidly permeating solutes, is the presence of "unstirred" layers of solution adjacent to the membrane. These unstirred layers can partially or completely control the rate of permeation and can thus mask the properties of the membrane itself. Since there is no discussion of this effect in the biological literature, I give a fairly detailed account here with particular reference to the problems connected with water transport. The section after this deals with the determination and meaning of a very important membrane parameter—the reflection coefficient for a solute. This parameter is particularly important for osmosis through leaky membranes.

Finally there are two sections on two rarely discussed, but often invoked, phenomena—electro-osmosis and polar permeability. I hope that the quantitative discussions in this article will help to clear away much of the dubious talk about these phenomena.

In many ways this article can be looked upon as a follow-up to Bennet-Clark's (1959) excellent chapter on water relations in Volume II of F. C. Steward's "Plant Physiology". I have tried to extend, make more quantitative and discuss in the light of modern physico-chemical theories his descriptions of various aspects of the water relations of single plant cells. Thus Bennet-Clark's chapter is essential reading for the descriptive basis of this article.

II. EQUILIBRIUM WATER RELATIONS OF PLANT CELLS

The water relations of vacuolated plant cells are usually expressed by the equation

$$S = \pi - P \tag{1}$$

where S is variously called the suction pressure or the diffusion pressure deficit (DPD), π is the osmotic pressure of the vacuolar contents, P is the hydrostatic pressure (excess over atmospheric) of the vacuolar contents and is called the turgor pressure or wall pressure. The use of this equation is implicitly restricted to cells in which the protoplasm is a thin envelope, with no significant elasticity, of a large central vacuole. The units of S, π and P are normally atmospheres.

This equation and its associated terminology have been used for a long time in plant physiology but changes now seem to be desirable. Ray (1960) has recently violently criticized the use of the expression diffusion pressure deficit. Two of his contentions are, I think, acceptable. He points out that the term is old fashioned and is not used outside plant physiology; physical chemists no longer speak of diffusion pressure as being the driving force of either osmosis or diffusion. His other acceptable criticism is that the term implies, or could be taken to imply, that the mechanism of osmosis is one of diffusion; he points out, and this will be discussed later, that osmotic water flow will be a bulk flow of a hydrodynamical nature if the membrane contains water-filled pores. His other criticisms of the term are not realistic. He points out for instance that, since the thermodynamic expression for DPD contains a term proportional to the logarithm of the water mole fraction, the DPD of a pure solute like glycerol is minus infinity and therefore, since the driving force on water is supposed to be proportional to DPD differences, there would be an infinite driving force across a membrane separating an aqueous solution from pure glycerol. He claims that this makes the term and concept of DPD absurd. But this is not so; the fact that chemical potential, and this is what DPD is, is logarithmically dependent on mole fraction and therefore becomes infinitely large, in a negative direction, when the mole fraction becomes zero is a familiar difficulty in physical chemistry. It does not lead the physical chemists to throw away the concept of chemical potential. In fact of course the difficulty arises out of a situation which does not correspond to physical reality: it is physically impossible to separate, say, an aqueous solution from pure glycerol by a membrane permeable to water in such a way that there is anywhere a plane on one side of which the water mole fraction is finite and on the other side it is zero. His other invalid criticism is based on the extension of equation (1) to actual water flow across a semi-permeable membrane. He indicates, quite correctly but for the wrong reasons, that it is only an approximation to assume the water flow caused by a DPD difference is proportional to that DPD difference. The answer to this is—it doesn't matter; the fact that one might write the water flow J, equal to a coefficient, k, times S does not imply that the coefficient, k, is a *constant* though the equation is most useful of course if k is reasonably constant. I have discussed Ray's criticisms in some detail because, in my opinion, his paper is a most important one and I shall be constantly referring to it.

Diffusion pressure deficit, S, "defined" for a vacuolated plant cell by equation (1), determines the direction in which water will move. Even if we had to deal with the vacuolated plant cell only there would be considerable incentive to change this terminology because it, and the equation (1), are ill-suited to deal with water flows when the protoplast is permeable to the solute used to cause the water flow. But we also have to deal with DPD's in protoplasm, in cell walls, in soils and in the atmosphere where the simple equation (1) is either invalid or simply not relevant. We can of course still continue to speak of DPD's in these other phases but neither soil scientists nor colloid chemists use this terminology. It would be much better, and now the time seems to be ripe, to get into step with other scientists and use a thermodynamic approach and terminology. The rapidly growing importance of irreversible thermodynamics which I hope this article will reveal also underlines the desirability of a change. The recent discussions at the Madrid UNESCO symposium on plant–water relationships in arid and semi-arid conditions (Slatyer and Taylor, 1960, 1961; Taylor and Slatyer, 1962) make acceptable suggestions which are considered below. The only objection to a change seems to be that botanical students do not get a good enough training in thermodynamics to be sufficiently at home with the proposed use of thermodynamic concepts and terms; the answer to this objection should be self-evident.

11§

As Slatyer and Taylor point out, the terms used by plant physiologists in discussing water relations are thermodynamic by nature. The
diffusion pressure deficit of the water in the vacuole of a plant cell is
simply related to the difference, with a negative sign, between the
partial molar free energies of the water in the vacuole and pure water
at the same temperature and at a pressure of one atmosphere. It is
given by

$$ -S = \frac{\Delta\mu_w}{\overline{V}_w} = P + \frac{RT}{\overline{V}_w}\ln a_w \simeq P + \frac{RT}{\overline{V}_w}\ln N_w \qquad (2) $$

In this equation $\Delta\mu_w$ is the difference between the partial *molar* free
energies (chemical potentials) of water in the vacuole and pure water
at the same temperature and at a pressure of one atmosphere. P is the
hydrostatic pressure in the vacuole exceeding one atmosphere; a_w and
N_w are the activity and the mole fraction of the water. R is the gas
constant, T the absolute temperature and $\overline{V}_w$ the partial molar volume
of water (18 cm³ mole⁻¹). In equation (2) I have thought it desirable
to use the chemical potential, i.e. the partial *molar* free energy, rather
than the partial *specific* free energy which Slatyer and Taylor recom-
mend. My reason for this choice is so as to bring the nomenclature
completely into line with the common physico-chemical usage. The
relationship between the two partial free energies is:

$$ \text{partial specific free energy} = \left(\frac{\partial G}{\partial m_w}\right)_{P,\,T,\,m_j} $$

$$ = \frac{1}{M_w}\left(\frac{\partial G}{\partial n_w}\right)_{P,\,T,\,n_j} = \frac{1}{M_w}\mu_w $$

where G is the Gibbs free energy and the differentiation with respect
to the mass of water (m_w) or the number of moles of water (n_w) is
carried out with all the other determining parameters—pressure (P),
temperature (T), other constituents (m_j or n_j)—held constant. M_w is
the molecular weight of water and it is, of course, approximately
numerically equal to $\overline{V}_w$, the molar volume of water.

Slatyer and Taylor propose that the partial specific free energy
difference of water be called the *water potential*. I shall use the same
term for the numerically very slightly different expression $\Delta\mu_w/\overline{V}_w$.
The two expressions differ in two ways: they are numerically very
slightly different because the density of water is not exactly unity at
all temperatures; and they are expressed in different physical units—
whereas partial specific free energy has the units of an energy per
unit mass, $\Delta\mu_w/\overline{V}_w$ has the units of an energy per unit volume. The
actual size of the units will probably be debated for some time. For
water potential as I have defined it ($\Delta\mu_w/\overline{V}_w$), the atmosphere or the

bar (10^6 dyn cm^{-2}) is quite a natural unit and I shall often use the atmosphere in this article. However it is likely that the best size of unit is the J cm^{-3}. This is equal to 10^7 dyn cm^{-2} or approximately 10 atm and is the natural unit of pressure to take when discussing the combined action of concentration, pressure and electric field gradients on the movements of solutes and solvent through membranes (see Spiegler, 1958).

In dealing with two aqueous solutions separated by a membrane, such as in the conventional vacuolated plant cell, equation (2) clearly has the same form as equation (1) for in this situation

$$\frac{\Delta \mu_w}{\overline{V}_w} = P + \frac{RT}{\overline{V}_w} \ln a_w = P - \pi \tag{3}$$

because the osmotic pressure, π, of a solution is given by

$$\pi = -\frac{RT}{\overline{V}_w} \ln a_w \simeq -\frac{RT}{\overline{V}_w} \ln N_w \simeq RTC_s \tag{4}$$

where C_s is the concentration of the (non-permeating) solutes. If we denote the water potential in the vacuole by M then we should replace equation (1) by

$$M = P - \pi \tag{5}$$

The driving "force" on water is then given by $(- dM/dx)$ as is usual with any potential.

The water potential depends on hydrostatic pressure and solvent mole fraction as shown in equation (3). It also depends, in a colloid phase such as protoplasm, cell wall or soil, on the relative amounts of water and solid matrix and of course on special properties of the solid matrix. This latter dependence can be included by adding to equation (3) a matric potential τ (Slatyer and Taylor, 1960, 1961; Taylor and Slatyer, 1962) which must be determined experimentally.

As Taylor and Slatyer (1962) point out, this matric potential may make quite a significant contribution to the water potential in the protoplasm and the cell wall. However it will not be discussed here for I am more concerned with the elucidation of certain principles of water transport in, essentially, highly vacuolated plant cells. If the matric potential be included, equation (5) should read

$$M = \frac{\Delta \mu_w}{\overline{V}_w} = P - \pi + \tau \tag{6}$$

III. Basic Theory of Transport Processes

The chief topic I wish to discuss in this article is the transport of water and, where relevant, solutes across membranes. In recent years a fairly satisfactory general approach to membrane transport has been developing: this is based on the thermodynamics of irreversible processes. I long debated whether or not to use this approach in an article of this kind addressed to botanists who are probably not very familiar with the subject of irreversible thermodynamics and who may be somewhat repelled by the apparent abstractness and by some of the terminology of irreversible thermodynamics. I decided that the general approach was so superior and its use disclosed so many errors of past methods, that it would be cowardly of me to avoid it and give the usual "half-correct" description of transport processes. I firmly believe that all workers on membrane transport, whether with animal or plant membranes, must in future have irreversible thermodynamics as their background theory. I therefore hope this article fulfils the important purpose of introducing this theory to plant physiologists and that they and others will not be unduly put off at the outset of this paper by its unfamiliar and perhaps "difficult" appearance. After the initial plunge into the Onsager equations and the like, familiar results will emerge though many of them will be slightly modified and some radically so.

It is not my intention to explain the theoretical foundations of the theory of the thermodynamics of irreversible processes; this can be found in the excellent texts of De Groot (1951) and Denbigh (1951). It is sufficient to state here that it is a theory about the relationships between fluxes or flows (of matter, energy, electrical charge, etc.) and the "forces" which cause these flows. Providing that the fluxes and forces are properly defined and that the system concerned is not too far from equilibrium, then the fluxes are linear functions of all the forces on the various components of the system.

We are concerned here with the flows of water (and solutes) across membranes in an isothermal system. In this article we shall also be concerned only with passive movement, i.e. movement under the influence of the physical "forces": pressure, electrical potential and concentration (or activity) gradients. "Active" transport, i.e. chemical coupling between the flows and metabolism, can be brought into the theory (see Kedem, 1961) but this is in its infancy and to try and include it, at this stage, in this article would lead to complications and obscurities because it is insufficiently worked out.

Consider the system shown in Fig. 1. A membrane separates two solutions of the same non-electrolyte. There are differences of concentration and pressure between the two sides which are conventionally called outside, on the left, and inside, on the right. For generality the membrane must be considered permeable to both solute and solvent. At this stage I am restricting the theory to non-electrolyte solutions.

When later discussing electro-osmosis, a more general theory must be used.

Let the solute and solvent fluxes across the membrane from left to right be denoted by ϕ_s and ϕ_w mole cm^{-2} sec^{-1}. Let the corresponding "forces" on these components be X_s and X_w. It can be shown from the basic theory of irreversible thermodynamics that if the fluxes are defined as above then the correct forces to take are the (negative) gradients of the chemical potential of the components. (If any of the

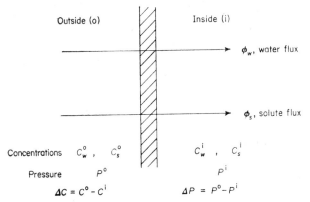

FIG. 1. Basic membrane system showing symbols and sign conventions used.

components were charged then the appropriate force would be the electro-chemical potential gradient.) The chemical potential gradient of an uncharged component i is given by

$$\frac{d\mu_i}{dx} = \overline{V}_i \frac{dP}{dx} + RT \frac{d}{dx} (\ln a_i) \tag{7}$$

where $d\mu_i/dx$ is the chemical potential gradient in the direction of flow, left to right perpendicular to the membrane, $\overline{V}_i$ is the molar volume in cm^3 mole^{-1} and a_i is the activity of component i, dP/dx is the pressure gradient in J cm^{-4}, R is the gas constant in J mole^{-1} deg^{-1} and T is the absolute temperature. The units of μ_i are J mole^{-1}. In biological systems, because of the non-existence of sufficiently accurate information, activity is replaced by mole fraction, N_i. The force X_i is given, then, by

$$X_i = -\frac{d\mu_i}{dx} = \overline{V}_i \left(-\frac{dP}{dx}\right) + RT \left(-\frac{d}{dx}(\ln N_i)\right) \tag{8}$$

(If i were a charged component a term $z_i F(-d\psi/dx)$ would have to be added to the right-hand side of equation (8), where z_i is the algebraic valency of the ion, F is the Faraday and $d\psi/dx$ is the electrical potential gradient.)

The theory of irreversible thermodynamics states that the fluxes are linear functions of *all* the forces. This means for our system of two perfectly well-stirred (an impossibility!) non-electrolyte solutions separated by a membrane, that

$$\left. \begin{aligned} \phi_s &= L_{ss}X_s + L_{sw}X_w \\[2mm] \phi_w &= L_{ws}X_s + L_{ww}X_w \end{aligned} \right\} \tag{9}$$

where the L's are the so-called Onsager coefficients relating fluxes to forces. Equations (9) are often referred to as the Onsager equations.

If we knew all the Onsager coefficients, as a function of x, we would know all about the passive transport processes in a particular system. In no biological system are all these coefficients known, but this knowledge must be the ultimate aim. The fundamental theory gives certain aids, the most important of which are the Onsager reciprocal relations :

$$L_{ij} = L_{ji} \text{ for } i \neq j \tag{10}$$

This symmetry relation for the "cross-coefficients", L_{ij}, decreases the number of Onsager coefficients to be determined, e.g. instead of four in equations (9) there are three. Other restrictions on the coefficients are that all the diagonal coefficients L_{ii} must be positive and whereas the cross-coefficients, L_{ij}, can be negative or positive, the determinant formed by the coefficients must be positive. This latter statement means for equations (9) that

$$L_{ss} L_{ww} \geqslant L_{sw}^2 \tag{11}$$

The great importance of formulating the transport fluxes by the method of equations (9) is that all possible parameters influencing passive flow of any component are correctly included. In particular this method clearly brings out that the flow of any component is, in principle, affected by the forces on all the other components and therefore points out the possible existence of interactions between the flows of the various components. This explicit bringing out of interactions, through the cross-coefficients L_{ij}, is in great contrast to the approaches used before. These previous approaches have normally used, implicitly, only the diagonal coefficients L_{ii}, i.e. they have assumed that all the cross-coefficients are zero. I am afraid that in the field of salt relations the diagonal coefficients have been usually used incorrectly, because it has not so far been clearly recognized that the correct driving force on an ion is the electrochemical potential gradient (see Dainty, 1962). (It is not necessarily incorrect to put all the cross-coefficients equal to zero but when it is done it should be done explicitly for clearly stated reasons.) Where the cross-coefficients have not been put equal to zero, e.g. in osmotic phenomena, this has also been done implicitly and hence has led to a great deal of misunderstanding about osmosis.

A more detailed discussion will now be given of our system of Fig. 1 in which two aqueous solutions of a non-electrolyte are separated by a membrane. Since this article is concerned with water relations more attention will be devoted to the water flow than to the non-electrolyte flow but it will soon be seen that the latter cannot be neglected. This subsequent discussion owes much to the papers of Staverman (1951, 1952), Kedem and Katchalsky (1958, 1961), Katchalsky (1961) and Spiegler (1958).

When discussing osmotic phenomena and non-electrolyte permeability the driving forces are normally taken to be the differences of hydrostatic pressure and osmotic pressure between the two sides of the membrane and equation (9) can be expressed, though not without modification, in these more familiar terms. Kedem and Katchalsky (1958) show how this is done for dilute, ideal, non-electrolyte solutions. The forces, X_i, on solute and water in the membrane are

$$X_s = - \bar{V}_s \frac{dP}{dx} - RT \frac{d}{dx} (\ln N_s) \tag{12}$$

$$X_w = - \bar{V}_w \frac{dP}{dx} - RT \frac{d}{dx} (\ln N_w) \tag{13}$$

where $\bar{V}_s$, $\bar{V}_w$ are the molar volumes of solute and water and N_s and N_w are the mole fractions of solute and water, respectively. For dilute solutions these equations simplify to

$$X_s = - \bar{V}_s \frac{dP}{dx} - RT \frac{d}{dx} (\ln C_s) \tag{14}$$

$$X_w = - \bar{V}_w \frac{dP}{dx} + \frac{RT}{C_w} \frac{dC_s}{dx} \tag{15}$$

where C_s, C_w are the solute and water concentrations at the point x in mole cm^{-3}. These expressions (14) and (15) are then inserted into equations (9) and a suitable integration is performed across the membrane thickness, d cm, assuming that a steady state exists, i.e. that the ϕ's are independent of x. This integration is fraught with a number of difficulties which are discussed in the papers of Kedem and Katchalsky (1961) and Dainty and Ginzburg (1963). Ignoring these difficulties, two equations similar in form to equations (9) can be derived but containing as "driving forces" the differences in hydrostatic $(\Delta P = P^0 - P^i)$

and osmotic ($RT\Delta C_s = RT\,(C_s{}^0 - C_s{}^i)$) pressures across the membrane; but the fluxes have changed and instead of ϕ_w and ϕ_s, we have J_v and J_D where

$$J_v = \text{volume flow} = \bar{V}_w\phi_w + \bar{V}_s\phi_s \tag{16}$$

$$J_D = \text{``exchange'' flow} = \frac{\phi_s}{C_s} - \frac{\phi_w}{C_w} \tag{17}$$

J_D, the exchange flow, is the relative velocity of the solute with respect to the water through the membrane. The equations connecting these new fluxes with their "conjugate" forces are (Kedem and Katchalsky, 1958)

$$J_v = L_P\,\Delta P + L_{PD}\,RT\Delta C_s \tag{18}$$

$$J_D = L_{DP}\,\Delta P + L_D\,RT\Delta C_s \tag{19}$$

and

$$L_{PD} = L_{DP} \tag{20}$$

As in equations (9), three independent Onsager coefficients are needed to specify completely the relations between fluxes and forces.

Examining equation (18) it can be seen that when there is no volume flow, i.e. $J_v = 0$,

$$\frac{\Delta P}{RT\Delta C_s} = -\frac{L_{PD}}{L_P} = \sigma \tag{21}$$

where

$$L_{PD} = -\sigma L_P \tag{22}$$

σ is called the reflection coefficient of the membrane for the particular solute and it gives the ratio of the hydrostatic pressure difference to the "thermodynamic" osmotic pressure difference when there is no net volume flow; it is the ratio of the apparent osmotic pressure to the theoretical osmotic pressure for a semipermeable membrane (Staverman, 1951). It can easily be shown that $\sigma = 1$ for a semipermeable membrane.

Equation (18) can thus be rewritten

$$J_v = L_P\,(\Delta P - \sigma\,RT\Delta C_s) \tag{23}$$

If in addition to the solute s to which the membrane may be permeable there also exists an osmotic pressure difference $\Delta\pi_{imp}$ due to other solutes to which the membrane is impermeable, then equation (23) becomes—with sufficient accuracy—

$$J_v = L_P\,[(\Delta P - \Delta\pi_{imp}) - \sigma\,RT\Delta C_s] \tag{24}$$

This is the correct equation to use in discussing changes in volume of plant cells in the presence of permeating solutes. It has to be supplemented by a suitable modification of equation (19). This is (Kedem and Katchalsky, 1958)

$$\phi_s = \omega RT\Delta C_s + J_v (1 - \sigma) \bar{C}_s \qquad (25)$$

where

$$\omega = \frac{L_P L_D - L^2_{PD}}{L_P} \bar{C}_s \qquad (26)$$

and $\bar{C}_s$ is a mean concentration of solute in the membrane. Note that ωRT is the conventional solute permeability coefficient, P_s, when there is no net flow of volume across the membrane. The three coefficients L_P, σ and ω, which replace the original L_{ss}, L_{sw}, L_{ww}, are a complete set of three independent coefficients needed for a correct description of osmotic phenomena in the presence of permeating solutes. It must be stressed that all three coefficients are completely necessary in order to specify the passive transport properties of a membrane. In the past attention has only been paid to L_P and ω; σ must now be given its due recognition.

Equations (24) and (25) are, then, our basic equations with their three coefficients L_P, σ and ω.

IV. The Mechanism of Osmosis

As has been noted before, for a truly semipermeable membrane $\omega = 0$ and $\sigma = 1$. Thus we get the familiar equation

$$J_v = (\text{in this case}) \ \bar{V}_w \phi_w = L_P (\Delta P - \Delta \pi) \qquad (27)$$

where $\Delta \pi$ here replaces the terms $\Delta \pi_{imp} + RT\Delta C_s$ in equation (24), with $\sigma = 1$. This equation, which has here been derived from the theory of irreversible thermodynamics, implies that an osmotic pressure difference, $\Delta \pi$, due to non-permeating solutes, produces the same kind of flow as a hydrostatic pressure difference. This is often called the Starling hypothesis by animal physiologists who use it to describe the flow of fluid across capillary walls and it has long been used by plant physiologists when discussing swelling and shrinking of plant cells. As discussed at the beginning of this article, plant physiologists accept the validity of this equation (27) when they use the concept of suction pressure or DPD.

The fact that the same coefficient L_P is used to describe flow of water through a semi-permeable membrane under the action of either a hydrostatic or an osmotic pressure difference has caused great conceptual difficulties among physiologists. First of all it is a fact; Mauro (1957), Robbins and Mauro (1960) have proved it fairly conclusively in work with artificial membranes. It seems to imply that somehow the osmotic pressure produces an equivalent hydrostatic pressure within the membrane; this is the real driving force and hence the coefficient relating flow to force must be L_P. Chinard (1952) had great difficulty in accepting this. He claimed that the driving force, being simply the water activity difference across the membrane, must produce flow by a diffusional mechanism. Others (Pappenheimer, 1953; Koefoed-Johnsen and Ussing, 1953; Durbin, Frank and Solomon, 1956) have accepted that osmotic pressure and hydrostatic pressure are equivalent in their effects and that both produce bulk flow of water through a semi-permeable membrane which contains water-filled pores. They contrast such bulk flow which is described by the parameter L_P with the diffusive exchange which would occur through the same membrane if the water on one side were simply labelled, with D_2O say, under conditions of osmotic equilibrium. By comparing the experimental value of L_P with the value of L_P calculated from measurements of tracer water exchange on the assumption that osmotic flow is a diffusional flow, i.e. by comparing L_P with $P_d \bar{V}_w / RT$ where P_d is the water permeability of the membrane in cm sec^{-1} determined in a tracer experiment, Solomon and his colleagues, in particular, claim to have proved the existence of bulk flow of water across certain biological membranes and hence the existence of water-filled pores; indeed they have estimated the diameter of these pores. I wish to return to a criticism of their experiments and those of others who have used this argument later in this article; at this point I would like to discuss the proposition that an osmotic pressure difference $\Delta\pi$ across a semipermeable membrane does indeed somehow produce an equivalent hydrostatic pressure in the pores of the membrane and hence the same kind of bulk flow. This discussion is based on the recent papers of Mauro (1960) and Ray (1960) together with some unpublished work of my own with Dr. P. Meares.

If it be assumed that the mechanism of the semipermeability of a porous membrane is the absolute exclusion of solute from the pores, then the pores are filled with water only. The basic equation of water flow, in the differential form, must then be—from equation (9)—

$$\phi_w = L_{ww}\left[-\bar{V}_w\frac{dP}{dx} - RT\frac{d}{dx}(\ln N_w)\right] \qquad (28)$$

if $\phi_w = 0$, then

$$\bar{V}_w P + RT \ln N_w = \text{constant} \qquad (29)$$

From equation (29) it can be immediately seen that a step in N_w at the solute mouth of the pore must be accompanied by a step in the pressure. Thus the water mole fraction and pressure profiles through the pore at zero water flow are as shown in Fig. 2.

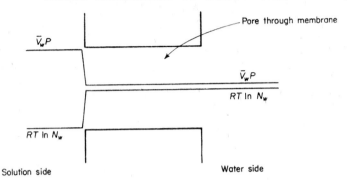

FIG. 2. Water mole fraction and pressure profiles through a water-filled pore at osmotic equilibrium. Actually the components of the chemical potential, $\bar{V}_w P$ and $RT \ln N_w$, have been drawn.

What happens during a flow situation (a) due to hydrostatic pressure alone when there is pure water on both sides of the membrane and (b) due to an osmotic pressure difference alone, when there is a solution of a non-permeating solute on one side and pure water on the other side of the membrane? The pure hydrostatic pressure case is straightforward and the pressure profile will be as shown in Fig. 3 where the positive pressure is applied on the right hand side. If the pores in the membrane are all right-cylinders of the same radius, there will be a linear fall in pressure.

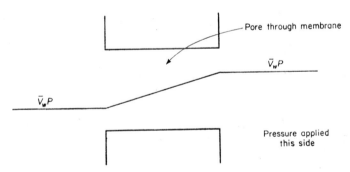

FIG. 3. Pressure profile through water-filled pore when an excess hydrostatic pressure is applied to the right-hand side and the solutions on the two sides have the same composition and concentration. Flow will take place from right to left down the hydrostatic pressure gradient.

When there is no hydrostatic pressure difference across the membrane, but only a water mole fraction difference, the pressure and water mole fraction profiles will be as shown in Fig. 4.

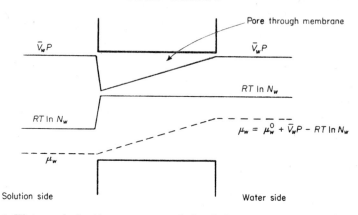

FIG. 4. Water mole fraction, pressure and chemical potential profiles through pore when there is pure water on the right-hand side and a solution of a non-permeating solute on the left-hand side; the pressures on both sides are equal. The water mole fraction profile is as shown because of the basic assumption that the pore is filled with pure water. The step in this profile at the solution end of the pore must be associated with an opposite step in the pressure profile, leading to a pressure profile as drawn. This is the same kind of pressure profile as that shown in Fig. 3 and indicates that a hydrostatic pressure gradient is the actual driving force on the water within the pore, causing it to flow from right to left.

It can clearly be seen from Fig. 4 that, within the pore, the pressure profile is exactly the same as in the pure hydrostatic pressure case of Fig. 3 and hence it can be said that an osmotic pressure difference produces water flow across a semipermeable membrane by exactly the same kind of mechanism as a hydrostatic pressure difference; in both cases hydrostatic pressure gradients are the operative forces within the pore.

Such pressure profiles have been given by Mauro (1960) and Ray (1960) and this demonstration of the existence of hydrostatic pressure gradients within the pore when only osmotic pressure differences exist in the bulk solutions is based on a thermodynamic argument: that the general chemical potential for water contains, as well as a term $RT \ln N_w$, a term $\bar{V}_w P$.

Ray (1960) has proceeded somewhat beyond the generalities of thermodynamics and gives a vivid and a very useful picture of the process of osmosis. He points out that there is a very steep concentration gradient at the solute end of the pore (see Fig. 4) and hence quite a rapid diffusion of water from the pore into the solution. This diffusion of water from within the pore aperture, if not compensated for by flow from the pure water, will result in a decrease in the density of the water just within the solution aperture and hence in a decrease of pressure there. This is the origin of the pressure gradient within the pore. Such a qualitative explanation has been given before by Koefoed-Johnsen and Ussing (1953). Ray (1960) and Dr. Meares and I have both, in different ways, converted this qualitative picture into quantitative theories.

Ray's approach is perhaps less fundamental in the sense that a molecular, kinetic, theory is not used. He assumes that the solute (or solvent) concentration gradient at the solution aperture of the pore extends over a distance approximately equal to the pore radius. He then calculates the rate of movement of water out of the pore down this very steep concentration gradient; since he correctly uses the chemical potential gradient as the driving force, the flux of water depends on the (unknown) pressure drop (ΔP) at the aperture as well as on the solute concentration difference. He next says that this water flux must be coming down the pore, being driven by the pressure drop (ΔP) across the whole length of the pore. Assuming a right-cylindrical pore to which Poiseuille's law applies, he calculates the relationship between the water flux and the pressure gradient within the pore. Equating the water fluxes calculated in these ways results in an equation for the drop in pressure (ΔP) at the solution aperture of the pore, which can easily be solved for ΔP. His result is

$$\Delta P = \frac{RTC_s}{1 + \dfrac{b}{l} + \dfrac{RTb^3}{8\,\eta D_w \overline{V}_w l}} \tag{30}$$

where C_s is the solute concentration at the solution side (pure water on the other side), b is the pore radius and l the pore length, η is the viscosity and D_w the diffusion coefficient of water in the pore.

This interesting formula indicates that the hydrostatic pressure produced by the solute concentration difference is equal to the osmotic pressure if the pores are very narrow compared with their length. Under these conditions there is diffusion equilibrium at the solution mouth of the pore while the flow is rate-controlled by the resistance of the whole length of the pore. Ray (1960) goes on to discuss this formula further and his important paper should be read for further details.

Dr. Meares and I have approached this problem from the point of view of the admittedly crude kinetic theory of liquids. An outline of our theory will be given here; the detailed argument will be published later. We consider two adjacent layers of molecules, one of pure water just inside the pore aperture and one of solution just outside the pore aperture. Each molecule oscillates within a confined volume and when, due to the random movements of its neighbours, a "hole" opens near by, it will jump into the hole, i.e. diffuse, and of course leave a vacancy behind it. Since the concentration of water is greater in the pure water layer than in the solution layer, there will be more jumps of water molecules from the pure water layer to the solution layer than *vice versa*. Thus there will be a net creation of vacancies in the pure water layer. These vacancies will of course be filled by water molecules jumping from further back in the pore, but the vacancies will persist for a short time and thus lead to a decrease in density of the pure water layer inside the pore. This decrease in density can be calculated in a relatively simple way from a kinetic theory of liquids. It can be related to a decrease in pressure by using the experimental value for the compressibility of water. This calculation on the basis of kinetic theory leads to a value for the drop in pressure just within the solution aperture of the pore of about one half of RTC_s, i.e. it gives the correct order of magnitude but not quite the right numerical factor. One could not expect the present crude kinetic theory of liquids to give accurate agreement with thermodynamic expressions and the result of our calculations is quite a satisfactory verification of the correctness of the approach.

It thus seems quite clear that osmotic flow of water through a membrane which contains pores is a bulk flow. The driving force can be looked upon as the hydrostatic pressure produced inside the pore by the markedly unbalanced diffusion of water molecules occurring at the solution end of the pore; it has now been demonstrated both theoretically and experimentally that this hydrostatic pressure is approximately equal to $RT\Delta C_s$. The fact that bulk flow occurs in an essentially diffusional situation should not cause surprise. Even the diffusion of, say, sucrose down a concentration gradient in an open vessel causes some bulk flow of the solution (see Hartley and Crank, 1949).

V. SWELLING AND SHRINKING OF PLANT CELLS

From equation (27), if no permeating solute is present, the flow of water across a membrane is given by

$$J_w = J_v = L_P (\Delta P - \Delta \pi) \tag{31}$$

It should be stressed that this equation is strictly valid—in the sense that L_P is a constant coefficient—only if the membrane is not too permeable to water and ΔP and $\Delta \pi$ are not too large. Ray (1960) discusses the validity of equation (31), or its equivalent, in a not very realistic way, taking very extreme cases in which the equation could not reasonably be expected to apply. For the normal discussion of vacuolated plant cells in an external aqueous solution of non-permeating solutes the equation can be taken as valid.

Using this equation Philip (1958) has discussed the swelling and shrinking of a turgid plant cell when transferred between various external, non-plasmolyzing, solutions of a non-permeating solute. He considers the cell as the usual classical cell osmometer, that is the volume, and any elastic properties, of the protoplast are neglected. The cell wall is assumed to be elastic only (not plastic) and to have an elastic modulus ϵ defined by the equation

$$P = \epsilon \left(\frac{V}{V_0} - 1 \right) = \epsilon v \tag{32}$$

where P is the excess hydrostatic pressure (above atmospheric) of the vacuole, V cm^3 is the cell volume at a pressure of P atm, V_0 is the cell volume when $P = 0$, i.e. at zero turgor pressure; v is defined in equation (32) as the relative departure of the cell volume from V_0. If the cell is in the process of swelling or shrinking after transfer from one solution to another, but still turgid of course, then the volume changes satisfy the following differential equation (a simple extension of equation (31))

$$-\frac{dV}{dt} = L_P A \, (P - \Delta\pi) \tag{33}$$

where A cm^2 is the surface area of the cell. To a first approximation, A can be taken as constant and equal to A_0, the area at zero turgor pressure, and $\Delta\pi$ is given by

$$\Delta\pi = \frac{\pi_0 V_0}{V} - \pi_e \tag{34}$$

where π_0 is the osmotic pressure of the vacuole at zero turgor pressure and π_e is the external osmotic pressure. After substituting (34) and (32) into (33) and making suitable adjustments (see Philip, 1958) the equation can be integrated to give

$$v = \frac{V}{V_0} - 1 = \frac{\pi_0}{\epsilon + \pi_0} \left[1 - \exp \left\{ - (\epsilon + \pi_0) \frac{L_P A_0}{V_0} t \right\} \right] \tag{35}$$

Actually this solution is for the special case of transfer from a solution of osmotic pressure π_0 to one of zero osmotic pressure, but the exponent which gives the rate of swelling or shrinking is the same no matter what the osmotic pressures of the original and final external solutions are, provided the cell is still turgid. The "half-time" for swelling or shrinking to a new equilibrium is given by

$$T_{\frac{1}{2}} = \frac{0 \cdot 693 \, V_0}{A_0 \, L_P (\epsilon + \pi_0)} \tag{36}$$

This is quite an important result; both the elastic properties of the cell wall and the amount of solute in the cell influence the rate of swelling or shrinking.

Philip (1958) goes on to discuss the rates of swelling and shrinking when the solute in the external solution is a permeating one. His analysis of this situation is, in principle, incorrect for he assumes that the movements of water and solute across the membrane are independent of each other, i.e. he implicitly assumes that the reflection coefficient is unity, or in other words that even a permeating solute exerts its full "thermodynamic" osmotic pressure. In effect he writes, when the fully turgid cell is changed from pure water to an external solution of a permeating solute of "osmotic pressure" $\pi_1 = RT(C_s)_1$ for the efflux of water, J_w,

$$J_w = L_P[\Delta P - \Delta\pi_{imp} + RT\Delta C_s] \tag{37}$$

where $\Delta\pi_{imp}$ ($= \pi_0 V_0/V$) is the osmotic pressure of the non-permeating vacuolar solutes and ΔC_s is the difference between the external and internal concentrations of the permeating solute. All the driving forces $-\Delta P$, $\Delta\pi_{imp}$, $RT\Delta C_s$—are functions of time. The concentration difference, ΔC_s, is related to the solute influx, ϕ_s, by

$$\phi_s = \omega RT\Delta C_s = P_s\Delta C_s, \text{ say,} \tag{38}$$

where P_s is the conventional solute permeability coefficient. After suitable manipulation of these equations Philip arrives at the following equation for the change of volume with time when a fully turgid cell is placed, at time zero, in a solution of a permeating solute of "osmotic pressure" π_1. (Actually he does not write down the analytical solution but presents it graphically.)

$$v = \frac{V}{V_0} - 1 = \frac{\pi_0}{\epsilon + \pi_0} + \frac{\pi_1 L_P}{L_P(\epsilon + \pi_0) - P_s}$$
$$\left[\exp\left\{-\frac{L_P A_0}{V_0}(\epsilon + \pi_0)t\right\} - \exp\left\{-\frac{P_s A_0}{V_0}t\right\}\right] \tag{39}$$

Figure 5, redrawn from Fig. 2 of Philip's paper, illustrates the behaviour of a turgid cell when placed in solutions of a permeating solute.

Though Philip's equation (39) is a good approximation when dealing with a slowly permeating solute, i.e. when

$$L_P(\epsilon + \pi_0) \gg P_s \tag{40}$$

it is not so good for describing the swelling and shrinking of cells in solutions of rapidly permeating solutes, such as in the experiments of

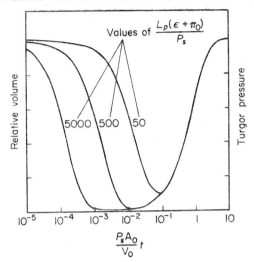

FIG. 5. Shrinking and subsequent swelling of a plant cell, initially fully turgid, when placed in a solution of a permeating solute. The curves are plotted according to equation (39). Note the logarithmic time scale.

Collander (1949, 1950) and Werth (1961). One important reason for the failure of equation (39) in these latter situations is that in the basic equations (37) and (38) the reflection coefficient, σ, is tacitly assumed to be unity. Equations (37) and (38) should be correctly written (modifications of equations (24) and (25)) as

$$\text{volume efflux} = J_v = L_P[\Delta P - \Delta\pi_{imp} + \sigma RT\Delta C_s] \qquad (41)$$

$$\text{solute influx} = \phi_s = P_s \Delta C_s - J_v(1 - \sigma)\bar{C}_s \qquad (42)$$

The manipulation and subsequent integration of equations (41) and (42) as they stand is difficult, one reason for this being the uncertainty of relating $\bar{C}_s$, the mean value of the concentration of the permeating solute in the appropriate part of the membrane, to ΔC_s. The effect of the second term on the right hand side of equation (42) is to slow down the entry of solute, but so far as I can judge this effect is small. If this retardation of the solute flow by the opposing water flow is neglected then the solution of equations (41) and (42) is similar to equation (39), except that instead of $\pi_1 = (RT\Delta C_s)_{t=0}$ we must write $\sigma\pi_1$.

This modification of the equation (39) describing the shrinking and subsequent swelling of a plant cell placed in a solution of a permeating solute has important consequences. Instead of the behaviour of the cell to various solutes being controlled, essentially, by the relative values of the solute permeability coefficients, P_s, (as compared with $L_P(\epsilon + \pi_0)$), another parameter—the reflection coefficient, σ—comes into play. For example the initial rate of shrinkage of the cell is given by

$$\left(\frac{dv}{dt}\right)_{t=0} = -\frac{\sigma\pi_1 L_P A_0}{V_0} \tag{43}$$

This is apparently a simple expression for comparing the values of σ for different solutes. Again, though the *time* to reach the minimum value of the cell volume is independent of σ, the *value* of this minimum volume depends on both σ and P_s.

Simple experiments of the kind done by Collander (1949, 1950) and Werth (1961) on the swelling and shrinking of plant cells in the presence of rapidly permeating solutes should clearly not be interpretated in terms of solute permeability coefficients only. It will be seen in the next section that even more caution than the previous paragraph indicates is needed, for there is an additional complicating factor—the so-called unstirred layers.

VI. UNSTIRRED LAYERS

The complexities of osmotic behaviour just described are made much worse by a further important factor when dealing with rapidly permeating solutes. This is the presence of so-called "unstirred layers" of solution adjoining the membrane on both sides; with most cells the outer layer is the more important (but see Dick, 1959). All the previous equations are based on the implicit assumption that the aqueous solutions on both sides of the membrane are so well-stirred that the concentrations at the membrane faces are the same as the bulk concentrations. It is however quite impossible to stir any solution right up to the membrane–solution interface and, with the usual rates of stirring used in biological experiments of the kind being discussed here, there is an effective unstirred layer of between 20 μ and 500 μ thick, the actual thickness depending partly on the size of the object used and partly on the rates of stirring (Bircumshaw and Riddiford, 1952; Dainty and Hope, 1959b). It should not be thought that the fluid throughout an unstirred layer is actually stationary, it is rather a region of slow laminar flow parallel to the membrane in which the only mechanism of transport is by diffusion; the layers are often called "Nernst diffusion layers". The thickness of the layer, δ, is also not the actual thickness but rather an operational thickness defined as the width of a uniform concentration gradient equal to that at the actual interface, i.e. it is defined by the equation

$$\left(\frac{\partial C}{\partial x}\right)_{\text{interface}} = \frac{C_b - C_{int}}{\delta} \tag{44}$$

where C_b is the bulk concentration of the solute and C_{int} is the concentration at the interface (see Fig. 6).

Whether these unstirred layers play an important part in membrane transport processes depends, essentially, on the permeability of the membrane itself to the particular molecule being transported. In considering this question it should be kept in mind that an unstirred layer can be looked upon as a membrane, in series with the actual membrane, with a permeability coefficient given by

$$P = D/\delta \qquad (45)$$

where D is the diffusion coefficient of the molecule in aqueous solution. Clearly, since $D \simeq 10^{-5}$ cm^2 sec^{-1} the permeability of the unstirred layer can be anything between 10^{-2} and 10^{-4} cm sec^{-1}. These figures are quite comparable with many quoted permeability coefficients of rapidly permeating solutes. Hence the possibility must always be considered that the transport of a rapidly penetrating solute across a membrane may be wholly or partially rate-controlled by the unstirred layers and not by the membrane at all.

The concentration profile, at some instant of time, during the permeation of a solute through a membrane, is illustrated in Fig. 6.

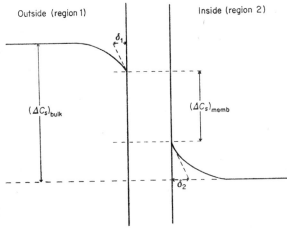

FIG. 6. Concentration profile for a permeating solute in the solutions adjacent to membrane. δ_1 and δ_2 are the thicknesses of the unstirred layers as defined by equation (44).

The apparent permeability of the membrane for a particular solute, as judged by the application of the usual Fick-type equation, is P, where

$$\phi_s = P(\Delta C_s)_{bulk} \qquad (46)$$

The "true" membrane permeability, P_t, is given by

$$\phi_s = P_t(\Delta C_s)_{memb} \tag{47}$$

and it can easily be shown that these two permeabilities are related by

$$\frac{1}{P} = \frac{1}{P_t} + \frac{\delta_1}{D_1} + \frac{\delta_2}{D_2} \tag{48}$$

where δ_1 and δ_2 are the thicknesses of the unstirred layers and D_1 and D_2 are the solute diffusion coefficients in the external and internal solutions respectively. The existence of these unstirred layers implies that in any phenomena depending on the difference of two surface concentrations possibly serious errors are made by using the difference of the bulk concentrations. A number of examples of this, some previously unpublished, follow.

The simplest example to consider is the straightforward determination of solute permeabilities by chemical or radioactive tracer methods. Such classical work is well-known in plant physiology from the papers of Collander on the permeability of the membranes of members of the Characeae (see Wartiovaara and Collander, 1960). In Collander's experiments there would be an external unstirred layer of up to at least 100 μ thick and an internal unstirred layer whose thickness is difficult to estimate (this is partly because of the protoplasmic streaming whose effect it is not easy to judge). Further these unstirred layers do not have a definite thickness independent of time but they will grow from some initial value up to a maximum. Thus the build-up in internal concentration, say, of an externally applied solute will be quite a complicated process since it is equivalent to solving equation (46) with a varying-with-time permeability coefficient P or, better, solving equation (47) when $(\Delta C_s)_{memb}$ is varying with time in a complicated way. Dr. N. Gilbert and I are trying to solve this "diffusion-with-permeation in cylindrical co-ordinates" problem by the use of an electronic computor. At the moment all that can be said is that any permeability coefficient greater than a few times 10^{-5} cm sec^{-1} is suspect and the bigger it is the more suspect it is. Collander has tried to correct his permeabilities, calculated from equation (46), for the unstirred layer effect by repeating the permeability experiments on dead cells and, in effect, saying that these latter experiments give a measure of $\delta_1/D_1 + \delta_2/D_2$ in equation (48). Unfortunately the effect of the unstirred layer is more complicated than this and, in any case, this method becomes very inaccurate when these two terms predominate, i.e. when the entry of solute is almost entirely rate-controlled by the unstirred layers.

Another example where, in my opinion, neglect of unstirred layers may have led to serious mistakes is in the comparisons that have been made between flow of water across a membrane due to a difference in osmotic pressure, and exchange of water across the membrane under conditions of osmotic equilibrium as measured by the permeation of a tracer such as DHO or THO for the water. If the osmotic flow of water proceeds by the mechanism of diffusion, as would occur if the membrane had no water-filled pores, then it can be shown that

$$L_P \frac{RT}{\overline{V}_w} = P_d \tag{49}$$

where P_d is the permeability coefficient of the membrane for water derived from the tracer experiment in which the water movement is purely by self-diffusion and L_P is the "hydraulic conductivity" of equation (27). The expression $L_P RT/\overline{V}_w$ is often referred to as P_{os}, the osmotic permeability coefficient for water, though strictly this is too loose an expression.

It was first noticed by Hevesy, Hofer and Krogh (1935), when studying the permeability of frog skin to water by the two methods—osmosis and tracer flux, that $L_P RT/\overline{V}_w$ was substantially greater than P_d. This kind of comparison has since been made many times: by Prescott and Zeuthen (1953) on various kinds of egg cells, and particularly by Solomon and his colleagues on red blood cells and other materials (Durbin, Frank and Solomon, 1956; Curran and Solomon, 1957; Curran, 1960; Paganelli and Solomon, 1957; Villegas, Barton and Solomon, 1958). Mauro (1957), Robbins and Mauro (1960) and Durbin (1960) have done similar experiments on artificial membranes. The result is always the same: $L_P RT/\overline{V}_w > P_d$. The accepted explanation of these results is that all the membranes contain water-filled pores so that the mechanism of osmosis is by bulk flow as has been earlier discussed in this article; thus there is no reason why $L_P RT/\overline{V}_w$ should be equal to P_d. If bulk flow does occur during osmosis then L_P will be greater than it would be without bulk flow; an additional bulk flow term has to be added to the diffusional term. Thus $L_P RT/\overline{V}_w > P_d$ implies water-filled pores. By assuming Poiseuille's Law type bulk flow down right-cylindrical pores, the diameter of these pores can be calculated from the ratio L_P/P_d (see Koefoed-Johnsen and Ussing, 1953; Prescott and Zeuthen, 1953; Durbin, Frank and Solomon, 1956).

A hint that this might not be the whole explanation of the discrepancy between $L_P RT/\overline{V}_w$ and P_d came from the experiments of Dainty and Hope (1959a). They tried to measure L_P and P_d for the membranes of *Chara australis* and found that though it was easy to measure L_P it was impossible to measure P_d because the exchange of D_2O between the internal and external water was rate-controlled by diffusion inside the cell, and, though this was not realized at the time, by diffusion in the external unstirred layer. Had we not realized that the membrane was not the rate-controlling factor in the exchange of D_2O and therefore calculated a value of P_d, we would have been making a gross underestimate of the true value of P_d. I would like to make the point quite strongly here that in all the published comparisons of $L_P RT/\bar{V}_w$ with P_d, P_d has always been underestimated and therefore the difference between $L_P RT/\bar{V}_w$ and P_d is never as great as has been claimed; in many cases there may be no difference and hence no proof of pores and in the remaining cases the pores are certainly narrower than has been estimated if it be accepted that Poiseuille's law can be applied.

It must be remembered that cell membranes are usually very permeable to water, for instance $L_P RT/\bar{V}_w$ for *Chara australis*, *Nitella translucens* and various red blood cells is of the order of 2×10^{-2} cm sec^{-1}. If P_d were of the same order of magnitude, an unstirred layer of only about 10 μ thick would have the same permeability; to put it another way if the true P_d were equal to $L_P RT/\bar{V}_w$ and there was an unstirred layer 10 μ thick outside the membrane and none inside, the measured P_d would be equal to $\frac{1}{2} L_P RT/\bar{V}_w$. It may be objected that the unstirred layer would affect L_P and P_d equally; but this is not so. The effect on P_d, which basically involves a measurement of the permeability of solute D_2O or DHO or THO, is as already has been described by equation (48). The effect on L_P is quite different. Because the net water flow comes out with a certain velocity J_v cm sec^{-1}, the impermeable solute is swept away from the face of the membrane by this water flow. This outward, convective, movement of the solute is opposed, within the unstirred layer, by inward diffusion down the concentration gradient so formed. The relevant equation for this "convection-opposed-by-diffusion" situation is:

$$J_v C_s + D_s \frac{dC_s}{dx} = 0 \tag{50}$$

the solution of which, for the plane case which is sufficiently accurate, subject to the boundary condition that $C_s = (C_s)_{bulk}$ at $x = \delta$, the unstirred layer boundary, is

$$C_s = (C_s)_{bulk} \exp\left(-J_v \delta/D_s\right) \tag{51}$$

In the already quoted example of a cell with $L_P RT/\bar{V}_w$ equal to 2×10^{-2} cm sec^{-1}, J_v would be about 10^{-4} cm sec^{-1} when there is an osmotic pressure difference of about 7 atm. If the solute were the commonly-used sucrose, then $D_s \simeq 5 \times 10^{-6}$ cm^2 sec^{-1} and if δ were 10 μ (which reduces P_d by a factor of two) then $C_s = (C_s)_{bulk} \exp\left(-\frac{1}{50}\right)$.

Thus the real driving force ($RT\Delta C_s$) is about 2% less than the assumed driving force and this means that the apparent L_P is only about 2% less than the real L_P. Thus with the usual driving forces used in a measurement of L_P, the value of L_P is not seriously affected by the unstirred layer—at least not compared with the effect on P_d.

Now 10 μ is extremely thin for an unstirred layer and this order of magnitude is only likely to be approached in the vigorous stirring around small objects which occurred in the experiments of Solomon and his colleagues on red blood cells. Even here, as already seen, the effect of a 10 μ layer is such as to cause a serious underestimate of P_d hence reasonable doubt as to whether these workers have proved their point about the existence of pores and the values they put on the radii of these supposed pores are particularly dubious. Even more doubtful are the values of P_d quoted by Prescott and Zeuthen (1953), for they used the Cartesian diver technique in which no stirring is allowable and the unstirred layers might be 500 μ thick, both inside and outside their rather large objects. Two unstirred layers of this thickness would have an equivalent permeability, for water, of about 2×10^{-4} cm sec^{-1} which is comparable with most of their measured P_d's. Another complication with the Cartesian diver method of measuring P_d, of which they do not seem aware, is that it is not at all clear exactly what buoyancy forces are acting on the egg when exchange with D_2O is taking place.

Ray (1960) collects together five different cases from the botanical literature where he considers it is possible to get, effectively, a comparison between $L_P RT/\overline{V}_w$ and P_d. I am afraid that in my opinion he has completely failed to substantiate in any single case his contention that $L_P RT/\overline{V}_w > P_d$ and hence that there are water-filled pores in plant-cell membranes. Two of his cases, B and C in his paper, are concerned with attempted comparisons in the single cells of *Nitellopsis obtusa* and *Nitella mucronata*. In both these cases the values of P_d, measured by the exchange of D_2O, have been grossly underestimated because of the unstirred layers. I do not think from my own experience (Dainty and Hope, 1959a), that a determination of P_d for these large cells is yet possible because the rate of exchange of water between inside and outside is almost completely rate-controlled by diffusion in the unstirred layers. There seems to be a better chance of making a comparison of $L_P RT/\overline{V}_w$ with P_d for higher plant cells because the permeability of their cell membranes to water seems to be much lower. But none of Ray's three comparisons (his cases, A, D and E) seem to be valid. In case A he takes the work of Ordin and Bonner (1956) who compared the half-time for diffusion of deuterium-labelled water into and out of *Avena* coleoptile sections with the half-time for establishment of osmotic equilibrium when the sections were transferred into or out of a 0·4 M mannitol solution. This is not necessarily a comparison of P_d with $L_P RT/\overline{V}_w$ at all. The exchange experiment is quite complex; the exchanging molecules must diffuse through external unstirred layers (not important here) and then perhaps through the "free space" of the tissue before crossing the actual membranes; thus the rate-controlling process might be diffusion within the free space—a kind of internal unstirred layer—and the half-time would then have no relation whatever to P_d. Similarly the half-time for the establishment of osmotic equilibrium could quite simply be almost entirely dependent on the rate of diffusion of mannitol in the free space. Similar remarks apply to his case D in which Thimann and Samuel's (1955) measurements of the exchange of tritium-labelled water in potato tissue discs are compared with the measurements of Falk, Hertz and Virgin (1958) on the progress of osmotic equilibration in strips of potato tissue. Even more dubious is his comparison (his case E) of Ordin and Kramer's (1956) measurement of the half-time for the diffusion of DHO out of root segments of *Vicia faba* with Brewig's (1937) measurement of steady osmotic water movement into *Vicia faba* root segments.

Thus, in my opinion, no valid comparison of $L_P RT/\overline{V}_w$ with P_d has yet been made with any plant cells or tissues and it will prove a very difficult matter. I do not dispute the idea that if it can be proved that $L_P RT/\overline{V}_w$ is substantially greater than P_d, then this will be proof that water moves across the membrane by some bulk-flow mechanism, and the simplest of such mechanisms to imagine is *via* water-filled pores.

Another possible manifestation of the effect of the unstirred layer is the occurrence of an apparent polar permeability of membranes to water (Kamiya and Tazawa, 1956; Dainty and Hope, 1959a). Here the observation is that the endosmotic permeability coefficient, $(L_P)_{en}$, is greater than the exosmotic permeability coefficient, $(L_P)_{ex}$. There are many possible explanations of this apparent polarity (see Bennet-Clark, 1959) but before discussing them, the first thing to ensure is that true permeability coefficients are being considered, i.e. one must insert in the following equation:

$$J_v = L_P \left(\Delta P - \Delta \pi \right) \tag{52}$$

the true driving force. If $\Delta \pi$ is written as $RT\Delta C$, then the correct ΔC must be inserted by allowing for the unstirred layer. In most experiments only the external unstirred layer is important and, since stirring is usually minimal, its extent might be quite large—up to 0·5 mm. Equation (51) is the relevant equation for discussing this effect and the two conditions (solute concentration profiles) of endosmosis and exosmosis are illustrated in Fig. 7 on the assumption that the external unstirred layer is the more important and that during an endosmosis experiment the external solution is pure water.

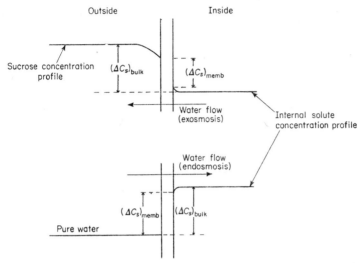

Fig. 7. Concentration profiles of outside solute (assumed sucrose) and inside solutes during exosmotic (upper diagram) and endosmotic (lower diagram) water flow. The unstirred layer effects are reasonably assumed to be small at the inner membrane surface. During endosmosis the external medium is taken to be pure water. Note the much greater discrepancy between $(\Delta C_s)_{memb}$ and $(\Delta C_s)_{bulk}$ during exosmosis.

It is clear that the driving force is overestimated during an exosmosis experiment but is reasonably correct in an endosmosis experiment. Thus $(L_P)_{ex}$ is underestimated as compared with $(L_P)_{en}$ and this is at least part of the discrepancy between the experimentally determined values of $(L_P)_{ex}$ and $(L_P)_{en}$. The possible magnitude of the discrepancy can be calculated from equation (51), which can be expressed as

$$(C_s)_{memb} = (C_s)_{bulk} \exp\left[- L_P(\Delta P - \Delta \pi)\, \delta/D_s\right] \qquad (53)$$

If $L_P(\Delta P - \Delta \pi)$ is 5×10^{-5} cm sec^{-1} ($L_P = 10^{-5}$ cm sec^{-1} atm^{-1} and $\Delta P - \Delta \pi = 5$ atm), $\delta = 200\ \mu$, D_s (sucrose) $= 5 \times 10^{-5}$ cm^2 sec^{-1}, then $(C_s)_{memb} = (C_s)_{bulk} \exp(-0.2)$; that is the effective driving force is more than 20% less during exosmosis than during endosmosis, when the apparent driving forces are the same. More detailed discussion of this effect will be found in a paper by Dainty (1963).

Another manifestation of the unstirred layer will be presented after the discussion of the determination of the reflection coefficient, but I hope that sufficient has been said to illustrate how important it might be when dealing with rapidly permeating substances and, in particular, when discussing water transport.

12

VII. The Reflection Coefficient of Membranes

I pointed out earlier that three parameters determine the passive properties of a membrane with respect to transport of water and a non-electrolyte solute; these are the hydraulic conductivity, L_P, the solute permeability, ω or P_s, and the reflection coefficient, σ. Previous research has been concentrated entirely on L_P and P_s, but σ may be equally important. In this section I shall be chiefly concerned with the information that may be gained about the structure of the membrane from a determination of σ.

Some time after we (Dainty and Hope, 1959a) had failed in our attempt to compare $L_P RT / \bar{V}_w$ and P_d for *Chara australis* and hence, as we thought, get some information about membrane structure, it occurred to Dr. Ginzburg and me that we might get some information from a determination of the reflection coefficient of the membrane for various solutes. This idea had independently occurred to Solomon and his co-workers and they have published a number of papers based on this idea (Goldstein and Solomon, 1960; Solomon, 1960; Whittembury, Sugino and Solomon, 1960; Villegas and Barnola, 1961). In their expression for σ, however, they have omitted what may be an important term and here I will outline, in a rather qualitative way, the theory of Dainty and Ginzburg (1963).

The reflection coefficient is defined, essentially, by equation (24)

$$J_v = L_P \left[(\Delta P - \Delta \pi_{imp}) - \sigma\, RT\Delta C_s \right] \qquad (24)$$

If J_v is equal to zero, then

$$\sigma = \frac{\Delta P - \Delta \pi_{imp}}{RT\Delta C_s} \qquad (54)$$

If it is remembered that $\Delta \pi_{imp}$ is the contribution to the osmotic pressure difference of the impermeable solutes, it can be seen that σ is the ratio of the apparent osmotic pressure to the osmotic pressure that would occur if the solutes s were truly non-penetrating. It is easier for discussion purposes to suppose that $\Delta \pi_{imp} = 0$, i.e. the only solute present is the permeating one s; then

$$\sigma = \frac{\Delta P}{RT\Delta C_s} \qquad (55)$$

A qualitative interpretation of σ can be seen by considering a membrane separating a solution of solute s from pure water. Suppose first that the membrane is impermeable to the solute, but permeable to the solvent (water); i.e. it is truly semipermeable. If the hydrostatic pressures are the same on the two sides (= 1 atm for convenience), then the chemical potential of water on the solution side is less than that on the pure water side (for this situation $\mu_w = \mu_w{}^0 + RT \ln N_w$); water will therefore flow from pure water into the solution. The chemical potential of the water on the solution side can be increased by applying hydrostatic pressure to this side (because of the term $\bar{V}_w P$ in the formula for the general chemical potential) and when it is increased to the value for pure water, the water flow will of course stop. The value required is quite simply given by

$$\bar{V}_w \Delta P = - RT \ln N_w \qquad (56)$$

where ΔP is equal to the hydrostatic pressure on the solution side minus the hydrostatic pressure on the pure water side and N_w is the mole fraction of water on the solution side. For a dilute solution, $\Delta P = RT \Delta C_s$, the familiar expression for the osmotic pressure of a solution.

Now suppose that the membrane is permeable to the solute as well as to the solvent and suppose, at first, that the solute and solvent do not interact as they go through the membrane, i.e. they pass independently of each other. If the same hydrostatic pressure difference is maintained, the chemical potential of water is still equal on both sides of the membrane and, since there is no interaction between solute and solvent in their passage across the membrane, there is still no net flux of water. But there is a solute flux, $\phi_s = P_s \Delta C_s$, and therefore a volume flow equal to $\bar{V}_s \phi_s = \bar{V}_s P_s \Delta C_s$. The apparent osmotic pressure is that hydrostatic pressure required to ensure no net *volume* flow, not no net water flow, so the hydrostatic pressure on the solution side must be decreased, i.e. the chemical potential of water must be lower than that of pure water, to allow a counterbalancing flow of water from the pure water side. Clearly the new hydrostatic pressure, ΔP_1 say, is found by equating the two volume flows, that is

$$L_P \left[- \Delta P_1 - \frac{RT}{\bar{V}_w} \ln N_w \right] = \bar{V}_s P_s \Delta C_s$$

i.e.

$$L_P \left[RT \Delta C_s - \Delta P_1 \right] = \bar{V}_s P_s \Delta C_s \qquad (57)$$

whence

$$\Delta P_1 = RT \Delta C_s \left[1 - \frac{P_s \bar{V}_s}{RT \, L_P} \right] \qquad (58)$$

Comparing (58) with (55) it can be seen that

$$\sigma = 1 - \frac{P_s \bar{V}_s}{RT \, L_P} \qquad (59)$$

This important expression for the reflection coefficient was first derived, by a rather different argument, by Kedem and Katchalsky (1958). It applies to the case when the solute and the solvent do *not* interact during their passage through the membrane, i.e. they take different pathways through the membrane. If the membrane is a biological one this case would probably apply when using a highly lipid soluble solute.

Finally, suppose that the membrane is permeable to both solute and solvent, but in their passage through the membrane they interact with each other, e.g. they both move through water-filled pores and exert a frictional drag on each other. The argument of the previous paragraph has to be modified because the solute and solvent no longer move solely under the influence of their own chemical potential gradients: in the general formulation of irreversible thermodynamics, the cross-coefficients are non-zero. As before, if the hydrostatic pressure on the solution side of the membrane has the "thermodynamic" value $RT'\Delta C_s$, which makes the chemical potential of the water on the two sides of the membrane equal, there is still a volume flow towards the water side due to the flow of solute. But in this case the volume flow is greater than in the previous case of no interaction between solute and water flows, for the solute flowing towards the pure water side drags water with it and hence increases the total volume flow. Thus the hydrostatic pressure has to be decreased even below the ΔP_1 of the previous case so as to produce no net volume flow. This means that, if there is direct interaction between solute and water flows, the reflection coefficient is less than when the solute and the water find pathways through the membrane independently of each other. In symbols, if solute and water interact,

$$\sigma < 1 - \frac{P_s \, \overline{V}_s}{RT \, L_P} \tag{60}$$

and this suggests an important experimental test for the existence of water-filled pores through the membrane: if an experimental determination of σ, P_s and L_P shows that σ is less than the value corresponding to independent passage, then this would suggest the existence of water-filled pores in the membrane through which the solute could also go.

The actual value of σ has been calculated for two special models by Kedem and Katchalsky (1961) and by Dainty and Ginzburg (1963). The more useful calculation for biological membranes is the latter. Since the calculation is complex only the final result is given here. It is assumed that the membrane is largely composed of lipid with a few water-filled pores (capillaries) through it. Both water and solute can go through both the lipid part of the membrane and through the pores. Those fractions of solute and water going through the lipid part are assumed to go through independently of each other, i.e. these fractions exert no frictional drag on each other. But the fractions which go through the water-filled pores do exert a frictional drag on each other. This model leads (Dainty and Ginzburg, 1963, see also Dainty, 1961) to the following formula for σ:

$$\sigma = 1 - \frac{P_s \overline{V}_s}{RT \, L_P} - \frac{K_s^c f_{sw}^c}{f_{sw}^c + f_{sm}^c} \tag{61}$$

In this formula, P_s is the permeability coefficient for the solute, L_P is the usual hydraulic conductivity for water, $\overline{V}_s$ is the molar volume of the solute, K_s^c is the partition coefficient for the solute between the water in the pores and the external solutions. f_{sw}^c and f_{sm}^c are frictional coefficients; f_{sw}^c is the frictional force, per mole of solute, between solute and water in the pores when the solute is moving at a speed of 1 cm sec^{-1} with respect to the water; f_{sm}^c is the frictional force, per mole of solute, between the solute in the pores and the solid wall of the pores when the solute is moving at a speed of 1 cm sec^{-1} relative to the solid parts of the membrane.

It is not necessary here to discuss the nature and value of the frictional coefficients. Some discussion will be found in Spiegler (1958), Mackay and Meares (1959), Kedem and Katchalsky (1961), Dainty and Ginzburg (1963). Frankly the kinetic theory of liquids is in such a poor state that no satisfactory theoretical calculation of these frictional coefficients can be given. All that is needed here is to point out that if there is any frictional interaction between the solute and the water as they travel through the pores, i.e. if there are water-filled pores, then

$$\sigma < 1 - \frac{P_s \overline{V}_s}{RT \, L_P} \tag{60}$$

and measurements of σ, P_s and L_P in a given system should answer the question as to the existence of pores without introducing any discussion of the actual mechanism of osmosis. Solomon and his co-workers have been content to show that $\sigma < 1$ for certain solutes and claim that this proves the existence of pores, but this procedure is not correct—in principle.

However in making these measurements we come up against the usual unstirred layer difficulties. The measurement of L_P is fairly straightforward and Dainty and Ginzburg (1963) used the method of transcellular osmosis (Kamiya and Tazawa, 1956; Dainty and Hope, 1959a) for *Chara australis* and *Nitella translucens*. The measurement of P_s for the solutes they used, various small molecular weight alcohols, was not easy—because of the effect of the unstirred layers—but the correction was not so hopelessly big as to make the value of P_s too unreliable. With values of P_s of from 2 to 5 times 10^{-4} cm sec^{-1} and L_P of the order of 10^{-5} cm sec^{-1} atm^{-1}, we could show that the second term in equation (61) only amounted to $0 \cdot 1$ at most. The real difficulty was the determination of σ. Two methods were used but the second of these two, the more sound one, is the only one which will be described here. Basically, by means of a trick, that concentration (C_s) of test solute was found which was equivalent—in its apparent osmotic pressure—to a certain concentration (C_{sucr}) of an impermeable solute (sucrose). It can then easily be shown that

$$\sigma = \frac{C_{sucr}}{C_s} \tag{62}$$

The trouble with this formula is that the denominator should really be $(\Delta C_s)_{memb}$ and not C_s, the bulk external concentration (the internal concentration being assumed zero). Even though the measurement of σ is theoretically made at time zero by an extrapolation procedure, the extrapolation is extremely dubious because of the unstirred layer. If we imagine an experiment actually being performed the procedure is as follows. The cell is sitting in a sucrose solution. The initiation of the experiment is the rapid removal of the sucrose solution and its replacement by a solution of the test, permeating, solute and the subsequent observation—effectively—of the changes in volume of the tested cell (e.g. *Nitella translucens*). However the removal of bulk sucrose solution does not remove all the solution adhering to the cell: besides that in the cell wall (10–20 μ thick) there is also a film about 20 μ thick surrounding the cell wall, i.e. a total thickness of about 40 μ of sucrose solution. When the test solution is added this 40 μ film of sucrose solution is essentially undisturbed at time zero—so no initial volume change would be expected. Then the sucrose starts diffusing away and the test solute starts diffusing inwards; the membrane therefore "sees" a rather complicated time-varying sum of the concentrations of sucrose and test solute and it is very difficult to decide what these are. In addition, of course, the test solute is diffusing through the membrane and hence $(\Delta C_s)_{memb}$ is lower than $(C_s)_{memb}$—and there is the sucrose to take into account. All-in-all it is very difficult to decide what the magnitude of these unstirred-layer corrections is and to get the correct value of σ.

There is a further difficulty in this problem of the determination of σ which is common to a great many investigations on the water relations of plant cells. This is the fact that the protoplast is not of negligible thickness; it too can probably swell and shrink when not in osmotic equilibrium and it is bounded by two membranes—the plasmalemma and the tonoplast. Though Briggs (1957) has discussed the effect of the swelling of the protoplast on the equilibrium water relations of vacuolated plant cells in various different kinds of external solutions, there has been little discussion of the way in which the properties of the protoplast might affect the rates of water movement. It is not known, for instance, whether the tonoplast is more or less permeable than the plasmalemma to water and it is time an attempt was made to find out. It is generally thought that the tonoplast is less permeable to most solutes than is the plasmalemma. MacRobbie and Dainty's (1958) results on ion transport in *Nitellopsis obtusa* indicate that in this plant cell the tonoplast is less permeable to K, Na and Cl ions than is the plasmalemma; but MacRobbie (1962) has shown that this situation is reversed in the fresh water *Nitella translucens*. Dainty and Ginzburg (1963) have recently shown that the tonoplast of *N. translucens* is less permeable to urea than is the plasmalemma.

If the tonoplast is less permeable to water, then it is likely that we, Dainty and Ginzburg (1963), were measuring σ for this membrane; but we don't know. In any case there would probably be swelling of the protoplast to take into account and, perhaps, some effective addition to the external unstirred layer.

VIII. Electro-Osmosis

Electro-osmosis and other electro-kinetic phenomena have often been invoked in plant physiology in an attempt to explain certain transport and equilibrium phenomena in which experiment and classical theory did not agree. One of these was the apparent discrepancy between the osmotic pressure of vacuolar contents determined (a) by plasmolytic measurements and (b) by extraction of the vacuolar sap (see Bennet-Clark, 1959, for latest discussion). This discrepancy suggested the occurrence of an active transport of water into the vacuole. Other phenomena have also suggested active transport of water and the mechanism frequently put forward for such a "water-pump" is electro-osmosis. Fensom (1957) and Spanner (1958) have also suggested that electro-osmosis is the driving mechanism in phloem transport. The discussions of electro-osmosis in the literature of plant physiology have rarely been quantitative and on those occasions where some quantitative approach has been used, it has almost certainly been used incorrectly. This section is an attempt to give a correct quantitative discussion of electro-osmosis and related phenomena, relevant to plant cells and tissues. A more detailed account will be found in a forthcoming paper by Dainty, Croghan and Fensom (1963).

Electro-kinetic phenomena are particularly well handled by the methods of irreversible thermodynamics; indeed this is the only way in which they can be clearly discussed (Staverman, 1952). We start, as usual, with general equations similar to equation (9) relating the fluxes of cation, anion and water to the corresponding forces on these components. Since there are three components there will be three equations and nine Onsager coefficients; three pairs of these coefficients are equal, from the Onsager reciprocal relations, thus reducing the number of independent coefficients from nine to six. The force X_i on any component is the gradient of the electro-chemical potential, i.e.

$$X_i = -\frac{d}{dx}[RT \ln a_i + \overline{V}_i P + z_i F \psi] \tag{63}$$

z_i is the algebraic valency, F the Faraday, ψ the electrical potential and the other symbols have been explained before. It is usual when considering electro-kinetic phenomena to assume that the concentrations of the components are the same on both sides of the membrane; this enables us to treat those aspects of the transport that are essentially electro-kinetic. Thus, dropping the activity term, we take X_i to be given by

$$X_i = z_i F \left(-\frac{d\psi}{dx}\right) + \overline{V}_i \left(-\frac{dP}{dx}\right) \tag{64}$$

The expression (64) for X_i is substituted into the general equations, then two new fluxes are worked out, viz.

$$J = \text{volume flow} = \phi_c \overline{V}_c + \phi_a \overline{V}_a + \phi_w \overline{V}_w$$

and

$$I = \text{current flow} = F(z_c \phi_c + z_a \phi_a),$$

and finally the resulting equations are integrated through the membrane assuming a steady state. The suffixes c, a and w refer to cation, anion and water respectively. This arithmetic leads to the following two equations

$$\left.\begin{array}{l} J = L_{PP}P + L_{PE}E \\[2mm] I = L_{EP}P + L_{EE}E \end{array}\right\} \tag{65}$$

in which P and E are the differences of pressure and electrical potential between the two sides of the membrane and the new Onsager coefficients are functions of the old ones in the general equations and other parameters. Since J and I are appropriate conjugate fluxes to the forces P and E the Onsager reciprocal relationship applies, i.e.

$$L_{PE} = L_{EP} \tag{66}$$

Also

$$L^2{}_{PE} \leqslant L_{PP} L_{EE} \tag{67}$$

These equations (65), (66) and (67) will be our basic equations. The units of J are cm^3 cm^{-2} sec^{-1}, of I are A cm^{-2}, of P are J cm^{-3} and of E are volts. L_{PP} can be described, as before, as the hydraulic conductivity of the membrane and L_{EE} is the electrical conductance in Ω^{-1} cm^{-2} of the membrane (we always assume that we are dealing with one cm^2 of the membrane). L_{PE} is the electro-kinetic term and this single cross-coefficient is involved in all electro-kinetic phenomena. For instance L_{PE} gives directly the volume flow under an applied potential difference when there are equal pressures on the two sides of the membrane; it also gives the current flow due to an applied pressure under zero electrical field conditions. L_{PE}/L_{EE} is numerically equal to the ratio of streaming potential to applied pressure and also, as is well known, it is equal to the ratio of electro-osmotic flow to applied current.

In the paper of Dainty, Croghan and Fensom (1963) detailed calculations of the values of the three Onsager coefficients are given for

TABLE I

Onsager coefficients for three electro-osmotic models

	Helmholtz-Smoluchowski model	Schmidt model	Frictional model
$\dfrac{l}{n\pi a^2}L_{PP}$	$\dfrac{a^2}{8\eta}10^7$	$\dfrac{a^2}{8\eta}10^7$	$\dfrac{1}{C_w f_{wm} + \dfrac{C_s f_{sw} f_{wm}}{f_{sw} + f_{sm}}}$
$\dfrac{l}{n\pi a^2}L_{PE}$	$\dfrac{\varepsilon\zeta}{4\pi\eta}\dfrac{1}{9\times10^4}$	$\dfrac{a^2}{8\eta}\bar{X}F\,10^7$	$\dfrac{F\left[\dfrac{C_s}{C_w}f_{sw} + C_s\bar{V}_s f_{wm}\right]}{f_{sm}\left[\dfrac{C_s}{C_w}f_{sw} + f_{wm}\right] + f_{sw}f_{wm}}$
$\dfrac{l}{n\pi a^2}L_{EE}$	λ	λ	$\dfrac{F^2\,C_s}{f_{wm} + \dfrac{f_{sw}f_{wm}}{\dfrac{C_s}{C_w}f_{sw} + f_{wm}}}$

Onsager coefficients for three electro-osmotic membrane models. The membrane is l cm thick, 1 cm^2 in area, and has n right-cylindrical pores each of radius a cm. η is the viscosity of the pore liquid in poises. ε is the dielectric constant of the pore liquid. ζ is the zeta-potential in volts in the double-layer for the Helmholtz-Smoluchowski model. λ is the specific conductivity in Ω^{-1} cm^{-1} of the pore liquid. $\bar{X}$ is the volume concentration in equiv cm^{-3} of the fixed charge for the Schmidt model. In the frictional model, suffix s refers to solute, either cation or anion depending on the sign of the fixed charge. C_s and C_w are the concentrations of mobile ion and water in mole cm^{-3} in the pore liquid. f_{sw}, f_{sm}, f_{wm} are so-called frictional coefficients, e.g. f_{sw} is the frictional force in J cm^{-1} exerted on one mole of solute when the relative velocity of the solute with respect to the water is 1 cm sec^{-1}.

12§

three membrane models, in each of which the membrane, l cm thick, is assumed to contain n right-cylindrical pores per cm^2 all of radius a cm. The three models are the conventional Helmholtz-Smoluchowski model (Overbeek, 1952), the Schmidt model (Schmidt, 1951) and a frictional model (Spiegler, 1958). The Helmholtz-Smoluchowski model applies to pores whose radii are much greater than the width of the electrical double layer at the pore wall surface and is least likely to be applicable to biological membranes. The Schmidt model applies to pores that are so narrow as completely to exclude all mobile ions of the same sign (co-ions) as the fixed charges on the walls of the pores; it also assumes that the mobile counter ions are uniformly distributed in the water in the pore. The frictional model, though capable of being more general, is here applied to narrow pores which exclude the co-ions; in this case it is a more sophisticated Schmidt model. Table I gives the Onsager coefficients for the three models.

The formulae for the Onsager coefficients in the frictional model are more complicated, but they give a better insight in many respects into what is happening. Though this particular frictional model is here developed only for very narrow pores which exclude co-ions, this is probably the relevant situation in most biological membranes. Furthermore electro-kinetic effects are likely to be most important with very narrow pores and this further justifies concentrating on this particular model. All previous discussions seem to have been based on the Helmholtz–Smoluchowski model.

It seems possible, qualitatively, to imagine that an electro-osmotic water pump could operate in a cell membrane so as to pump water up a chemical potential gradient. This could lead for instance to a higher turgor pressure inside a vacuolated cell than that given by simple considerations of osmotic equilibrium. It is of course necessary to supply a steady electrical current across the membrane but some supply of current can be envisaged though only by an electrogenic ion pump (see Dainty, 1962). This idea must now be examined quantitatively.

The excess hydrostatic pressure, at the steady state of zero net volume flow, created inside the vacuole of a plant cell by the operation of an inwardly-directed electro-osmotic pump is given, from equation (65) when $J = 0$, numerically by

$$\frac{P}{E} = \frac{L_{PE}}{L_{PP}} \tag{68}$$

For our three models we have:

Helmholtz-Smoluchowski:

$$P/E = 2\ \epsilon\zeta/\pi a^2\ 9 \times 10^{11}\ \text{J cm}^{-3}\ \text{V}^{-1} \tag{69}$$

Schmidt:

$$P/E = \overline{X}F \quad \text{J cm}^{-3} \text{ V}^{-1} \tag{70}$$

Frictional:

$$P/E \simeq C_s F f_{sw}/(f_{sw} + f_{sm}) \quad \text{J cm}^{-3} \text{ V}^{-1} \tag{71}$$

Taking $a = 10^{-5}$ cm (limit of applicability of Helmholtz-Smoluchowski model), $\epsilon = 80$, $\zeta = 0 \cdot 1$ V, $\overline{X} = C_s = 10^{-3}$ mole cm^{-3}, $F = 10^5$ C mole^{-1}, we get for the three models $P/E = 1/20$ atm, 10^3 atm, $< 10^3$ atm, respectively. Thus except in the Helmholtz-Smoluchowski case, the applicability of which is dubious anyway, quite substantial pressures could be produced by quite small ($\sim 0 \cdot 1$ V) driving potentials.

But this conclusion must be seriously qualified. There is first of all the problem of supplying the electrical current through the charged pores to produce the electro-osmosis. This can only be provided by a so-called electrogenic ion pump operating in the lipid parts of the membrane; the existence of ion pumps is now quite well established in plant cells (MacRobbie and Dainty, 1958) but it seems most likely that these are neutral, i.e. not charge carrying, and therefore could not supply the currents required for electro-osmosis (Dainty, 1962). Secondly there is the question of the power required to supply such currents. I shall not discuss this question at this stage except to say that energetically speaking electro-osmosis is inherently a rather inefficient mechanism of transport. Thirdly, and perhaps most important, the possibly high turgor pressures that could be produced according to the Schmidt or frictional models according to equations (70) and (71), presuppose that the charged pores through which electro-osmosis takes place are the only channels of water movement. If this is not so, then the conclusions based on equations (70) and (71) must be modified; the discussion of the next paragraph will show how profound the modification may be.

Instead of using equations (70) and (71) I shall use a somewhat different approach based on the maximum possible efficiency of an electro-osmotic process. It can be shown (Dainty, Croghan and Fensom, 1963) that the maximum ratio of moles of water transported to moles of ion in electro-osmosis is given by the ratio of the molar concentrations of water and ions. Thus

$$\frac{J_w}{I} \leqslant \frac{C_w}{C_s} \frac{\overline{V}_w}{F} \simeq \frac{1}{FC_s} \text{ and therefore } \frac{J_w}{E} \leqslant \frac{L_{EE}}{FC_s} \tag{72}$$

This gives us the maximum volume of water transported, due to an electrical potential difference, E. The only plant cells for which we know the value of the membrane conductance, L_{EE}, are one or two of the Characean internodal cells; for these cells, L_{EE}, is approximately equal to $5 \times 10^{-5} \, \Omega^{-1} \, cm^{-2}$ under physiological conditions (Hope and Walker, 1960, 1961; Dainty, Johnston and Williams, 1961). Therefore for these cells, from (72),

$$J_w \leqslant \frac{5 \times 10^{-10}}{C_s} E \tag{73}$$

taking $F = 10^5 \, C \, mole^{-1}$.

Suppose this electro-osmotic water movement gives rise to an extra pressure P atmospheres. This extra pressure will cause an outward flow of water which can be calculated from the known hydraulic conductivity (L_P or L_{PP}) of these cells: approximately $10^{-5} \, cm \, sec^{-1} \, atm^{-1}$ (Kamiya and Tazawa, 1956; Dainty and Hope, 1959a). (It can be shown that this outward flow is *via* channels other than the electro-osmotically used charged pores.) This extra pressure P, then, will produce an outward flow of water equal to $10^{-5} \, P \, cm \, sec^{-1}$. When the cell is in a state of flux equilibrium for water, the inward electro-osmotic flow will be balanced by the outward, pressure driven, flow of water. Thus

$$\frac{5 \times 10^{-10}}{C_s} E \geqslant 10^{-5} \, P$$

i.e.

$$\frac{P}{E} \leqslant \frac{5 \times 10^{-5}}{C_s} \tag{74}$$

A reasonable value for C_s, the counter-ion concentration in the pore, is about $10^{-3} \, mole \, cm^{-3}$, giving $P \leqslant 5 \times 10^{-2} \, E$ atm. For $E = 0 \cdot 1$ V— a maximum possible value—$P \leqslant 5 \times 10^{-3}$ atm which is negligible.

It would therefore appear that the reason why an electro-osmotic, or indeed any other water pump, can make no serious contribution to the turgor pressure is that the passive permeability of the cell membranes to water is so high that any pumping activity is "short-circuited" by the passive water movements. Our hypothetical electro-osmotic pump is very inefficient because almost all the passive water flow occurs through channels other than the charged pores through which electro-osmosis takes place.

It will be objected that I have chosen as an example a special cell type. In fact I had no choice because it is the only kind of cell for which we know both the membrane conductance and the water permeability. *If* the usual parenchyma cell has a much lower permeability to water and at the same time a membrane conductance of the same order of magnitude as *Nitella*, then a more substantial electro-osmotic counter pressure could be built up; but it will be seen from my general arguments that it is unlikely to be a significant contribution to the turgor pressure. It is clearly urgent to measure the membrane conductance of higher plant cells and to improve the measurements of water permeability.

Fensom (1957) and Spanner (1958) have both suggested that an electro-osmotic pump may be the operative mechanism at the sieve plates of the phloem and so achieve transport of water and neutral solutes across the sieve plates. I would like to point out here that whatever other virtues an electro-osmotic pump may have it certainly does not economize in power as compared with, say, pressure driven flow. Qualitatively it can be said that electro-osmosis achieves water, and therefore neutral-solute, transport by the frictional drag between the ions and the water. Since there will always be some slip between the ions and the water—by the very nature of the frictional drag— extra dissipation of energy will always occur in electro-osmosis as compared with a pressure driven mechanism. Clearly the more the co-ions are excluded from the pores, the more efficient will be the electro-osmotic mechanism, i.e. the Schmidt or our special frictional model should give the highest electro-osmotic efficiencies.

A simple quantitative argument, without any appeal to a model, demonstrates the fact that to achieve the same volume flow electro-osmotic power is greater than "pressure-volume" power. From the fundamental equations (65) it can easily be seen that:

$$\text{electro-osmotic power} = W_{eo} = (IE)_{P=0} = L_{EE}E^2 \qquad (75)$$

$$\text{pressure-volume power} = W_V = (PJ)_{E=0} = L_{PP}P^2 \qquad (76)$$

For the same value of J in each case the ratio of the voltage to the pressure is given by

$$\frac{E}{P} = \frac{L_{PP}}{L_{PE}} \qquad (77)$$

Therefore

$$\frac{W_{eo}}{W_{PV}} = \frac{L_{EE}}{L_{PP}} \frac{E^2}{P^2} = \frac{L_{EE} L_{PP}}{L^2_{PE}} \qquad (78)$$

But a fundamental law of irreversible thermodynamics is that the entropy of the system must increase during the flow process and this is epitomized by equation (67). Therefore $W_{eo} \geqslant W_{PV}$. Dainty, Croghan and Fensom (1963) discuss these considerations in more detail and for specific models.

One interesting consequence of the existence of electro-osmosis, though not directly concerned with water transport, is that if it occurs the membrane conductance, L_{EE}, is greater than would be calculated from a measurement of the flux of ions across the membrane when this latter is determined by using isotopic tracers during conditions of flux equilibrium. (This situation is very reminiscent of the distinction between osmotic and diffusional movements of water.) The qualitative reason for this increased conductance is that the bulk movement of water during electro-osmosis increases the velocity of the ions, i.e. the ionic mobility, and hence increases the conductance. This is discussed in detail by Dainty, Croghan and Fensom (1963) and the magnitude of the increased conductance is quantitatively worked out and it is shown how it depends on the various frictional coefficients in the special frictional model. The great interest of this calculation is that wherever the conductance of a cell membrane has been measured and compared with the conductance calculated from measurements of all the ion fluxes, large discrepancies have appeared. These discrepancies have been observed with muscle cells (Hodgkin and Keynes, 1955) and in plant cells (Hope and Walker, 1960, 1961; MacRobbie and Dainty, 1958; MacRobbie, 1962; Dainty, Johnston and Williams, 1961) and the usual explanation has up to now been that the ions move in "files" through the membrane. This explanation gets a bit far-fetched with plant cells where files of at least 10 ions are needed. The explanation by electro-osmosis is a much more natural one.

It has occasionally been suggested that some of the electrical potentials observed across cell membranes and in tissues may be streaming potentials. A little quantitative thinking is also required here before they are invoked. From equation (65) it can be seen that the streaming potential is given numerically by

$$\left(\frac{J}{I}\right)_{P=0} = \left(\frac{E}{P}\right)_{I=0} = \frac{L_{PE}}{L_{EE}} \tag{79}$$

As already mentioned, equation (72), the maximum value of $(J/I)_{P=0}$ is given by $1/FC_s$, where C_s is the counter-ion concentration in the pores. Thus the maximum streaming potential is given by

$$E = \frac{P}{FC_s} \tag{80}$$

If $P = 1$ J cm^{-3} (10 atm) and $C_s = 10^{-3}$ equiv cm^{-3} then the maximum possible streaming potential is 10 mV for the large driving pressure difference of 10 atm. These facts should be kept in mind when streaming potentials are invoked without any quantitative calculations.

IX. POLARITY OF WATER MOVEMENT

It is well-known that certain biological membranes are more permeable to water when it moves in one direction than when it moves in the opposite direction, under apparently the same driving force, of course. Such plant membranes as seed coats and the outer coats of many fruits, e.g. the tomato, are more permeable to inward than to outward water movement. This kind of polarity is also exhibited by many insect cuticles and the large amount of excellent work on insect cuticles has recently been very well discussed by Beament (1961), the chief worker in this field. Bennet-Clark (1959) has recently summarized the position with respect to plant membranes.

It is first necessary to be clear that there is nothing particularly mysterious about the polar permeability to water, i.e. the different rates of movement of water inwards and outwards across a membrane when the same chemical potential difference of water is applied across the membrane. Quite simply, the membrane itself is different in the two cases. The fact that an insect, a seed or a fruit loses less water per second when exposed to water vapour at a certain lower chemical potential than the inside water chemical potential as compared with the water it gains per second when exposed to water vapour at a chemical potential higher by the same amount, may be simply due to the fact that the limiting membrane is less hydrated in the first case than in the second; it is well-known (Hartley, 1957) that in many polymers the diffusion coefficient of water is very strongly dependent on the degree of hydration of the polymer. Even if the water permeability were more correctly measured in the two directions by using certain chemical potentials of water on the two sides and then reversing the membrane without changing the chemical potentials of water so that the average degree of hydration might be expected to be the same, the same explanation would apply if the membrane were a complex one made up of a hydratable membrane and a non-hydratable, rather impermeable, membrane in series. For when the hydratable membrane was in contact with the solution of lower chemical potential (of water) it would be less hydrated and therefore less permeable.

Beament (1957) does not consider that these are possible explanations for the polar permeabilities of insect cuticles. He is more inclined to ascribe the polarity to a valve mechanism in the highly oriented wax layer and gives a very feasible picture of such a valve. However, the fact remains that such polar permeability, which is clearly of great survival value in the examples mentioned, is explainable in simple physico-chemical terms and we can now proceed to analyse it without being bothered by some faint suspicion of magic being associated with the phenomenon.

Brauner (see Bennet-Clark, 1959) favours an electro-osmotic interpretation of the polar permeability of seed coats. He finds that there is a potential difference between two aqueous salt solutions of equal concentrations when separated by the seed coat of *Aesculus hippo-*

campus and suggests that this electrical asymmetry is associated with an electro-osmosis which will add to the water flow in one direction and subtract from it in the other. This is an interesting suggestion but I feel intrinsically unlikely; it should however be much more intensively investigated. There are many things which should be done. For instance it is unlikely that a steady-state electrical potential difference across such impermeable membranes will be established except after a very long time and the time to establish such a potential will almost certainly be different depending on the orientation of the membrane. This time to reach a steady state will always be an important factor in the apparent polar permeability of the rather impermeable membranes I have so far been discussing. Another consequence of an electro-osmotic explanation to look for is the associated ionic permeability which must be present.

I have so far discussed this problem of polar permeability on the assumption that the true permeability had been determined. In fact this is rarely so because the true driving force—right at the membrane faces—is different from the apparent driving force as given by the difference between the chemical potentials of the water in bulk. This discrepancy between true and apparent driving forces, which of course is due to unstirred layers, is usually recognized when one of the phases bathing the membrane is gaseous, as with insects, seeds, fruits, etc., transpiring or gaining water to or from damp air (see Beament, 1961); but it rarely or never seems to be recognized when the phases are liquid. As I have shown earlier, unstirred layers in liquids cannot necessarily be neglected.

An example of the influence of unstirred layers in producing a certain degree of apparent polar permeability to water arises in the studies of water permeability in *Nitella flexilis* by Kamiya and Tazawa (1956) and in *Chara australis* by Dainty and Hope (1959a), both by the method of transcellular osmosis. This is a very valuable method of measuring the permeability of the membranes of these large cylindrical cells to water when exposed to an osmotic pressure difference which does not plasmolyze them. The cell is mounted in a double chamber, the volume of liquid in one or both chambers being measurable by the movement of an air bubble in an attached capillary tube (see Fig. 8). Usually the cell is mounted symmetrically so that equal areas are in each chamber. The experiment is started by having water in each chamber and then, when any drifting of the air bubble has ceased, the water on one side is replaced by a suitable solution of a non-permeating solute such as sucrose. This leads to a flow of water from the dilute side (water), through the cell, and out into the concentrated side (sucrose solution). The initial rate of flow is given, in the symmetrical case, by

$$J_w \simeq \frac{A}{2} L_P \pi_0 \tag{81}$$

where A cm² is the exposed area of the cell in either chamber and π_0 is the osmotic pressure of the sucrose solution used to cause the trans-cellular osmosis. This is a very nice way of determining L_P.

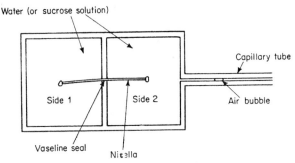

FIG. 8. Diagram of double chamber system used to measure trans-cellular osmosis in *Nitella*.

It can easily be shown (Kamiya and Tazawa, 1956) that if the cell is put into the double chamber asymmetrically, with an area A_1 cm² in one chamber and A_2 cm² in the other chamber, and if L_P does not depend on whether the water is moving into the cell (endosmosis) or out of the cell (exosmosis), the initial rate of flow is given by

$$J_w \simeq \frac{A_1 A_2}{A_1 + A_2} L_P \pi_0 \tag{82}$$

According to this formula the rate of flow does not depend on whether the short or the long end is in the sucrose solution. However the experimental observations (Kamiya and Tazawa, 1956; Dainty and Hope, 1959a; Dainty and Ginzburg, 1963) are quite clear: when exosmosis occurs at the short end, the transcellular flow is smaller than when it takes place at the long end.

Kamiya and Tazawa (1956) say that this result indicates that the exosmotic permeability coefficient is less than the endosmotic permeability coefficient.

In symbols

$$(L_P)_{ex} < (L_P)_{en} \tag{83}$$

This follows quite simply from the modification to equation (82) required if these two permeability coefficients are different. If the short end corresponds to A_1 and the long end corresponds to A_2 and if the sucrose solution surrounds the short end of the cell (case I), then the flow is

$$(J_w)_I \simeq \frac{(L_P)_{en} (L_P)_{ex} A_1 A_2}{(L_P)_{en} A_2 + (L_P)_{ex} A_1} \pi_0 \tag{84}$$

If the sucrose solution surrounds the long end of the cell (case II), then the flow is

$$(J_w)_{II} \simeq \frac{(L_P)_{en} (L_P)_{ex} A_1 A_2}{(L_P)_{en} A_1 + (L_P)_{ex} A_2} \pi_0 \tag{85}$$

The experimental observation is that $(J_w)_I < (J_w)_{II}$. Since $A_1 < A_2$ this implies that $(L_P)_{ex} < (L_P)_{en}$, i.e. the water permeability is greater when water is moving into the cell than when it is moving out.

However when unstirred layers are taken into account this conclusion is not necessarily justified. There are three unstirred layers to consider: one outside the cell in the sucrose solution and two inside the cell—one at the exosmosis end and the other at the endosmosis end. The effect of all these unstirred layers is to decrease the driving "force" below the bulk value π_0; and the driving force is decreased more in case I (sucrose at the short end) than in case II, because the velocity of the water leaving the cell in case I is greater than in case II. Thus one would expect $(J_w)_I < (J_w)_{II}$ without involving any concept of polar permeability. I have given a quantitative theory of the effect of unstirred layers in transcellular osmosis (Dainty, 1963). The conclusions reached were that the apparent polarity observed by Dainty and Hope (1959a) could be explained by the effect of unstirred layers but that the effects observed by Kamiya and Tazawa (1956) were rather too large to be entirely due to these layers. Recently Dainty and Ginzburg (1963) have done some further experiments and they have concluded that in addition to the unstirred-layer effect, which must always be corrected for, there seems to be a "true" polarity of permeability which, we think, is probably due to the dehydration of the membrane by the plasmolytica used to produce exosmosis; during endosmosis the membrane will be much more hydrated than during exosmosis.

REFERENCES

Beament, J. W. L. (1961). *Biol. Rev.* **36**, 281–320.
Bennet-Clark, T. A. (1959). "Water relations of cells". *In* "Plant Physiology" (F. C. Steward, ed.), Vol. II, pp. 105–191. Academic Press, New York and London.
Bircumshaw, L. L. and Riddiford, A. C. (1952). *Quart. Rev. chem. Soc., Lond.* **6**, 157–185.
Brewig, A. (1937). *Z. Bot.* **31**, 481–540.
Briggs, G. E. (1957). *New Phytol.* **56**, 258–260.

Carr, D. J. and Gaff, D. F. (1962). *In* "Plant-Water Relationships in Arid and Semi-Arid Conditions", pp. 117–125. UNESCO, Paris.

Chinard, F. P. (1952). *Amer. J. Physiol.* **171**, 578–586.

Collander, R. (1949). *Physiol. Plant.* **2**, 300–311.

Collander, R. (1950). *Physiol. Plant.* **3**, 45–57.

Curran, P. F. (1960). *J. gen. Physiol.* **43**, 1137–1148.

Curran, P. F. and Solomon, A. K. (1957). *J. gen. Physiol.* **41**, 143–168.

Dainty, J. (1961). *In* "Membrane Transport and Metabolism" (A. Kleinzeller and A. Kotyk, eds.), pp. 109–110. Academic Press, New York and London.

Dainty, J. (1962). *Annu. Rev. Pl. Physiol.* **13**, 379–402.

Dainty, J. (1963). *Protoplasma* (in press).

Dainty, J., Croghan, P. C. and Fensom, D. S. (1963). To be published.

Dainty, J. and Ginzburg, B–Z. (1963). To be published.

Dainty, J. and Hope, A. B. (1959a). *Australian J. Biol. Sci.* **12**, 136–145.

Dainty, J. and Hope, A. B. (1959b). *Australian J. Biol. Sci.* **12**, 395–411.

Dainty, J., Johnston, R. J. and Williams, E. J. (1961). Abstracts of contributed papers to International Biophysics Congress, Stockholm, p. 163.

De Groot, S. R. (1951). "Thermodynamics of irreversible processes", 242 pp. North Holland, Amsterdam.

Denbigh, K. G. (1951) "The thermodynamics of the steady state", 103 pp. Methuen, London.

Dick, D. A. T. (1959). *Int. Rev. Cytol.* **8**, 387–448.

Durbin, R. P. (1960). *J. gen. Physiol.* **44**, 315–326.

Durbin, R. P., Frank, H. and Solomon, A. K. (1956). *J. gen. Physiol.* **39**, 535–551.

Falk, S., Hertz, C. H. and Virgin, H. I. (1958). *Physiol. Plant.* **11**, 802–817.

Fensom, D. S. (1957). *Canad. J. Res. (Bot.)* **35**, 573–582.

Gaff, D. F. and Carr, D. J. (1961). *Australian J. Biol. Sci.* **14**, 299–311.

Goldstein, D. A. and Solomon, A. K. (1960). *J. gen. Physiol.* **44**, 1–17.

Hartley, G. S. (1957). *Trans. Faraday Soc.* **53**, 1148.

Hartley, G. S. and Crank, J. (1949). *Trans. Faraday Soc.* **45**, 801–818.

Hevesy, G., Hofer, E. and Krogh, A. (1935). *Skand. Arch. Physiol.* **72**, 199–214.

Hodgkin, A. L. and Keynes, R. D. (1955). *J. Physiol.* **128**, 61–68.

Hope, A. B. and Walker, N. A. (1960). *Australian J. Biol. Sci.* **13**, 277–291.

Hope, A. B. and Walker, N. A. (1961). *Australian J. Biol. Sci.* **14**, 26–44.

Kamiya, N. and Tazawa, M. (1956). *Protoplasma* **46**, 394–422.

Katchalsky, A. (1961). *In* "Membrane Transport and Metabolism" (A. Kleinzeller and A. Kotyk, eds.), pp. 69–86. Academic Press, New York and London.

Kedem, O. (1961). *In* "Membrane Transport and Metabolism" (A. Kleinzeller and A. Kotyk, eds.), pp. 87–93. Academic Press, New York and London.

Kedem, O. and Katchalsky, A. (1958). *Biochim. biophys. Acta* **27**, 229–246.

Kedem, O. and Katchalsky, A. (1961). *J. gen. Physiol.* **45**, 143–179.

Koefoed–Johnsen, V. and Ussing, H. H. (1953). *Acta Physiol. Scand.* **28**, 60–76.

Mackay, D. and Meares, P. (1959). *Trans. Faraday Soc.* **55**, 1221–1238.

MacRobbie, Enid, A. C. (1962). *J. gen. Physiol.* **45**, 861–878.

MacRobbie, Enid, A. C. and Dainty, J. (1958). *J. gen. Physiol.* **42**, 335–353.

Mauro, A. (1957). *Science* **126**, 252–253.

Mauro, A. (1960). *Circulation* **21**, 845–854.

Myers, G. M. P. (1951). *J. exp. Bot.* **2**, 129–144.

Ordin, L. and Bonner, J. (1956). *Plant Physiol.* **31**, 53–57.

Ordin, L. and Kramer, P. J. (1956). *Plant Physiol.* **31**, 468–471.

Overbeek, J. T. G. (1952). *In* "Colloid Science" (H. R. Kruyt, ed.), Vol. I, pp. 115–244. Elsevier, Amsterdam.

Paganelli, C. V. and Solomon, A. K. (1957). *J. gen. Physiol.* **41**, 259–277.

Pappenheimer, J. R. (1953). *Physiol.Rev.* **33**, 387–423.

Philip, J. R. (1958). *Plant Physiol.* **33**, 264–271.

Prescott, D. M. and Zeuthen, E. (1953). *Acta physiol. scand.* **28**, 77–94.

Ray, P. M. (1960). *Plant Physiol.* **35**, 783–795.

Robbins, E. and Mauro, A. (1960). *J. gen. Physiol.* **43**, 523–532.

Schmidt, G. (1951). *Z. Elektrochem.* **55**, 229–237.

Slatyer, R. O. and Taylor, S. A. (1960). *Nature, Lond.* **187**, 922–924.

Slatyer, R. O. and Taylor, S. A. (1961). *Nature, Lond.* **189**, 207–209.

Solomon, A. K. (1960). *J. gen. Physiol.* **43**, Suppl. 1–15.

Spanner, D. C. (1958). *J. exp. Bot.* **9**, 332–342.

Spiegler, K. S. (1958). *Trans. Faraday Soc.* **54**, 1408–1428.

Stadelmann, E. J. (1962). *In* "Physiology and Biochemistry of Algae". Academic Press, New York and London. (in press).

Staverman, A. J. (1951). *Rev. trav. chim.* **70**, 344–352.

Staverman, A. J. (1952). *Rev. trav. chim.* **71**, 623–633.

Taylor, S. A. and Slatyer, R. O. (1962). *In* "Plant-Water Relationships in Arid and Semi-Arid Conditions", pp. 339–349. UNESCO, Paris.

Thimann, K. V. and Samuel, E. W. (1955). *Proc. Nat. Acad. Sci., Wash.* **41**, 1029–1033.

Villegas, R. and Barnola, F. V. (1961). *J. gen. Physiol.* **44**, 963–977.

Villegas, R., Barton, T. C. and Solomon, A. K. (1958). *J. gen. Physiol.* **42**, 355–369.

Wartiovaara, V. and Collander, R. (1960). "Permeabilitätstheorien". *Protoplasmatologia, II, C8d*, pp. 98. Springer-Verlag, Wien.

Weatherley, P. E. (1962). *In* "The Water Relations of Plants" (F. H. Whitehead, ed.). Blackwell, Oxford. (in press).

Werth, W. (1961). *Protoplasma*, **53**, 457–503.

Whittembury, G., Sugino, N. and Solomon, A. K. (1960). *Nature, Lond.* **187**, 699–701.

Electron Paramagnetic Resonance in Photosynthetic Studies*

G. M. ANDROES

*Physics Department, American University of Beirut,
Beirut, Lebanon*

I. INTRODUCTION

The material of which living organisms are made is largely organic. Even the inorganic constituents, apart from the sodium chloride, are most often found very closely associated with, if not a component part of, organic substances in living organisms. The structure and change of structure of such materials, that is, organic substances in general, is determined by the electronic configuration of the atoms of which they are made, and by changes in these electronic configurations. Therefore, any method of observation which permits us to look closely into the nature of the electronic configurations and the changes of these electronic configurations is likely to become a major tool in biological studies.

* The preparation of this paper was sponsored by the U.S. Atomic Energy Commission.

A method of observing both static and, in some cases, changing electronic configurations has recently (that is, in the last decade or so) risen to prominence. This method has been important particularly in the study of the electronic structure of inorganic materials and is now becoming of importance in the examination of both the statics and dynamics of organic and biological materials. It depends upon the fact that a spinning electron is magnetic as are, in fact, a good many other nuclei. Therefore such an electron, when placed in an external magnetic field, will interact with that field in a characteristic way which can be determined by suitable configuration of the magnet and electromagnetic radiation, which we will discuss later.

Most of the electrons in organic materials however, although they individually do have a spin and a magnetic moment, exist in the organic material in pairs in which the magnetic moments are oppositely directed and are closely coupled so that the individual electrons cannot interact in an as yet observable fashion with an external magnetic field. However, many important chemical transformations which occur in biological materials involve at some stage of their occurrence the uncoupling of these paired electrons in some way or another. Indeed, when photosynthetic materials are illuminated certain paired electrons are uncoupled, and observations can be made on the resulting unpaired electrons. Such observations open the way to new and perhaps unique insights into the mechanism of photosynthesis. In what follows we shall be concerned primarily with the observations which have been made on the photo-induced, unpaired electrons in photosynthetic materials, and with the correlations which have been made between the resonance observations and biological parameters.

Electrons of the type which have been most studied in photosynthetic materials using the techniques to be discussed below are only one of the types of unpaired electrons susceptible to such study. Before proceeding with our discussion we shall briefly mention the various kinds of electrons which can be studied using electron paramagnetic resonance techniques.

Materials which contain electrons in any of the following conditions might be susceptible to study by electron paramagnetic resonance methods, and many of them have been observed in biological materials.

Free electrons. Such electrons as are found as conduction electrons in a metal are indeed free, and under certain special conditions their spin life times are sufficiently long so that they may be observed by these methods (e.g. Feher and Kip, 1955; and Wagoner, 1960).

Not-quite-free electrons. Such electrons as are found in the narrow conduction bands in semiconductors are also susceptible to observation. These electrons are produced by being raised from bound, or coupled, states into their conduction, or free, states, usually by thermal energy or light (e.g. Singer and Kommandeur, 1961).

Trapped "free" electrons. These are electrons which are not paired but which are physically trapped in a solid lattice (either ionic, atomic or molecular) and are definitely readily observable by this method (e.g. Fletcher *et al.*, 1954; Portis, 1953).

Unpaired electrons from paired electrons. These unpaired electrons, created by the fission of a paired electron bond which generally involves the separation of atoms as well, are also clearly observable by these methods. Such (ordered) molecules may appear as chemical reaction intermediates (e.g., Yamazaki *et al.*, 1960) or as broken bonds created by high energy radiation (e.g. Shields and Gordy, 1959).

Unpaired electrons in even molecules. These are unpaired electrons from molecules containing an even number of electrons in which two or more have been uncoupled to form what is called a state of higher multiplicity. (When only one electron pair is uncoupled to give two electrons with parallel spin moments, it is called a triplet state.) For example, see Hutchinson and Mangum, 1958.

Unpaired electrons from transition metal ions. These electrons in the transition metal ions such as iron, cobalt, nickel and copper are generally observable by the method of electron spin resonance. It is clear that these unpaired electrons, existing as they do in the d-orbitals of transition elements, are quite common in biological systems. Unfortunately, they are not always observed, at least under the conditions which are currently known to us. However, much important information is beginning to appear from studies where they have been observed (e.g. Beinert and Lee, 1961).

In what follows we shall discuss the nature of the electron paramagnetic resonance (EPR) experiment by which these various types of unpaired electrons are observed, how this type of experiment has been applied to photosynthetic systems, and the conclusions concerning photosynthesis that can be drawn from the experiments.

In Section II we will describe the parameters associated with the observation of unpaired electrons. Here the emphasis will be placed on those aspects of the observation which become important when the sample material is polycrystalline or amorphous. Some of the resonance parameters are not, as yet, generally useful in making statements about electronic configurations or environment in photosynthetic systems. We will try to indicate these.

In Section III we will list the types of photosynthetic materials on which resonance experiments have been performed. Then in varying degrees of detail we will discuss some of the experiments which have been performed. Our treatment will not be encyclopaedic, but, rather, will deal with those experiments that seem to us to be most pertinent.

Section IV will be concerned with the conclusions one can draw from the experiments discussed in Section III. Mention will be made of some experiments which may be profitably carried out in the future.

Finally, a brief appendix is attached in which several terms are defined, and various experimental details are described.

II. The Magnetic Resonance Experiment

The magnetic resonance experiment has been the subject of several text-books. These are listed at the head of the list of references. Texts on nuclear as well as electron resonance have been included, since many of the concepts involved in the two types of experiment are the same. A glance at any of the listed texts will show that the subject, when treated in detail, can be quite complicated. The treatment here is considerably simplified; we hope without the sacrifice of accuracy.

A. ZEEMAN ENERGY LEVELS

If an atomic or molecular system with spin angular momentum of magnitude $\hbar[S(S + 1)]^{1/2}$ is placed in a magnetic field, H, the component of angular momentum along the direction of the field may assume only the values

$$S\hbar, (S - 1)\hbar, \ldots, -S\hbar.$$

Each of these $2S + 1$ orientations will have a different energy. The energy of these states, the Zeeman energy, may be written

$$\mathscr{H} = g\beta\mathbf{H} \cdot \mathbf{S} = g\beta H_z m_s$$

if H is in the z direction and $m_s = S, S - 1, \ldots, -S$. In this expression $\beta (= e\hbar/2mc)$ is the Bohr magneton, and g represents the effective size of the magnetic moment being acted upon by the magnetic field. It is equivalent to the spectroscopic splitting factor

$$g = 1 + \frac{J(J + 1) + S(S + 1) - L(L + 1)}{2J(J + 1)}$$

of the free atom (White, 1934). According to the basic quantum mechanical principle, transitions between the magnetic states can be induced by providing a quantum of energy, $h\nu$, of the appropriate size.

Except under special circumstances, transitions may be induced only between adjacent levels: i.e. $\Delta m_s = \pm 1$. Thus, the condition for inducing the transitions (the resonance condition) is $h\nu = g\beta H_z$ when H is 3300 G and $g = 2 \cdot 00$, ν is $9 \cdot 5$ kMc/s ($\lambda \approx 3 \cdot 2$ cm).

B. RESONANT ENERGY ABSORPTION

When the resonance condition is satisfied, transitions from m_s to $m_s + 1$ and from $m_s + 1$ to m_s are induced with equal probability by the high frequency (microwave) magnetic field. This in itself leads to no energy absorption. A net energy absorption arises from two facts: (1) The spin system, in thermal equilibrium, has a few more spins in the state m_s than in the state $m_s + 1$. The slightly greater number of spins in the states of lower energy means that slightly more transitions up than down will be induced by the microwave field. (2) There are alternative routes by which spins excited to the state $m_s + 1$ can be returned to m_s; the routes through which thermal equilibrium would be re-established should the microwave field be suddenly removed. These processes are called thermal relaxation processes.

The margin of operation for observation of the net energy absorbed is small. In Table I we have listed values for the Boltzman factor

TABLE I

Boltzman Constant and Fractional Spin Unbalance for an $S = 1/2$ System at Three Temperatures

	$300° K$	$77° K$	$4·2° K$
$\dfrac{g\beta H}{kT}$	$1·48 \times 10^{-3}$	$5·76 \times 10^{-3}$	$1·05 \times 10^{-1}$
$\dfrac{n\uparrow - n\downarrow}{n\uparrow + n\downarrow}$	$7·4 \times 10^{-4}$	$2·9 \times 10^{-3}$	$5·0 \times 10^{-2}$

$$g = 2·00; \ \beta = 0·927 \times 10^{-20} \ \text{erg/G}, \ H = 3300 \ \text{G}$$

determining the population distribution, and the fractional spin unbalance for an $S = 1/2$ system at three different temperatures. It is seen that even at $4·2°$K, of 100 spins there are only five more in the ground than in the excited state. The first person to observe these small energy absorptions was Zavoisky (1945) in the U.S.S.R. He used solutions containing transition metal ions. In the United States Cummerow and Halliday (1946) made the first observation of this type. The field has been developing rapidly ever since.

We would now like to answer the following two questions. What are the measurable parameters associated with this resonance absorption and, which of these, if any, might be useful in the solution of problems concerning photosynthesis?

C. AREA UNDER CURVE

The reaction of a system of free electrons to an applied magnetic field is depicted in Fig. 1. In a resonance experiment (holding ν constant)

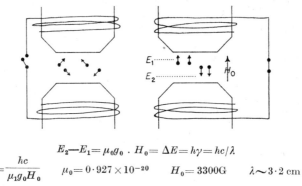

$$E_2 - E_1 = \mu_0 g_0 \cdot H_0 = \Delta E = h\gamma = hc/\lambda$$

$$\lambda = \frac{hc}{\mu_1 g_0 H_0} \qquad \mu_0 = 0 \cdot 927 \times 10^{-20} \qquad H_0 = 3300\text{G} \qquad \lambda \sim 3 \cdot 2 \text{ cm}$$

Fig. 1. Diagrammatic representation of the behaviour of a population of free electrons in an external magnetic field.

each electron would absorb energy at exactly the same value of H_0 (H_0 and H_z are used interchangeably to identify the large applied field. If its direction in space is important the z axis is specified). When unpaired electrons exist inside an actual sample the resonance condition is not satisfied at one unique applied field, but over a range of values. This is because a range of local magnetic fields is contributed by the sample itself. These add vectorially to H_0 to produce the effective resonance field at a particular electron (see below). In the simplest cases the net effect is to produce a resonance curve as shown in Fig. 2.

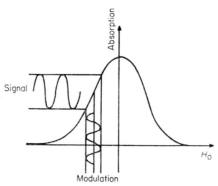

Fig. 2. Field modulation applied to a resonance absorption, showing the origin of the derivative signal. The "modulation" is a small alternating magnetic field which is superposed on the large magnetic field, H_0. Because of the modulation the total magnetic field moves in and out of the absorption curve sinusoidally in time. As a result the microwave power incident on the detector (the "signal") also varies sinusoidally with time, and greater spectrometer sensitivity can thereby be achieved.

As the resonance condition is traversed, energy is absorbed from the microwave field causing an electrical unbalance in the spectrometer.

This unbalance is proportional to the amplitude of the absorption curve (Fig. 2) for a particular value of H_0. Recording the unbalance of the spectrometer as a function of H_0 will thus yield the absorption curve.

The sensitivity of the spectrometer can be considerably enhanced by causing the resonance absorption to unbalance the spectrometer sinusoidally at a frequency ν_m. One then looks specifically at the spectrometer unbalance occurring at frequency ν_m, any other frequency variation in unbalance being ignored. This scheme is accomplished, usually, by modulating H_0 at frequency ν_m as H_0 is also being slowly swept through the resonance condition (Fig. 2). The unbalance that is recorded using this scheme is not proportional to the height of the absorption curve at the mean value of H_0 at a given point, but is proportional to the difference in height of the absorption curve between the extremes of the modulation envelope. The resultant sinusoidal unbalance is designated "signal" in Fig. 2. This, then is the signal leaving the detector (Fig. 3). Its magnitude is proportional to the first derivative of the absorption curve.

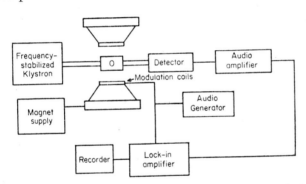

Fig. 3. Transmission spectrometer employing field modulation and phase sensitive detection. The microwave magnetic field, H_1, in the cavity is perpendicular to H_0, the applied magnetic field.

In Fig. 3 the lock-in amplifier analyses the detected and amplified signal for only those components varying at frequency ν_m. The magnitude of these components is recorded, and yields the first derivative of the absorption curve. This is the curve commonly observed in published work. (See Appendix A for a more detailed description of the operation of the spectrometer.)

The integrated area under the absorption curve (the second integral of the derivative curve) is proportional to the number of magnetic species absorbing microwave power. For a voltage-sensitive microwave detector and Lorentzian line,

$$\text{Area} \propto H_1 Q_L \chi_0 (1 + \tfrac{1}{4}\gamma^2 H_1^2 T_1 T_2)^{-1/2}$$

where H_1 is the magnitude of the microwave magnetic field at the site of the sample, Q_L is the quality factor for the microwave cavity in place in the spectrometer, $\gamma = \omega_0/H_0$, T_1 is the spin lattice relaxation time (discussed below) and, for our purposes, T_2 defines the width of the absorption line (see the Appendix for a brief discussion of detector types, line shape functions, Q_L and T_2). $X_0 = N_0 g^2 \beta^2 S(S+1)/3kT$ is the static magnetic susceptibility. Here N_0 is the number of spins per unit volume and T is the absolute temperature.

The constant of proportionality between area and N_0 may be evaluated in terms of the gains, power, time constant, etc., of the spectrometer. It is easier and more accurate, however, to relate the area under the curve of an unknown to the area under the curve of a sample with a known number of spins.

In view of the form of the above factor, relating area to N_0, several facts should be observed. (1) The same microwave power level should be applied to both samples, and this power level should be such that both spin populations remain in thermal equilibrium with the lattice ($\frac{1}{4}\gamma^2 H_1^2 T_1 T_2 \ll 1$). (2) The two samples should affect the electrical properties of the microwave cavity in the same way (giving the same Q_L). (3) Most microwave spectrometers obtain the necessary sensitivity by modulating H_0 and employing a phase sensitive detection system (Fig. 3). This modulation scheme broadens the recorded absorption line. If a narrow line is compared with a wide one and the same modulation amplitude is used on both, the narrow line will be relatively more broadened than the wide one.

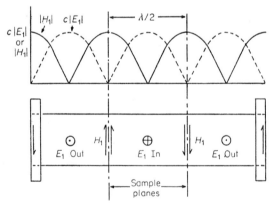

FIG. 4. Top: The magnitude of the electric, – – – –, and magnetic, ————, microwave fields as a function of distance along the centre line of the cavity. C is related to the dielectric constant of the medium, and is such that $C \mid E_1 \mid_{max} = \mid H_1 \mid_{max}$. Bottom : The cavity (a rectangular transmission type, TE_{103}) showing the sample positions and the field directions at the positions of the maxima in E_1 and H_1. H_1 and E_1 reverse directions sinusoidally in time.

A double cavity, first used by Kohnlein and Muller (1961), automatically takes points (1) and (2) into consideration. This cavity, illustrated in Fig. 4, allows the simultaneous placement of both samples on equivalent planes of the stationary microwave pattern of the cavity. Overlapping resonance lines are separated by using steel shims or Helmholz coils to provide an auxiliary field at one sample site. We have used this type of cavity to advantage with samples whose dielectric loss properties vary greatly.

In some situations one wants to know the spin concentration only as a function of time or as a function of some other external variable. Then it is necessary only to follow some resonance parameter which is proportional to the area under the curve. Assuming the line shape and width do not change, the amplitude of the resonance is such a parameter.

D. LINE SHAPES

1. Resolved structure

A second measurable parameter is the structure of the resonant absorption. This is capable of providing quite detailed information about the environment of the observed magnetic species. Obviously, this is the sort of thing an experimenter using biological systems would delight in. Unfortunately, it is seldom observed in such samples.

The principal interaction which will produce structure in a resonance absorption is called the hyperfine or "contact" interaction. The name results from the fact that the electronic wave function must be in "contact" with the nucleus (i.e. the wave function must have s character) or the interaction is zero. The energy of this interaction may be written

$$\mathscr{H}_i = A \mathbf{I} \cdot \mathbf{S} = A m_I m_s$$

where A is a constant representing the strength of the interaction and $\mathbf{I}$ and $\mathbf{S}$ are the spin quantum numbers of the nuclear and electronic systems, respectively. The last form of the equation results from the fact that both spin moments are quantized along the same axis,

$$[m_I = I, I - 1, \ldots, -(I - 1), -I].$$

This interaction is isotropic in space. Its effect is to split each electronic energy level into $2I + 1$ levels. Hence, the resonance absorption is split into $2I + 1$ equally spaced, equally intense lines.

Several cases may be distinguished.

a. The electronic wave function is confined to one atom and is interacting with the nucleus of that atom. An example of this case appears in Fig. 5. The state of the $3d$ electrons contains some s character so that the Mn^{55} nucleus is seen. $I_{Mn}^{55} = 5/2$ so that a six-line spectrum results.

b. The electronic wave function is delocalized so that the interaction is equally distributed among several nuclei. In this case $I_t = \Sigma_i I_i$ and $2I_t + 1$ equally spaced lines result. However, now a statistical effect enters. A given value of $(m_I)_t$ may possibly be achieved with several

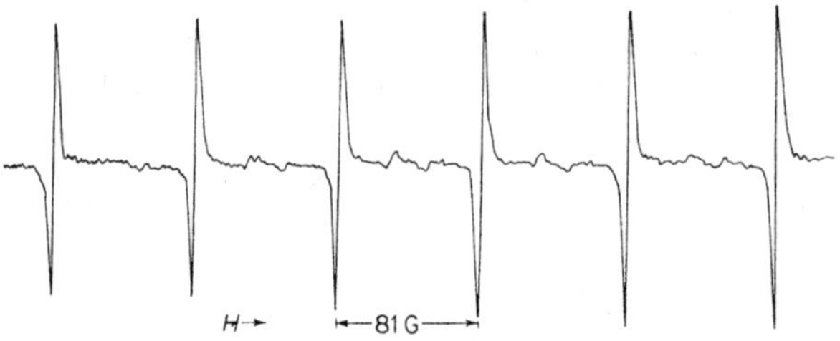

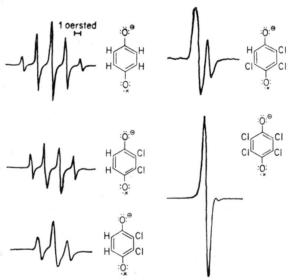

$H \rightarrow$ $\leftarrow$—81 G—$\rightarrow$

FIG. 5. The EPR spectrum of Mn^{++} (at 200 molar ppm) in MgO.

different nuclear configurations. For example, the states with $(m_I)_t = \pm I_t$ can be achieved in only one way (all nuclear spins parallel), while the states with $(m_I)_t \approx 0$ can be made in several ways. The probability of these states (and thus the line intensities) follows the binomial coefficients. This type of behaviour is shown for a series of halogenated semiquinones in Fig. 6.

In general, when the electron interacts with several different types of nuclei (the same element but in non-equivalent molecular sites, or

1 oersted

FIG. 6. The EPR spectra of a series of chlorine-substituted semiquinones.
(After Wertz and Vivo, 1955.)

different elements in otherwise equivalent molecular sites as in Fig. 6, or a combination of both) each type will contribute its own splitting to the resonance pattern. Presumably in Fig. 6 the splitting produced by the interaction with the chlorine is too small to be observed.

c. The electron is coupled primarily to one nucleus, but is delocalized enough to interact with neighbouring nuclei: a combination of (1) and (2). An example of this situation is found in copper etioporphyrin-II (Roberts and Koski, 1960). The Cu^{++} electrons are coupled most strongly to the Cu^{63} nucleus ($I = 3/2$). This coupling produces a basic four-line spectrum. The Cu^{++} electrons interact also, but to a lesser extent, with the four neighbouring N^{14} nuclei of the porphyrin ring. This interaction is equally distributed among the nitrogens, and splits each of the four lines due to Cu^{63} into nine (i.e. $2I_t + 1 = 9$). Effects, such as incomplete rotational averaging in solution, prevent the complete resolution of the spectrum.

There are three nuclei of wide biological occurrence that have no nuclear moments and will thus not produce this type of structure. These are listed in Table II with several other nuclei from which splitting might be expected.

TABLE II

Nuclei Capable of (Having Magnetic Moments) and Incapable of (Lacking Magnetic Moments) Producing Structure in EPR Absorption

Structure Producing Nuclei			Nuclei without Magnetic Moments		
Isotope	Spin	% Abundance	Isotope	Spin	% Abundance
H^1	1/2	99·9	C^{12}	0	98·9
N^{14}	1	99·6	O^{16}	0	99·8
P^{31}	1/2	100	S^{32}	0	95·1
Cl^{35}	3/2	75·4			
Mn^{55}	5/2	100			
Cu^{63}	3/2	69·1			

As mentioned above, resonance lines with well resolved structures are seldom seen in samples of biological materials. We shall next inquire why structures are sometimes unresolved, and what other parameters characteristic of such lines can be measured.

2. No resolved structure

There are several mechanisms which might operate singly or in combination to obliterate structures. In extreme cases the entire line might become unobservable.

a. Dipole-dipole broadening. The effect could be electron-electron (between two unpaired electrons on the same or neighbouring molecules) or electron-nuclear (with nuclei on the same molecule). It results from the fact that the electrons and nuclei involved have magnetic moments associated with them. The field of a magnetic moment is anisotropic in space and can be averaged out by the tumbling action of a molecule in solution. In amorphous or multicrystalline solids a broadening results. The order of magnitude of the effect is approximated by the expression $h \approx \mu r^{-3}$, where h is the local field produced at distance r from a spin with magnetic moment μ. For an electron, $h \approx 80$ G when $r = 5$ Å. If this type of broadening is a problem, magnetic dilution of the sample (increasing the mean unpaired spin separation) is an obvious solution.

The line shape resulting from this type of broadening is approximately Gaussian. The resonance may be further classified as homogeneous; that is, every magnetic species being observed is equivalent. A spin found resonating in the low field wing of the resonance line may at a slightly later time be found resonating in the high field wing due to a change in its local environment.

b. Inhomogeneous broadening. In distinction to the definition just given for a homogeneous line, a spin found resonating in the low field wing of an inhomogeneous line will always be found in that wing because its local environment will never change sufficiently (in the time of the experiment) to move it to any other section of the resonance. This type of broadening can be brought about, of course, by putting the sample in an inhomogeneous external magnetic field where the variation in applied field is large compared to the local fields found in the sample. An entirely analogous situation may arise inside some samples. In these

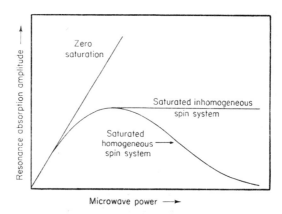

Fig. 7. Idealized power saturation behaviour.

cases the conditions inside the sample are such that the observed resonance is only the envelope of a large number of overlapping narrower lines. This is depicted on the left side of Fig. 11. The classic case of this kind is F centres in alkali-halide crystals (halogen atom vacancies occupied by a single electron). Here, there is a Gaussian distribution of types of magnetic sites. In the other extreme, an unresolved five or seven line structure displays some of the characteristics of an inhomogeneous system.

Figure 7 shows the idealized behaviour of homogeneous and inhomogeneous systems as a function of microwave power. At very low levels, the increase in signal amplitude is approximately linear with power. In this region thermal processes are fast compared with the induced processes, and thermal equilibrium is maintained. As the microwave power is further increased, the induced transitions begin to catch up with the thermal relaxation processes. Saturation begins to set in as the spin system departs from thermal equilibrium; the energy levels are becoming more and more equally populated. The behaviour of the signal amplitude of a homogeneous line is shown in Fig. 7. In addition, the line width of the homogeneous system increases as the microwave power is increased. For example, the width of an homogeneous Lorentzian line (between points of maximum slope) increases as $T_2^{-1}(1 + \frac{1}{4}\gamma^2 H_1^2 T_1 T_2)^{1/2}$. The saturation behaviour of an inhomogeneous system is distinctly different. Although each narrow line making up the envelope follows the homogeneous behaviour just described, the envelope as a whole does not; the broadening of any one component line is small compared to the width of the envelope; each individual component saturates in approximately the same way so the shape of the envelope does not change. Thus, under saturating conditions this resonance does not broaden, and its amplitude becomes independent of microwave power (Portis, 1953).

c. Exchange narrowing. This may be thought of as resulting from the actual physical exchange of unpaired electrons among different magnetic environments in the sample. As the exchange becomes rapid, the electron will see an effective magnetic field which is some average of the local fields of the various sites. Because of the averaging effect the resolved structure will collapse, with a single narrowed line resulting from sufficiently rapid exchange. The single line tends toward Lorentzian in shape.

Two cases will be mentioned. (1) The unpaired electron is exchanging with other unpaired electrons. After each exchange a particular electron is still unpaired. The different local environments then result from different nuclear configurations (the nuclear configuration is "constant" as compared with the electron relaxation times). (2) A particular un-

13

paired electron becomes paired after an exchange. This case is well illustrated in Fig. 8. In the top picture, 1 in 50 fluoranils has reacted to become a semiquinone radical (at 10^{-4} M concentration), in the middle picture only 1 in 1000 fluoranils has become a radical, while at the bottom only 1 in 9000 has reacted to become a semiquinone radical. The effect of exchange between the fluoranil semiquinone radical and the unreacted fluoranil is clear.

d. g-Value anisotropy. As noted above, the factor *g* appearing in the

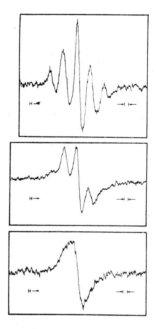

Fig. 8. The effect of electron exchange on the EPR spectrum of 10^{-4} M fluoranil semiquinone in 90% tetrahydrofuran-10% acetonitrile. The spectra were produced by reacting 0·005 M NaI with 0·005 M (top), 0·100 M (middle) and (bottom) 0·900 M fluoranil at $-75°$C. The scale magnitude of one *g* is indicated. (Eastman *et al.*, 1962.)

resonance equation represents the effective size of the magnetic moment being acted upon by the magnetic field. In crystal structures of less than cubic symmetry, the value of *g* can vary as the direction of the magnetic field is varied with respect to the crystal. Why this is so will be mentioned briefly below. The point we wish to make here is that in amorphous or polycrystalline materials an anisotropic *g*-value can produce quite broad asymmetric absorptions.

As an example of such an effect we can take the resonance of the copper in the protein complex ceruloplasmin (Fig. 9). The extremes in *g*-value are probably associated with the asymmetry of the molecular field around the copper. In a polycrystalline material there will be contributions from all g-values between extremes. Assuming random orientation in space and tetragonal symmetry (Malmstrom and Vanngard, 1960) the probability of H being parallel to the symmetry plane is twice that of its being perpendicular to this plane. The parallel configurations will therefore contribute more to the resonance absorption and it becomes asymmetric. The spectrum resembles that of a frozen solution of the copper-histidine complex.

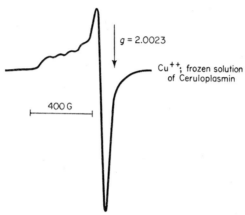

FIG. 9. The EPR spectrum of Cu^{++} in solid ceruloplasmin. $T = 77°$K. Magnetic field increases from left to right. (After Malmstrom and Vanngard, 1960.)

This broadening mechanism differs from the others discussed in that its effect is directly proportional to the applied field. Working in two different applied fields will distinguish this type of asymmetry from other possibilities.

e. Lifetime broadening (relaxation broadening). As the lifetime, T_1, of a spin state becomes very short, the energy of that state should, according to the uncertainty principle, become correspondingly uncertain. In the extreme of very short T_1's ($\approx 10^{-11}$ seconds) the resonance may go unobserved because of its width. The remedy in this situation is to lengthen T_1, usually by lowering the temperature. Although this effect can be important for conduction electrons, and for transition metal ions under certain circumstances, it is not generally important for free radicals.

E. g-VALUES

Whether or not structure is resolved one can measure the g-value for the resonance. Experimentally this requires measuring, simultaneously, the strength of the magnetic field at which resonance occurs and the frequency of the microwave source.

The value of g differs from the "free spin" value $(2 \cdot 0023)$ because the electronic state has some degree of orbital angular momentum associated with it. The electron, as a magnetic moment, interacts with the magnetic field produced by the electron, as an electric charge, moving in an orbit. This interaction is called the spin-orbit interaction, and it

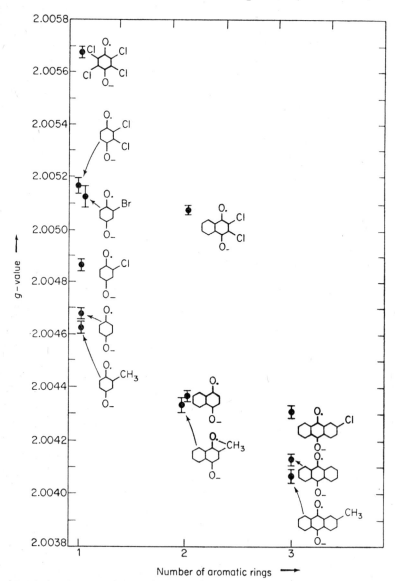

FIG. 10. g-value as a function of the number of aromatic rings for some substituted benzosemiquinones. (After Blois et al., 1961.)

is characterized by the relation $\mathscr{H}_{so} = \lambda \mathbf{L} \cdot \mathbf{S}$ where $\mathbf{L}$ is the orbital-angular momentum quantum number, and λ is the spin-orbit coupling constant. This perturbation mixes states of higher energy and perhaps different symmetry properties with the ground state. When the Zeeman energy levels are calculated using the new (admixed) wave functions it is found that their separation has been altered by the perturbation. Also, their separation may be anisotropic with respect to the magnetic field direction because the perturbation has mixed in states of different symmetry. The effect, from the point of view of the resonance experiment, is to shift the g-value. The magnitude of the shift in g-value (from $2 \cdot 0023$) may be roughly approximated as $| \Delta g | \approx \lambda / \Delta$ where Δ is the energy separation between the ground and the first excited state.

$| \Delta g |$ is very small for organic free radicals. Blois et al. (1961), using sophisticated methods, have been able to establish some correlation between g-value and molecular structure or chemical substituents on a given structure. Some of their results are shown in Fig. 10. Most laboratories are not equipped for such refined measurements.

Values of $| \Delta g |$ as high as 2 or 4 have been observed for transition metal ions (Ingram and Bennett, 1955; Bennett et al., 1957; Gibson and Ingram, 1957). For a particular ion $| \Delta g |$ will depend very strongly on the strength of the molecular or crystal fields, and on the symmetry of the site.

g-Values as aids to identification can be useful if they are supported with independent data. These data could be in the form of spectrophotometric observations, chemical analyses, etc.

F. THERMAL RELAXATION

The concept of thermal relaxation processes has been mentioned several times. Any process which changes magnetic to thermal lattice energy is such a process. A time characteristic of such a process can be defined in the following way. A spin system with a unique temperature T_s is not in thermal equilibrium with the lattice at temperature $T_L (<T_s)$. At a certain time the perturbation keeping the spin system at T_s is removed. The spin system then returns exponentially to thermal equilibrium (i.e. to the lattice temperature) according to the expression $\exp(-t/T_1)$. T_1 is the spin-lattice relaxation time. This time is characteristic of the environment of the spin system, and will limit its rate of absorption of microwave power.

T_1 may be measured by continuous or transient methods. The amplitude of the resonance signal as a function of applied microwave power can be determined. As noted in Fig. 7, the shape of this curve is dependent upon whether the spin system is homogeneous or inhomogeneous. But given the type of system, the shape of the curve is determined by the magnitude of the product $H_1{}^2T_1T_2$. T_1 can thus be determined if T_2 and H_1 are known (Portis, 1953). The transient method makes more direct use of the above definition of T_1. In this method the spectrometer is adjusted so that the amplitude of the resonance line is continuously observed. A short, intense burst of microwave power drives the system from equilibrium. The return to equilibrium is followed. See, for example, Liefson and Jeffries (1961).

G. MISCELLANEOUS MEASUREMENTS

Even though a resonance line displays no resolved structure, there is one technique available for deciding which nuclei are coupled to the resonating electrons. The line must be inhomogeneously broadened in the sense that it is made up of unresolved components, and the relaxation times must be long. The technique, called "electron-nuclear

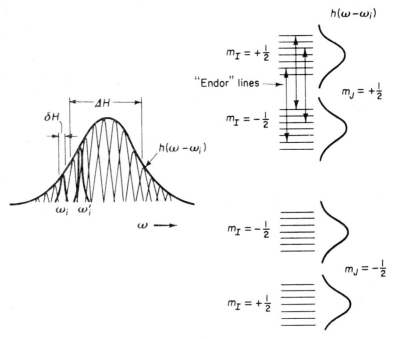

Fig. 11. Characteristics of an inhomogeneously broadened line. At left: The observed electron spin resonance line shape. One observes the envelope of many narrow resonance lines, each with slightly different resonant frequency, ω_i. At right: The energy level system which produces the observed resonance. Double resonance effects are observed in the electron resonance ($\Delta m_s = \Delta m_j = \pm 1$) when the nuclear transitions ($\Delta m_I = \pm 1$) are induced. (After Feher, 1958.)

double resonance" (ENDOR), is due to Feher (1959) and is depicted in Fig. 11 for $S = \frac{1}{2}, I = \frac{1}{2}$.

As in the transient determination of T_1, the spectrometer is adjusted so that the electron spin resonance amplitude is continuously observed. This resonance is partially saturated. Then, while observing the amplitude of the electron resonance, a second high frequency magnetic field is applied to the sample. This frequency is in the range of nuclear transitions (0 to 100 Mc/s). Assume that while this frequency is being swept, one passes through the resonance of a nucleus coupled to the electron via the contact interaction, $A\mathbf{I} \cdot \mathbf{S}$. Inducing the nuclear transitions changes the population of the corresponding electronic levels. The latter change is reflected in the resonance amplitude being observed. ENDOR effects will be observed when

$$h\nu_I = \tfrac{1}{2}A \pm \gamma_I H_0 (A \gg \gamma_I H_0).$$

The $\pm$ sign comes from the fact that the top and bottom electronic levels are not split quite equally by the nuclei. Thus, one is able to solve for both A and γ_I. γ_I identifies the nucleus since in the absence of the unpaired electron the nucleus would resonant at frequency $h\nu_I = \gamma_I H_0$.

An example of the ENDOR technique is shown in Fig. 12 for phosphorus impurities in silicon. γ_I for P^{31} at about 3000 G is ≈ 6 Mc/s. Approximately twice this frequency is indeed the separation of the resonance peaks in Fig. 12a. Thus, the extra electrons added to the

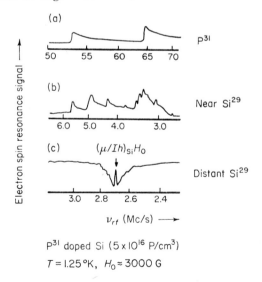

FIG. 12. Variation of the amplitude of the EPR of phosphorus impurities in silicon as a function of the frequency of the "nuclear-transition" radio frequency field. (After Feher, 1959.)

silicon lattice by the phosphorus impurities are most strongly coupled to the phosphorus nucleii themselves (highest value of the coupling constant A), but also interact in a complicated way with near Si^{29} and with many distant Si^{29} nuclei ($A \approx 0$). None of this information can be obtained from the EPR line itself, a broad symmetric line as on the left in Fig. 11.

Only one application of this type of experiment has been made to organic systems (Cole et al., 1961), but it should be generally applicable to inhomogeneously broadened lines.

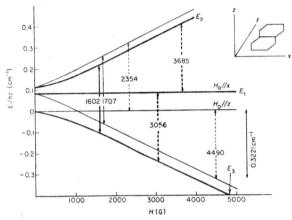

FIG. 13. Relative energies of the components of the lowest triplet state of naphthalene as a function of magnetic field. Heavy lines: Magnetic field along x axis. Thin lines: Magnetic field along z axis. The dashed transitions indicate those observed by Hutchinson and Mangum (1958), the solid transitions those observed by van der Waals and de Groot (1959). (After van der Waals and de Groot, 1959.)

Long-lived triplet states ($S = 1$) have been observed using EPR techniques. These states were first observed (Hutchinson and Mangum, 1958) using magnetically dilute, single crystals of naphthalene in durene (1,2,4,5-tetramethylbenzene). The transitions observed in the single crystal are depicted by the dashed lines in Fig. 13. Recently van der Waals and de Groot (1959, 1960) have discovered that the $\Delta m_s = \pm 2$ triplet transitions can be observed in glasses containing the excited molecules. This may make this technique applicable to biological samples in certain instances, although it should be noted that the probability of this transition is an order of magnitude less than that for $\Delta m_s = \pm 1$. The $\Delta m_s = \pm 2$ transitions observed in naphthalene-durene crystals are also shown in Fig. 13 as well as the anisotropy of the energy level system as H_0 is varied with respect to the naphthalene molecules. The $\Delta m_s = \pm 2$, $g = 4$ resonances observed in naphthalene-containing glasses are shown in Fig. 14. The $\Delta m_s = \pm 2$ transitions are allowed for H_1, both parallel and perpendicular to H_0.

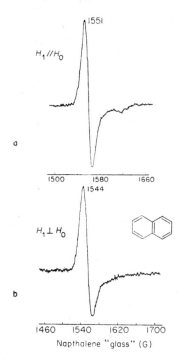

FIG. 14. The EPR spectra of $\Delta m_s = \pm 2$ transitions of naphthalene in a rigid glass. (a) H_1 parallel to H_0. (b) H_1 perpendicular to H_0. $T = 77°$K. (After van der Waals and de Groot, 1960.)

H. CONTROL OF EXTERNAL FACTORS

Photosynthetic materials seem to combine most of the features that would seem to obliterate the information carried in an EPR spectrum. Thus, it is not surprising that the measurable parameters of the resonances found in these materials yield, in and of themselves, little information concerning photosynthesis.

The obvious thing to do, of course, is to vary the physicochemical and "biological" environments of the sample material while making observations on these resonance parameters. For example, whether the observed radicals are physical or chemical intermediates can be determined from the temperature dependence of the resonance. At sufficiently low temperatures all chemical reactivity should cease. One might also include as parameters in a temperature study the rise and decay times of the resonance as the light is turned on and off. These can be followed by adjusting the spectrometer to resonance conditions and observing the response to the changing light conditions. If the rise and decay times are faster than the response time of the spectrometer then dispersion techniques, in which the incident light intensity is amplitude modulated at successively higher frequencies (Melville and Burnett, 1953), can be employed to advantage. From such kinetic studies should come information concerning the number of different types of unpaired electrons (biological pools) contributing to the observed resonance. The ambient atmosphere may be important in stabilizing the observed radicals. Pigments important in the production of unpaired electrons can be determined from an action spectrum for unpaired spin production.

Concerning the "biological" environments, working with mutant types of photosynthetic species offers many advantages. The correlation of parameters associated with unpaired spin production with quantities long considered a measure of photosynthesis, e.g. oxygen evolution, quantum yield, carbon dioxide fixation, etc. should prove enlightening. Simplified systems, obtained from selective extractions, fragmentations, or by using selective inhibitors, should provide useful information. In the extreme of simplification model systems might be employed.

Most of the approaches just listed have been employed to a greater or lesser extent. In the following section we shall be concerned with several of the experiments in which they are involved.

III. Experimental Results

A. TYPES OF SYSTEMS THAT HAVE BEEN STUDIED

It might have been noted that most of the functional dependences suggested for determination in the preceding paragraph were oriented more towards identifying the observed radical species than towards making remarks concerning photosynthesis. Indeed, to date this has been the case. Investigators have had to use their knowledge of photosynthesis to try to establish the position of the unpaired electron spins in the photosynthetic cycle rather than the reverse, i.e. making new remarks concerning photosynthesis on the basis of the EPR observations. Still, some statements concerning photosynthesis are emerging.

The first EPR observations on photo-induced unpaired electrons in photosynthetic materials were made on relatively complete systems (Commoner et al., 1956; Sogo et al., 1957). In the earliest work the physiological condition of the samples was poor; in most cases they were dried. The spectrum observed consisted of a single line approximately 10 to 15 G wide, with g-value in the region expected for most organic free radicals. There was no evidence of structure. In short, these resonances yielded a niggardly amount of information.

Since that time the development has been in two directions; one, towards working with complete photosynthetic organisms in relatively good physiological condition, and the other towards simplifying the system. Both approaches are important. It is important to know that the pools of unpaired electrons exist in real functioning photosynthetic systems. On the other hand, it has usually been the case that if one is to understand a complicated situation one must start by first understanding simplified approximations to that situation. In some senses the "simplified" photosynthetic systems used to date are still much more complicated than one would wish.

The types of photosynthetic materials which have been employed in in the various EPR investigations are as follows. 1. Extracts from photosynthetic systems. 2. "Crystals" composed of pigment molecules. The crystalline property of long range order is probably not too well met by these samples. 3. Very small fragments of photosynthetic systems. 4. Large fragments of photosynthetic systems. 5. Whole or complete photosynthetic systems.

B. RESULTS

Under the correct conditions photo-induced EPR signals can be observed in each of these classes of materials. Where g-values have been measured they are not (at present) significantly different. The line widths vary from class to class, and there are some marked differences in the overall shape of the observed resonances. Perhaps the largest variation is found in the kinetics of the rise and decay of the photosignal. The times involved vary from milliseconds to hours. We shall present the results of the resonance experiments in the order of the class distinctions just made.

1. Extracts

Anderson (1960) has observed photo-induced EPR signals in acetone extracts of spinach chloroplasts which contain essentially all of the pigment molecules in the chloroplast. The extract was evaporated to dryness and tested *in vacuo* for the presence of a photo-induced EPR signal. No signal, either in the light or in the dark, was observed. However, when small amounts of water vapour were admitted to the system, narrow (≈ 4 G wide, $g = 2 \cdot 00$) photo-induced signals were observed. Their rise and decay times were of the order of seconds and minutes, respectively. The measured rates seem to be dependent on the presence of as yet unknown components in the extract. The equilibrium amplitude of the induced photo signal is a function of the water vapour pressure (Fig. 15).

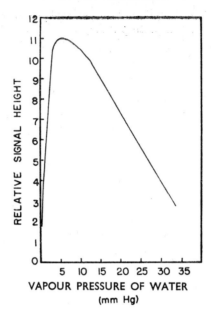

FIG. 15. The light induced EPR amplitude in an acetone extract of spinach chloroplasts as a function of the pressure of water vapour present. The extract was evaporated to dryness and evacuated before admission of water vapour.

2. Crystals

A dependence of the photo-induced EPR signal on the presence of water vapour has also been observed by Holmogorov and Terenin (1961) in a recent study on "crystalline" chlorophyll $(a + b)$. In their experiment the sample space was first evacuated, then water vapour was admitted. They observed two overlapping resonance lines (Fig. 16). One $(g = 2 \cdot 0035$, half-width $= 11$ G; "a" in Fig. 16) is always present. The second $(g = 2 \cdot 0030$, half-width $= 7$ G) is light-induced in the presence of water vapour. The presence of p-benzoquinone (at 2×10^{-2} mm Hg pressure) was also effective in stabilizing in light-induced free radical. The water affects much faster rise and decay kinetics than does the p-benzoquinone.

The role of the water is not entirely clear. By analogy with the effect of the benzoquinone it would appear that it might act as an electron acceptor, a common role for benzoquinone. However, it is difficult to visualize water in this role. Perhaps it is worth noting that the presence of water vapour aids in the initial formation of the chlorophyll crystals (Jacobs and Holt, 1954). Thus, water-induced structural modifications may be responsible for the formation and stabilization of the light-induced unpaired electrons.

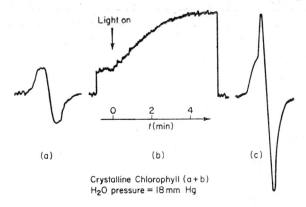

FIG. 16. Light induced EPR in "crystalline" chlorophyll $(a + b)$ *in vacuo* in the presence of water vapour. (a) dark, (b) response to light, (c) superimposed dark and light signals. (After Holmogorov and Terenin, 1961.)

3. Very small fragments

EPR experiments have been performed on systems composed of very small fragments obtained from both the aerobic and anaerobic growing photosynthetic systems (Androes *et al.*, 1962). These fragments are obtained in both cases by rupturing the cell structure in intense sonic fields, and then using centrifugation techniques to obtain the desired particle size fraction.

The particles obtained from the purple bacterium *Rhodospirillum rubrum* are called chromatophores (Frenkel and Hickman, 1959). They are spherical, have diameters of approximately 200 Å, and contain all of the light absorbing pigments of the whole bacterium. There are two indications that the local pigment environment is left essentially unaltered by the method of preparation. The particles will perform cyclic photophosphorylation reactions, and the optical absorption spectrum is left unaltered by the mechanical disruption of the cell.

The sample of chromatophores is placed in an appropriately buffered aqueous medium (Frenkel and Hickman, 1959) and inserted into the EPR cavity. (See the Appendix for the method of inserting water into the microwave field.) The observed photo-induced resonance absorptions seem to be the same in g-value, width and shape as those observed in the whole bacterium (see Fig. 19). The rise and decay times of the signals are somewhat altered as might be expected when the terminations of the energy transfer system have been removed. In particular, at room temperature the decay scheme consists of a fast and a slow component with time constants of about 2 sec and 20 sec, respectively. The observed decay in the whole cell in aqueous suspension is instrument limited. That is, it is completed in less than 1 sec, the time constant of the spectrometer. When the chromatophore sample is prepared as a dried film on a slide, the rise and decay times eventually become instrument limited as the temperature is reduced to $-150°C$. This is also true of dried films of the whole bacteria (see Fig. 20 and the discussion connected with it).

Where the spin concentration is photo-induced it is of some interest to determine the action spectrum of the equilibrium spin concentration as a function of the wavelength of the incident light. In previous attempts to obtain such a spectrum (Sogo et al., 1961) the sample has been infinitely thick compared to the distance the active light penetrates into it. Self-absorption effects were pronounced, shifting the maximum in the action spectrum to the long wavelength side of the

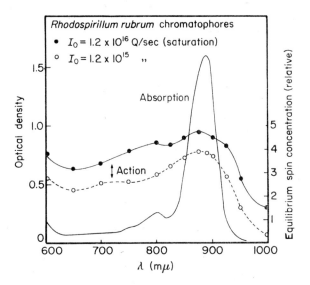

Fig. 17. Action spectrum of *Rhodospirillum rubrum* chromatophores. The action spectra were performed on different samples on different days. They cannot be compared in absolute magnitude since this varied unpredictably from sample to sample. However, the shape of either curve is reproducible (*e.g.* the top action spectrum always peaked at ~800 mμ and at ~880 mμ). For the action spectra the monochromator half-intensity band width was 15 mμ.

chlorophyll absorption maximum. The chromatophore samples are non-scattering, and the sample used here was only slightly coloured to the eye (the O.D. recorded in Fig. 17 represents a sample of the same thickness, but a factor of $3·3$ greater in concentration than those used in the spin determination). As seen in Fig. 17, the action spectrum now peaks at the absorption maximum of the bacteriochlorophyll.

This spectrum still suffers from self-absorption effects, but is nearer the true situation. It is evident that the bacteriochlorophyll is the principal pigment responsible for spin production.

The other type of small fragment can be obtained from spinach leaves (Park and Pon, 1961). The preparative procedure gives one fragments of the lamellar structure of the chloroplast. The smallest fragments are oblate spheres with principal dimensions of 100 and 200 Å. Larger fragments in the preparation appear to be made up of agglomerations, still in sheet form, of the smallest particles. These particles, or the agglomerations of them, will perform the Hill reaction, and when combined with the water-soluble protein leached out in the preparative procedure they will fix carbon dioxide. Again, the optical absorption spectrum is unaltered by the preparative procedure. These particles have been called quantasomes (Park, 1962). We have made the suspending solution 10% in methanol (Milner et al., 1950) to enhance the chemical stability of the quantasome preparation during the long periods of illumination to which it was subjected.

It is interesting to note that with a modulation amplitude of only one G and a signal-to-noise ratio of about 20, only one resonance line has been observed in these particles. This resonance line is approximately 10 G wide, has an asymmetry similar to that shown in Fig. 19 for *R. rubrum*, and has relatively fast kinetics. This contrasts with the observations on whole chloroplasts. In the whole chloroplast two overlapping resonance lines appear (see the following section). The resonance occurring in the quantasomes corresponds to the narrow resonance observed in the whole chloroplasts.

At the time of writing, the action spectrum for the production of unpaired spins in the quantasomes has not been completely determined. When the quantasomes are in aqueous suspension at room temperature the kinetics of the photo-induced signal are similar to those for the signals induced in the chromatophores in equivalent circumstances. However, when the quantasomes are subject to certain conditions the decay time for the photo-induced signal becomes very long (at least hours). These conditions are (1) when the quantasomes are thoroughly dried, and (2) when they remain in water, but their temperature is reduced to less than 0°C. Changing the ambient atmosphere from air to nitrogen does not effect the results. Thus, it seems that in these particles diffusing water molecules, or perhaps molecular species carried by the water, must interact with the site of an unpaired electron to bring about its return to the diamagnetic state.

An attempt was made to substitute deuterons for all of the exchangeable protons in both the chromatophores and the quantasomes. This was done by substituting D_2O for H_2O in the preparative procedure.

No effect on the resonance line shape was observed. Since a relatively complete exchange of water is probable, the observed radical species either has no protons in its environment or the protons in its environment are non-exchangeable. Replacing H by D in the immediate environment of an unpaired electron should narrow the resulting resonance line. (In this connection see also Sec. III, B.5.)

4. Large fragments

Because of the extremely small dimensions of the particles discussed in the previous section we list the chloroplast as a "large" fragment of a photosynthetic system. This fragment has been studied widely, and its photosynthetic abilities and characteristics are well known. In the EPR experiment it behaves when in aqueous suspension essentially as do the whole green algae, so it will be discussed in connection with them (Sec. III, B.5).

Much larger leaf fragments have been employed in EPR experiments. In particular, eucalyptus leaves give small light-induced signals (Sogo *et al.*, 1957). These signals rise rapidly when the leaf is illuminated (< one second), but remain for hours after the light has been turned off.

Bubnov *et al.* (1960) report observations on photo-induced EPR in the leaves of the cereals *Triticum vulgare*, *Hordeum vulgare* and *Avena sativa*, and on the variegated leaves of the decorative plant *Sancheria*. In the cereal leaves a doublet (splitting = $1 \cdot 8$ G, $g = 2 \cdot 004$) with fast rise and decay times was observed. In *Sancheria* a broad, complex spectrum was observed. The exact form of this spectrum was different in the chlorophyll and the non-chlorophyll containing parts of the leaves. These observations have not been independently corroborated. Using model photochemical reactions as a basis for reasoning, this group has attributed the photo-induced doublet in the cereal leaves to an oxidized form of ascorbic acid.

5. Whole systems

Systems of this type have been studied while essentially dry and, more recently, while in aqueous suspension (i.e. in relatively good physiological condition).

The purple bacterium *Rhodopseudomonas spheroides* also exists in a blue-green mutant form in which the carotenoid pigments are absent (Griffiths *et al.*, 1955). The EPR spectra of the wild and this mutant form of the bacterium were compared in the dry state. Few differences were noted. The mutant is subject to photo-killing when exposed to both light and oxygen (Dworkin, 1958). In the experiments performed the mutant and the wild type both reacted to light-plus-oxygen in a parallel way. The photo-induced resonance amplitude was monitored as a function of time after exposure of the sample, in the form of a thin film, to oxygen. Both resonance amplitudes increased to an asymptotic value of approximately three times their initial values after about 2 h exposure to oxygen in the light. Thus, oxygen appears to stabilize the radicals formed in these bacteria. (In this connection see also Section IV below.)

One difference in behaviour which was noted was the way in which the two resonances power saturated. This is shown in Fig. 18. The wild

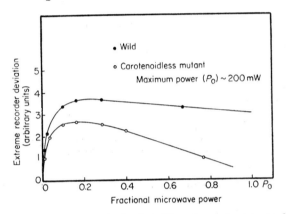

Fig. 18. Power saturation of the EPR in the wild and a mutant type of *Rhodopseudomonas spheroides*. Dried films at room temperature.

type seems to saturate more like an inhomogeneously broadened system than does the mutant, and to have a somewhat shorter relaxation time. This would suggest that the mutant has fewer varieties of unpaired electrons participating in the production of this signal.

Dried films are a natural choice for the form of the sample if the material is to be studied below zero degrees centigrade. Such studies have been carried out on whole *Rhodospirillum rubrum* (Fig. 19). The principal points to note are that photo-induced signals can be produced at these low temperatures, that the decay scheme contains multiple decay times, and that the shorter decay times predominate as the temperature is reduced.

This behaviour leads one to believe that at a given temperature one is observing free radicals in several biological pools. If this is so, the observed resonance should be of the inhomogeneously broadened type, i.e. it should be the envelope of several narrower overlapping resonance lines. There is additional evidence that this is so. The line width does not increase with applied microwave power (one criterion), but the saturation is not ideally inhomogeneous as in Fig. 7. Rather, its satura-

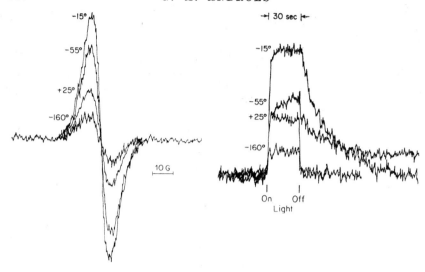

Fig. 19. Shape, amplitude and rise and decay kinetics of the EPR signal in *Rhodospirillum rubrum* as a function of temperature. Dried film.

tion behaviour is similar to that observed in *R. spheroides* (upper curve in Fig. 18).

The derivative of the photo-induced resonance which is observed in *R. rubrum* is shown in Fig. 20. It is seen that, disregarding the slight asymmetry that is observed, the curve is more nearly Gaussian than Lorentzian. Note that with the modulation amplitude at only 0·6 G there is no evidence of resolved structure.

One resonance with easily resolved structure which is observed in chloroplasts and green algae is that of ionic, or cubically bound, Mn^{++}. It is not observed in purple bacteria. This resonance does not seem to be photosensitive. As shown in Fig. 5 the structure consists of six lines. The Mn^{++} concentration in the samples is as high as 10^{-6} to 10^{-5} M.

In contrast to the fast low temperature kinetics just cited for *R. rubrum*, the decay of the photo-induced signal in chloroplasts and algae becomes very long as the temperature is reduced (Sogo *et al.*, 1959). The slowing of the kinetics for these materials is not abrupt at the freezing point of water as it appears to be in the quantasome particles mentioned above.

In green algae and chloroplasts in aqueous suspension two overlapping photosensitive resonance lines can be resolved with amodulation amplitude of the order of 1G (Commoner *et al.*, 1957; Commoner, 1961; Allen *et al.*, 1961; and Weaver, 1961). The resolved structure is made up of a narrower line (≈ 10 G wide) with $g = 2·002$ and a broader

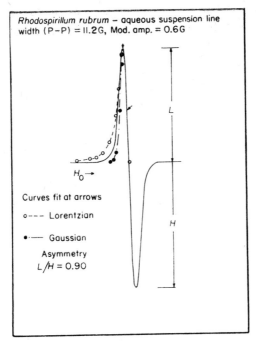

Rhodospirillum rubrum – aqueous suspension line width (P–P) = 11.2G, Mod. amp. = 0.6G

$H_0 \rightarrow$

L

H

Curves fit at arrows

o--- Lorentzian

•—— Gaussian

Asymmetry

$L/H = 0.90$

FIG. 20. Line shape analysis of the EPR spectrum of *Rhodospirillum rubrum* in aqueous suspension at room temperature.

line (of the order of 20 G wide) with $g = 2 \cdot 005$. The narrow line is fast rising and decaying (a decay time of 50 to 60 msec is reported by Allen *et al.*, 1961) while the broad line takes many minutes to decay. Commoner *et al.* (1957) have partially resolved this broad resonance into five lines each separated by 6G.

Some insight into the origins of these two signals has been obtained through the use of mutants. Allen *et al.* (1961) have observed the two overlapping resonance lines in several green algae. Some of their results for *Chlorella pyrenoidosa* are reproduced in Fig. 21. The dependence of the relative signal amplitude on the wavelength of the light incident on the sample made it probable that the two resonance lines were resulting from the absorption of light by two different pigment systems. The pigment system absorbing the light which was producing the broader resonance was implicated with chlorophyll *b* by using a *Chlorella* mutant in which this pigment was absent. In this latter case the broad resonance failed to appear.

The magnetic environment partially responsible for the width of both the resonances observed in green algae has been determined. Commoner (1961) reports EPR observations on *C. pyrenoidosa* cultured in 99·9%

D_2O growth medium by Chorney et al. (1960). Both resonances observed in *C. pyrenoidosa* were significantly narrowed by substitution of *D* for *H*. This narrowing results from the much smaller magnetic moment of the deuteron.

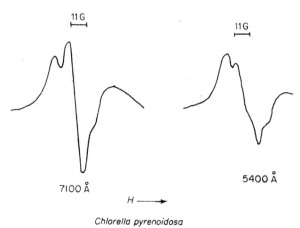

FIG. 21. Light induced EPR signals in *C. pyrenoidosa* at two different wavelengths of illumination. (After Allen *et al.*, 1961.)

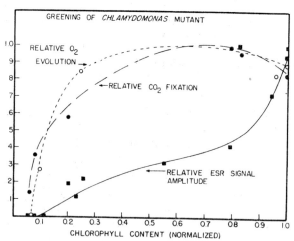

FIG. 22. The EPR amplitude, O_2 evolution rate and $C^{14}O_2$ fixation rate as a function of chlorophyll ($a + b$) content during the greening of a yellow mutant of *Chlamydomonas reinhardi*.

It has been shown that few long-lived radicals exist in etiolated as compared with fully greened leaves (Commoner *et al.*, 1954). We are attempting to inquire further into the relationship between the unpaired spin concentration and the chlorophyll content of the sample material. This study is being carried out on a yellow (chlorophyll-less) mutant of *Chlamydomonas reinhardi* supplied to us by Dr. Ruth Sager.

In this experiment six flasks of cell cultures were grown in the dark until the total cell volume per flask was sufficient for the needs of the experiment. When this condition was fulfilled, all six flasks were exposed to uniform illumination. At this time the total chlorophyll content of the cells was quite low. On being exposed to light, though, the cells start to green, regaining their total chlorophyll compliment in 15 h. One flask of cells was harvested immediately after exposure to light. The others were harvested at intervals of 3 h thereafter. The uniform illumination in which the cell cultures were greening ($\sim$700 ft. candles) was approximately 5 to 10 times smaller than the illumination to which they were exposed in the later parts of the experiment. The later parts consisted of several independent measurements as follows. On each cell sample, a chlorophyll content, an oxygen evolution rate, a carbon dioxide fixation rate, and an equilibrium EPR signal amplitude (using white light in each case) were determined. The oxygen evolution and carbon dioxide fixation experiments were performed on very dilute suspensions so that variation in observed rates can be attributed to physiological changes in the cell, and not to the shielding effects of an optically dense suspension. The EPR experiments were performed on very dense aqueous suspensions so that, excepting possibly the first one or two measurements at zero time and at 3 h, all of the light is absorbed by the sample. All of these measurements were normalized to the same volume of wet packed cells.

Our preliminary results are plotted in Fig. 22. The EPR amplitude recorded is the composite of two overlapping lines. These lines have been resolved in older cultures, but instrument sensitivity has prevented our resolving them during the greening period. The observed line shape and the concentration ratio of chlorophyll *a*/chlorophyll *b* remain constant during the greening. The equilibrium EPR amplitude, the

TABLE III

Chlorophyll Concentration as a Function of Time in a Greening Chlamydomonas *Mutant*

	Time (hours)					
	0	3	6	9	12	15
Relative chlorophyll ($a + b$) concentration	0·05	0·08	0·20	0·70	0·84	1·00*

* Corresponds to 2·7 mg chlorophyll ($a + b$)/ml wet packed cells.

rate of $C^{14}O_2$ fixation and the rate of oxygen evolution are all normalized so that the maximum value observed in each variable is equal to one. These are plotted against chlorophyll $(a + b)$ content, also normalized to one at maximum value. The chlorophyll $(a + b)$ concentration as a function of time is given in Table III.

There are several clear results. (1) A photo-induced EPR signal grows in as the chlorophyll content increases. (2) The relationship between chlorophyll content and EPR amplitude is not linear, the larger part of the signal growing with the last 20% of chlorophyll synthesized. (3) The rate of photosynthesis, as measured by rate of oxygen evolution, and the rate of turnover of the carbon cycle, as measured by the rate of $C^{14}O_2$ fixation, are maximal long before the EPR signal starts its steepest rise.

IV. CONCLUSIONS

One sequence in the photosynthetic cycle which at present is among the least understood is that series of steps by which the electromagnetic energy of the absorbed light quantum is converted to chemical energy. In recent years a type of spectrophotometry has been developed which allows one to follow spectral changes which take place shortly after the illumination of a photosynthetic system, presumably the steps of quantum conversion. These techniques have been developed especially by Witt et al. (1961) and to a lesser degree by Kok (1959). Witt et al. have been able to make tentative identification of several compounds which undergo electronic transitions in times as short as 10^{-6} seconds after illumination of the system. Some of these changes are temperature independent.

When photo-induced EPR was first observed in photosynthetic systems it was hoped that a method for following the quantum conversion process was at hand. Response times in the EPR experiment are not nearly as rapid as can be obtained in the spectrophotometric experiment, but upon onset of illumination the resonance appeared as rapidly as the EPR spectrometer could follow them, and a portion of the observed resonance is temperature independent. The EPR approach to quantum conversion will probably prove to be valuable, but several years have elapsed and the sites of the observed radicals have still not been identified. There is, as yet, insufficient evidence to place them unequivocally in the main pathway of quantum conversion, as distinct from putting them in some less important side reaction. What then, can be said of the photo-induced EPR signals which are observed?

Chlorophyll is definitely the sensitizing pigment for their formation. This is evident from the absence of the photo-signal in etiolated leaves and chlorophyll-less algae mutants, and their appearance with the sub-

sequent greening of these materials (Sec. III, B.5). Apparently caro-
tenoids, the only other pigments always present in photosynthetic
systems, are not involved in a fundamental way (Sec. III, B.5).

Given the presence of chlorophyll, the formation of photo-induced
signals, as such, requires little else of the living organism. The structure
into which the chlorophyll is incorporated in the chromatophore or
quantasome as it exists there is not required (extracts, Sec. III, B.1);
nor is the presence of the pigment molecules associated with chlorophyll
in the structure of the photosynthetic unit required (chlorophyll
crystals, Sec. III, B.2). It should be remarked, however, that the presence of water or possibly some electron acceptor is required in both
the experiments just cited. The condition of "life" is obviously not
required. To be sure, these signals differ in many respects from those
observed in photosynthetic systems in good physiological condition,
but they do suggest that at least some of the radicals contributing to
the resonances in "living" photosynthetic systems might be of a
physical nature. That is, their formation should be relatively independent of temperature, which it would not be if they were the intermediates in some chemical reaction. This is, indeed, the case (Fig. 19).
In the purple bacteria it appears that at a given temperature two or
more radicals are contributing to the observed resonance (multiple
decay times; inhomogeneous line behaviour, Sec. III, B.5), and that as
the temperature is lowered various routes by which these radicals are
utilized become blocked. At the lowest temperatures studied they apparently disappear by reversal of the mechanism by which they were formed.

It should be emphasized in this connection that some of the faster
optical density changes observed by Witt et al. (1961) in chloroplasts
and algae are temperature independent at least down to −150°C.
Others occur, but are irreversible at this temperature. Irreversible
optical density changes have also been observed to occur in the bacterium *Chromatium* at 77°K by Chance and Nishimura (1960). However,
reversible changes have been observed in the chromatophores of
R. spheroides by Arnold and Clayton (1960) at 1°K. Witt et al. correlate
the changes they observe with the oxidation of chlorophyll a and a
cytochrome. Chance and Nishimura identify the change they observe
with the oxidation of a cytochrome. The unpaired electrons which have
been observed in the above systems may be associated with these molecular components.

Of the two resonances observed in green algae and chloroplasts, the
narrow one with the faster kinetics seems to be more closely associated
with the quantum conversion act. This is shown by the fact that the
broad resonance is absent in the quantasomes although present in the
chloroplasts before the final stages of sample preparation. Thus, while

this broad signal might result from light absorbed by the auxiliary pigments in these systems (e.g. chlorophyll *b*, Allen *et al.*, 1961) its position in space seems to lie outside the basic photosynthetic unit. The narrow resonance in aerobic systems seems to correspond to the narrow single resonance observed in anaerobic systems.

The fundamental difference between the arobic and the anaerobic photosynthetic systems is reflected in the EPR experiment in several ways. Neither the six-line spectrum due to Mn^{++} ion nor the second broader resonance with slow kinetics is observed in the anaerobic systems. In addition, the low temperature decay kinetics are markedly different, the photosignals in the aerobic systems being produced irreversibly at sufficiently low temperatures. Finally, there is apparently a complicated dependence of the amplitude of the photo-induced EPR signal in whole anaerobic photosynthetic systems on oxygen concentration, light intensity and time (see Sec. III, B.5 and below). This dependence is absent in aerobic photosynthetic systems, but it must be remembered that these systems liberate oxygen during photosynthesis

We have placed the resonance observed in *R. rubrum* and the narrow resonance observed in green algae and chloroplasts in what may prove to be the basic photosynthetic units for these two different types of systems. Some of the radical species that can be produced at room temperature can be produced at liquid nitrogen temperatures, and may be correlated with the molecules undergoing electronic transitions at these temperatures. Is there evidence for placing the observed radical species directly in the photosynthetic pathway?

Several experimental cases lead one to conclude that when photosynthesis is limited or inhibited in any way the photo-induced EPR becomes considerably easier to see. If the radicals exist in several biological pools then the resonance will become easier to see when the kinetics of the system are altered in such a way as to make these pool sizes larger. Thus, the photosynthetic systems in the best physiological condition require the highest light intensities to attain a given EPR amplitude. When the systems are dried or fragmented, the kinetics slow down and a given EPR amplitude is obtained with a much lower light intensity. It was remarked in Sec. III, B.5 that the presence of oxygen enhanced the amplitude of the EPR observed in *R. spheroides*. This effect is much more marked as well as being considerably more complicated in *R. rubrum* and *R. spheroides* when they are in good physiological condition (in aqueous suspension in a closed system) where it is known that traces of oxygen markedly inhibit anaerobic photosynthetic processes (van Niel, 1941; Clayton, 1955). A photo-induced EPR is observed in oxygen-free samples of these bacteria. When small amounts of oxygen are admitted the resonance amplitude is generally enhanced, but the amplitude behaves in a complicated way as a function of light intensity, oxygen concentration and time. This behaviour has not, as yet, been completely elucidated. (It might be remarked here that the resonance observed in the *R. rubrum* chromatophores displays none of the complicated behaviour observed in the whole cells.)

In the study of the greening of the chlorophyll-less mutant of *C. reinhardi*, described above, one has a case in which the photo-induced EPR becomes visible only after photosynthesis, as measured, for example, by oxygen evolution, has become limited. In the early stages of greening one might expect the low chlorophyll content to be the factor limiting photosynthesis. This is the case in the early stages of the greening of etiolated barley leaves (Smith, 1954). After a short induction period in which chlorophyll *a* is synthesized and the oxygen evolution rate is zero, there follows a period in which the rate of oxygen evolution is linear with chlorophyll *a* content. Following this period, the rate of photosynthesis, again as measured by the rate of oxygen evolution in etiolated leaves, is limited by other factors (Blaauw-Jansen *et al.*, 1950). A state exists in which the rate at which energy is absorbed is greater than the rate at which it can be used in subsequent reactions. The number of unpaired electrons in the biological pools presumably lying between the point of energy absorption and the limiting reaction, is thus enhanced.

These pools of unpaired electrons are certainly quite near the point of energy absorption (they can be produced in the smallest photosynthetic units which have so far been removed from plant materials and purple bacteria, and at temperatures as low as 77°K) and are quite possibly in the path of energy flow into the photosynthetic cycle. The facts are consistent with, but, of course, do not prove, the latter assertion.

The statement regarding photosynthesis which emerges from these considerations is that one or more of the steps following the absorption of the light quantum is of a physical nature. That is, it, or they, are essentially independent of temperature. This is no new statement; it was made soon after the original observation of photo-induced EPR in photosynthetic systems, and has been made more recently in connection with the temperature independent changes in absorption spectra. When made in connection with the absorption changes this assertion has some validity because of the additional information available about the optical absorption spectra of the molecular content of the photosynthetic apparatus. Here, molecular species thought to lie in the photosynthetic pathway have been tentatively identified with the observed changes in absorption. When made in connection with the EPR observations it has been based on the assumption that the observed radical species lie in the photosynthetic pathway. Although the validity of this assumption has been hard to prove, the growing body of information concerning the radicals is consistent with it. A physical description of the quantum conversion process has been given by Calvin (1961).

Before the EPR experiment can be used to make more concrete and far reaching statements about photosynthesis, the sites of the unpaired electrons ought to be identified. This problem might be approached in several ways.

One possible approach is the study of a wide variety of mutant organisms in which specific known molecules were absent or altered. Success in this approach depends on selecting the right mutant.

A complete understanding of the difference between the photo-induced EPR in aerobic and anaerobic photosynthetic systems would aid in the identification of the sites. Additional correlations with the behaviour of the observed optical density changes in these systems could well supplement this approach.

An approach more on the physical side is that of electron-nuclear-double resonance. As has been indicated (Sec. III, B.5) the photo-induced resonances display some behaviour characteristic of inhomo-geneous spin systems. Thus, a double resonance experiment may provide information concerning the radical site. The success of this experiment will depend on the unpaired electrons being coupled isotropically (i.e. through the contact interaction) to some uncommon nucleus in the system. Finding that the unpaired electrons are coupled only to protons will probably not yield a great deal of information about the site.

Some attention has been given to photo-induced EPR in model systems (Tollin et al., 1960; Kearns et al., 1960; Kearns and Calvin, 1961). These systems, whose form is based on the ultrastructure exhibited in electron micrographs of photosynthetic materials, are composed of lamellar layers of electron-donating and electron-accepting molecules. Several parallels with real photosynthetic systems have been found. The models have the ability to separate positive and negative charge (oxidizing and reducing power) when illuminated. EPR is observed in them. The amplitude of the resonance is photosensitive. That is, upon illumination of the system the amplitude increases or decreases, depending critically upon the electron donating and accepting properties of the molecules involved. The photo-induced paramagnetism is essentially temperature independent. As these systems become better understood and more subtle donor-acceptor combinations are used (e.g. molecules extracted from photosynthetic systems) the conditions under which the unpaired electrons are produced will be so defined as to help identify the sites in the real systems. However, this is certainly a long range project since the sites of the unpaired electrons in the lamellar systems are also, as yet, not known with certainty.

V. Appendix

A. THE EPR SPECTROMETER

We shall briefly describe the steps necessary to obtain a recorded resonance curve using a transmission type of spectrometer, as shown in Fig. 3. The procedure can be divided into several steps. (1) After the klystron (Fig. 3) has been turned on and allowed to come to thermal equilibrium its frequency is adjusted to the resonant frequency of the sample cavity (with the sample in place). This condition is determined by monitoring the amount of microwave power incident on the detector. When the klystron frequency is far removed from the resonant frequency of the cavity no power reaches the detector. As the klystron frequency approaches that of the cavity power begins to be coupled from the waveguide into the cavity and from the cavity into the waveguide leading to the detector. The power incident on the detector is maximum when the cavity frequency and that of the klystron coincide. It is important that the cavity frequency and the klystron frequency remain coincident during an experiment. The spectrometer loses sensitivity rapidly as these two frequencies drift apart. For this reason most spectrometers have an automatic frequency control (AFC) circuit (not shown in Fig. 3) which senses the drift of the klystron frequency from that of the cavity, and sends a correction signal back to the klystron, correcting its frequency back to the resonant frequency of the cavity. This circuitry is made operative at this point. (2) The microwave power supplied to the cavity by the klystron is then adjusted to the desired value using a calibrated attenuator between the klystron and the cavity (not shown in Fig. 3). (3) The modulation amplitude is adjusted. As can be seen in Fig. 2, the signal leaving the detector goes from zero at zero modulation amplitude to a maximum value when the modulation amplitude is on the order of the width of the absorption curve (between points of maximum slope). If resonance curves which are undistorted by modulation broadening are desired, the modulation amplitude should be small compared to the line width. (4) The recorder is then turned on, and a workable noise level is achieved by adjusting the gain and time constant of the lock-in amplifier. Most of the noise in the spectrometer is generated in the klystron and detector. Increasing the amplifier gain increases the recorded noise, while increasing the time constant decreases it. The time constant cannot be increased indefinitely, however. The longer the time constant the more slowly must one traverse the resonance condition. General spectrometer stability with time and the time available to do the experiment limit the convenient time constants to $\gtrsim 10$ sec. (5) The magnetic field, H_0, is adjusted to approximately the value required for resonance (as determined by the resonance equation), and then is swept (linearly with time) through the resonance. Since the recorder chart also moves linearly with time an absorption curve derivative will be recorded on a scale which is linear with magnetic field.

B. DETECTORS

Two types of detecting elements are used in spectrometers to remove the resonance information from the microwaves incident upon them. One of these is sensitive to the incident microwave power (e.g. bolometers), the other is sensitive to the microwave electric field (e.g. crystal diodes). The first gives a signal proportional to $H_1{}^2$ (H_1 is the magnitude of the microwave magnetic field at the site of the sample), the second a signal proportional to H_1. Thus, the exact shape of the recorded resonance curve and the power saturation curves will depend on the particular type of detector employed.

C. LINE SHAPE FUNCTIONS

Two resonance line shapes result from the solution of problems concerning nuclear and electronic paramagnetism. These are the Gaussian and the Lorentzian line shapes. The line shape observed in most actual resonance experiments falls somewhere between these two forms. They may be defined as follows:
Gaussian

$$g(\omega - \omega_0) = T_2\pi^{-1} \exp[-(\omega - \omega_0)^2 T_2{}^2 \pi^{-1}]$$

Lorentzian

$$g(\omega - \omega_0) = T_2\pi^{-1} [1 + (\omega - \omega_0)^2 T_2{}^2]^{-1}$$

D. T_2

The parameter T_2 appearing in these expressions is related to the width of the resonance line. If one determines the distance between points of maximum slope (the peak-to-peak separation in the derivative representation of the absorption line) one finds for
Gaussian

$$\Delta\omega_{p-p} = (2\pi)^{1/2} T_2{}^{-1}$$

Lorentzian

$$\Delta\omega_{p-p} = 2(3^{-1/2} T_2{}^{-1})$$

As a time characteristic of the resonating system T_2 may be thought of as follows. Magnetic moments precess around the direction of a field H_0 at a frequency given by the resonance condition. If two magnetic moments in the sample are precessing in phase at a given time, then at time T_2 later they will be out of phase, say, by π radians. This results from slightly different precessional frequencies at the different magnetic sites (i.e. different effective values of H_0) and this, of course, fixes the line width.

E. Q

The cavity Q enters as an important parameter in the spectrometer sensitivity. The cavity serves to concentrate the microwave field (i.e. to increase the magnitude of H_1). This it does by establishing a standing microwave pattern in the cavity. There are energy losses in the cavity walls and in the devices by which the cavity is coupled into the spectrometer, as well as in the sample itself. The loaded cavity Q is defined as the ratio of the energy stored to the energy dissipated, and is a measure of how efficiently the sample concentrates H_1. In general, the greater the efficiency the greater the sensitivity.

F. INSERTING WATER INTO CAVITY

The energy losses in the sample are usually very much smaller than those in the cavity walls and coupling devices. However, when highly polar substances such as water are inserted into the cavity this situation may change. The dipole moment of the polar substance interacts strongly with the microwave electric field, and energy dissipation is enhanced. This can drastically reduce spectrometer sensitivity. If the polar substance could be placed in a region in the cavity of zero electric field, then the electric interaction would be zero and the sensitivity would remain unimpaired. This can be partially achieved. Reference to Fig. 4 shows that in the plane of maximum H_1 (where we would like to place the sample) the field E_1 is zero. A thin planar cell (~ 0.25 mm thick) positioned in this plane allows the insertion of water into the cavity with little loss in spectrometer sensitivity.

G. ILLUMINATING THE SAMPLE

Light is admitted into the cavity usually in one of two ways. (1) The hole in the cavity wall is at the end of a short section of tubing (of the same dimensions as the hole) which to the microwaves looks like a waveguide beyond cut-off. This terminology refers to the fact that microwaves of a given wavelength, λ_g, will not travel down waveguides which are too small. The limiting dimensions are related to λ_g. The tube, then, through which light is admitted to the cavity is much too small for waveguide propagation, and the standing pave pattern in the cavity is not seriously altered. (2) The hole in the wall is composed of a system of narrow slits oriented so as not to interrupt the flow of current in the cavity walls, and each with width very much smaller than λ_g. In this case, the system of slits looks like a solid wall to the microwaves, but the wavelength of the light is very much smaller than the width of the slits, so it passes easily through.

REFERENCES

NMR

Abragam, A. (1961). "The Principles of Nuclear Magnetism", Oxford University Press, London.
Andrew, E. R. (1955). "Nuclear Magnetic Resonance", Cambridge University Press, London.

EPR

Ingram, D. J. E. (1955). "Spectroscopy at Radio and Microwave Frequencies", Butterworths Scientific Publications, London.
Ingram, D. J. E. (1958). "Free Radicals as Studied by Electron Spin Resonance" Butterworths Scientific Publications, London.

Allen, M. B., Piette, L. and Murchio, J. C. (1961). *Biochem. Biophys. Res. Comm.* 4, 271.
Anderson, A. F. H. (1960). Unpublished results.
Androes, G. M., Singleton, M. F. and Calvin, M. (1962). To be published.
Arnold, W. and Clayton, R. K. (1960). *Proc. Nat. Acad. Sci., Wash.* 46, 769.
Beinert, H. and Lee, W. (1961). *Biochem. Biophys. Res. Comm.* 5, 40.
Bennett, J. E., Gibson, J. F. and Ingram, D .J. E. (1957). *Proc. Roy. Soc.* A 240, 67.
Blaauw-Jansen, G., Komen, J. G. and Thomas, J. B. (1950). *Biochim. Biophys. Acta,* 5, 179.
Blois, M. S., Brown, H. W. and Maling, J. E. (1961). In "Free Radicals in Biological Systems" (M. S. Blois *et al.*, eds.), p. 117, Academic Press, New York.
Bubnov, N. N., Krasnovskii, A. A., Umrikhina, A. V., Tsepalov, V. F. and Shliapintokh, V. Ia. (1960). *Biofizika,* 5, 121.
Calvin, M. (1961). *J. Theoret. Biol.* 1, 258.
Chance, B. and Nishimura, M. (1960). *Proc. Nat. Acad. Sci., Wash.* 46, 19.
Chorney, W., Scully, N. J., Crespi, H. L. and Katz, J. J. (1960). *Biochim. Biophys.* Acta, 37, 280.
Clayton, R. K. (1955). *Archiv Mikrobiol.* 22, 180.
Cole, T., Heller, C. and Lambe, J. (1961). *J. Chem. Phys.* 34, 1447.
Commoner, B. (1961). In "Light and Life" (W. D. McElroy and B. Glass, eds.), p. 356, Johns Hopkins Press, Baltimore.
Commoner, B., Townsend, J. and Pake, G. E. (1954). *Nature, Lond.* 174, 689.
Commoner, B., Heise, J. J. and Townsend, J. (1956). *Proc. Nat. Acad. Sci., Wash.* 42, 710.
Commoner, B., Heise, J. J., Lippincott, B. B., Norberg, R. E., Passoneau, J. V. and Townsend, J. (1957). *Science* 126, 57.
Cummerow, R. L. and Halliday, D. (1946). *Phys. Rev.* 70, 433.
Dworkin, M. (1958). *J. gen Physiol.* 41, 1099.
Eastman, J., Androes, G. and Calvin, M. (1962). *Nature, Lond.,* in press.
Feher, G. (1958). *In* "Proceedings of Kamerlingh Onnes Conference on Low Temperature Physics, Leiden", p. 83, supplement to *Physica,* Sept. 1958.
Feher, G. (1959). *Phys. Rev.* 114, 1219.

Feher, G. and Kip, A. F. (1955). *Phys. Rev.* **98**, 337.
Fletcher, R. C., Yager, W. A., Pearson, G. L. and Merritt, F. R. (1954). *Phys. Rev.* **95**, 844.
Frenkel, A. W. and Hickman, D. D. (1959). *J. Biophys. Biochem. Cytology* **6**, 285.
Gibson, J. F. and Ingram, D. J. E. (1957). *Nature, Lond.* **180**, 29.
Griffiths, M., Sistrom, W., Cohen-Bazire, G. and Stanier, R. Y. (1955). *Nature, Lond.* **176**, 1211
Holmogorov, V. and Terenin, A. (1961). *Naturwissenschaften*, **48**, 158.
Hutchinson, C. A. and Mangum, B. W. (1958). *J. Chem. Phys.* **29**, 952.
Ingram, D. J. E. and Bennett, J. E. (1955). *Discussions, Faraday Soc.* **19**, 140.
Jacobs, E. and Holt, A. S. (1954). *Amer. J. Bot.* **41**, 710.
Kearns, D. R. and Calvin, M. (1961). *J. Amer. Chem. Soc.* **83**, 2110
Kearns, D. R., Tollin, G. and Calvin, M. (1960). *J. Chem. Phys.* **32**, 1020.
Kohnlein, W. and Muller, A. (1961). In "Free Radicals in Biological Systems" (M. S. Blois *et al.*, eds.), p. 113, Academic Press, New York.
Kok, B. (1959). *Plant. Physiol.* **34**, 184.
Leifson, O. S. and Jeffries, C. D. (1961). *Phys. Rev.* **112**, 1781.
Malmstrom, B. G. and Vanngard, T. (1960). *J. Mol. Biol.* **2**, 118.
Melville, H. W. and Burnett, G. M. (1953). In "Techniques of Organic Chemistry" (S. L. Freiss and A. Weissberger, eds.), Vol. VIII, p. 138, Interscience Publishers, New York.
Milner, H. W., French, C. S., Koenig, M. L. G. and Lawrence, N. S. (1950). *Arch. Biochem.* **28**, 193.
Park, R. B. (1962). *J. Chem. Ed.*, in press.
Park, R. B. and Pon, N. G. (1961). *J. Mol. Biol.* **3**, 1.
Portis, A. M. (1953). *Phys. Rev.* **91**, 1071.
Roberts, E. M. and Koski, W. S. (1960). *J. Amer. Chem. Soc.* **82**, 3006.
Shields, H. and Gordy, W. (1959). *Proc. Nat. Acad. Sci., Wash.* **45**, 269.
Singer, L. S. and Kommandeur, J. (1961). *J. Chem. Phys.* **34**, 133.
Smith, J. H. C. (1954). *Plant Physiol.* **29**, 143.
Sogo, P., Pon, N. G. and Calvin, M. (1957). *Proc. Nat. Acad. Sci., Wash.* **43**, 387.
Sogo, P., Jost, M. and Calvin, M. (1959). *Rad. Res. Suppl.* **1**, 511.
Sogo, P., Carter, L. A. and Calvin, M. (1961). In "Free Radicals in Biological Systems" (M. S. Blois *et al.*, eds.), p. 311, Academic Press, New York.
Tollin, G., Kearns, D. R. and Calvin, M. (1960). *J. Chem. Phys.* **32**, 1013.
van der Waals, T. H. and de Groot, M. S. (1959). *Mol. Phys.* **2**, 333.
van der Waals, T. H. and de Groot, M. S. (1960). *Mol. Phys.* **3**, 190.
van Niel, C. B. (1941). *Adv. Enzymol.* **1**, 263.
Wagoner, G. (1960). *Phys. Rev.* **118**, 647.
Weaver, H. E. (1961). Private Communication.
Wertz, J. E. and Vivo, J. (1955). *J. Chem. Phys.* **23**, 2441.
White, H. (1934). "Introduction to Atomic Spectra", p. 215, McGraw-Hill, New York.
Witt, H. T., Muller, A. and Rumberg, B. (1961). *Nature, Lond.* **192**, 967. This is the latest of several relevant papers by Witt and co-workers. References to previous works are contained in the reference given.
Yamazaki, I., Mason, H. S. and Piette, L. H. (1960). *J. Biol. Chem.* **235**, 2444.
Zavoisky, E. (1945). *J. Phys. U.S.S.R.* **9**, 211.

Author Index

Page numbers in ordinary figures are text references ; page number in italic figures are bibliographical references.

A

Abragam, A., *368*
Adams, K. F., *206*
Afanasieva, T. P., 247, 256, 257, 259, 260, 262, *276*
Afzelius, B. M., 156, 168, *204*
Agardh, J. G., 189, *204*
Allen, M. B., 356, 357, 358, 362, *368*
Alt, K. S., 7, 9, 68
Ananiev, A. R., 14, 16, *68*
Anderson, A. F. H., 349, *368*
Andrew, E. R., *368*
Andrews, H. N., 7, 8, 9, 13, 14, 15, 16, 26, 28, 30, 34, *68*, *70*
Andrews, H. N., Jr., 199, *204*
Andreyeva, T. F., 213, *277*
Androes, G. M., 340, 351, *368*
Andrup, O., 203, *204*
Arisz, W. H., 214, *274*
Arnold, C. A., 22, 30, *68*
Arnold, W., 361, *368*
Aronoff, S., 250, 251, 254, 256, *275*, *277*
Artüz, S., 199, *204*
Asada, K., 266, 274, *274*
Ask-Upmark, E., 203, *204*
Axelrod, D. I., 9, 61, *68*

B

Bachmann, E., 75, *99*
Bachofen, R., 267, 272, *277*
Bailey, I. W., 169, *204*
Balme, B. E., 199, *204*
Bancroft, N., 36, *71*
Banks, H. P., 9, 19, *68*
Barbesino, M., 263, *275*
Barghoorn, E. S., 29, *71*
Barnola, F. V., 308, *326*
Barrier, G., 222, 249, *274*
Barton, T. C., 303, *326*
Bassham, J. A., 213, *274*
Battaglia, F., 224, 227, 228, 256, *274*
Bauer, L., 261, *274*
Baxter, R. W., 17, 26, 28, 31, *68*
Beament, J. W. L., 321, 322, *324*

Beinert, H., 329, *368*
Belikov, I. F., 265, 266, 270, 273, *275*
Bennet-Clarke, T. A., 75, 84, 86, 92, *99*, 280, 282, 306, 313, 321, *324*
Bennet, J. E., 343, *368*, *369*
Benson, A., 213, *275*
Bentall, R., 199, *207*
Berglund, B., 152, 179, *205*
Beug, H. J., 184, *204*
Biddulph, O., 246, 250, 251, 254, 256, *275*
Bieleski, R. L., 224, 225, 227, 228, *275*
Bischoff, G. W., 165, *204*
Blaauw-Jansen, G., 363, *368*
Blois, M. S., 342, 343, *368*
Böszörmenyi, L., 231, *277*
Bolchovitina, N. A., 199, *204*, *205*
Bonner, J., 86, *99*, 305, *325*
Bose, A., 9, *69*
Bosemark, N.-O., 79, *99*
Bremekamp, C. E. D., 175, *205*
Brewig, A., 306, *324*
Briggs, G. E., 313, *324*
Brovchenko, M. I., 215, 217, 218, 219, 221, 222, 230, 231, 232, 245, 246, 247, 248, 252, 256, *275*, *276*
Brown, H. W., 342, 343, *368*
Brown, J. C., 74, 83, *99*, *100*
Bubenschchikova, N. K., 250, 252, 253, 269, *276*
Bubnov, N. N., 354, *368*
Buchanan, J., 217, 222, *275*
Burcumshaw, L. L., 300, *324*
Burley, J., 252, *255*
Burnett, G. M., 348, *369*
Burström, H., 76, 77, 83, 84, 85, 86, 87, 90, 91, 92, 93, 95, 96, 97, *99*
Busse, M., 97, *99*

C

Caldwell, J., 274, *275*
Calvin, M., 73, 89, 95, *99*, 213, *274*, *275*, *277*, 340, 348, 351, 352, 354, 356, 364, *368*, *369*

14

Subject Index